Robots in Academic Libraries:

Advancements in Library Automation

Edward Iglesias
Central Connecticut State University, USA

A volume in the Advances in Library and Information Science (ALIS) Book Series

Managing Director:	Lindsay Johnston
Editorial Director:	Joel Gamon
Book Production Manager:	Jennifer Yoder
Publishing Systems Analyst:	Adrienne Freeland
Development Editor:	Myla Merkel
Assistant Acquisitions Editor:	Kayla Wolfe
Typesetter:	Lisandro Gonzalez
Cover Design:	Jason Mull

Published in the United States of America by
Information Science Reference (an imprint of IGI Global)
701 E. Chocolate Avenue
Hershey PA 17033
Tel: 717-533-8845
Fax: 717-533-8661
E-mail: cust@igi-global.com
Web site: http://www.igi-global.com

Library of Congress Cataloging-in-Publication Data

Robots in academic libraries : advancements in library automation / Edward Iglesias, editor.
   pages cm
  Summary: "This book provides an overview on the current state of library automation, addresses the need for changing personnel to accommodate these changes, and assesses the future for academic libraries as a whole"--Provided by publisher.
  Includes bibliographical references and index.
  ISBN 978-1-4666-3938-6 (hardcover) -- ISBN 978-1-4666-3939-3 (ebook) -- ISBN (invalid) 978-1-4666-3940-9 (print & perpetual access) 1. Academic libraries--Automation. 2. Academic libraries--Effect of technological innovations on. 3. Academic librarians--Effect of automation on. 4. Integrated library systems (Computer systems) 5. Libraries--Automation--Case studies. I. Iglesias, Edward G., 1966-
  Z675.U5R615 2013
  025.1'977--dc23
                              2012049782

This book is published in the IGI Global book series Advances in Library and Information Science (ALIS) Book Series (ISSN: 2326-4136; eISSN: 2326-4144)

British Cataloguing in Publication Data
A Cataloguing in Publication record for this book is available from the British Library.

All work contributed to this book is new, previously-unpublished material. The views expressed in this book are those of the authors, but not necessarily of the publisher.

Advances in Library and Information Science (ALIS) Book Series

ISSN: 2326-4136
EISSN: 2326-4144

MISSION

The **Advances in Library and Information Science (ALIS) Book Series** is comprised of high quality, research-oriented publications on the continuing developments and trends affecting the public, school, and academic fields, as well as specialized libraries and librarians globally. These discussions on professional and organizational considerations in library and information resource development and management assist in showcasing the latest methodologies and tools in the field.

The **ALIS Book Series** aims to expand the body of library science literature by covering a wide range of topics affecting the profession and field at large. The series also seeks to provide readers with an essential resource for uncovering the latest research in library and information science management, development, and technologies.

COVERAGE

- Academic libraries in the digital age
- Blogging in libraries
- Cataloging and classification
- Collection development
- Community outreach
- Digital literacy
- Ethical practices in libraries
- Green libraries
- Librarian education
- Mobile library services
- Remote access technologies
- University libraries in developing countries

IGI Global is currently accepting manuscripts for publication within this series. To submit a proposal for a volume in this series, please contact our Acquisition Editors at Acquisitions@igi-global.com or visit: http://www.igi-global.com/publish/.

Titles in this Series

For a list of additional titles in this series, please visit: www.igi-global.com

Challenges of Academic Library Management in Developing Countries
S. Thanuskodi (Annamalai University, India)
Information Science Reference • copyright 2013 • 348pp • H/C (ISBN: 9781466640702) • US $175.00 (our price)

Robots in Academic Libraries Advancements in Library Automation
Edward Iglesias (Central Connecticut State University, USA)
Information Science Reference • copyright 2013 • 341pp • H/C (ISBN: 9781466639386) • US $175.00 (our price)

Advancing Library Education Technological Innovation and Instructional Design
Ari Sigal (Catawba Valley Community College, USA)
Information Science Reference • copyright 2013 • 339pp • H/C (ISBN: 9781466636880) • US $175.00 (our price)

Recent Developments in the Design, Construction, and Evaluation of Digital Libraries Case Studies
Colleen Cool (Graduate School of Library and Information Studies, Queens College, USA) and Kwong Bor Ng
(Queens College, CUNY, USA)
Information Science Reference • copyright 2013 • 275pp • H/C (ISBN: 9781466629912) • US $175.00 (our price)

Design, Development, and Management of Resources for Digital Library Services
Tariq Ashraf (University of Delhi, India) and Puja Anand Gulati (University of Delhi, India)
Information Science Reference • copyright 2013 • 438pp • H/C (ISBN: 9781466625006) • US $175.00 (our price)

Public Law Librarianship Objectives, Challenges, and Solutions
Laurie Selwyn (Law Librarian [Ret.], USA) and Virginia Eldridge (Grayson County, Texas Law Library, USA)
Information Science Reference • copyright 2013 • 341pp • H/C (ISBN: 9781466621848) • US $175.00 (our price)

Library Collection Development for Professional Programs Trends and Best Practices
Sara Holder (McGill University, Canada)
Information Science Reference • copyright 2013 • 504pp • H/C (ISBN: 9781466618978) • US $175.00 (our price)

Library Automation and OPAC 2.0 Information Access and Services in the 2.0 Landscape
Jesus Tramullas (University of Zaragoza, Spain) and Piedad Garrido (University of Zaragoza, Spain)
Information Science Reference • copyright 2013 • 409pp • H/C (ISBN: 9781466619128) • US $175.00 (our price)

Remote Access Technologies for Library Collections Tools for Library Users and Managers
Diane M. Fulkerson (University of South Florida Polytechnic Library, USA)
Information Science Reference • copyright 2012 • 232pp • H/C (ISBN: 9781466602342) • US $175.00 (our price)

DISSEMINATOR OF KNOWLEDGE

www.igi-global.com

701 E. Chocolate Ave., Hershey, PA 17033
Order online at www.igi-global.com or call 717-533-8845 x100
To place a standing order for titles released in this series, contact: cust@igi-global.com
Mon-Fri 8:00 am - 5:00 pm (est) or fax 24 hours a day 717-533-8661

To my wife, Paula Quenoy, for her never ending patience and grace.

Editorial Advisory Board

Table of Contents

Detailed Table of Contents

Chapter 1
    Edward Iglesias, Central Connecticut State University, USA

Library automation is considered in terms of technological directionality citing sources from various disciplines including the work of various theorists in the field. A brief history of library automation is followed by a look at library organizational structure and how it might be affected by technology in the future just as it has been by technology in the past. Finally, with a strong nod to pioneering economic theorists Brynjolfsson & McAfee there is a discussion on how Artificial Intelligence will affect library jobs and organization in the future. This chapter looks at the history of library automation within the context of technological directionality. Much has been written about the history and evolution of libraries, but less as to the eventual consequences of automation. The author seeks to correct this by looking at how current workflows and departments will be impacted by the use of Artificial Intelligence in automated processes to take over work formerly done by trained library professionals. For the purposes of this chapter, these AIs and automated processes are referred to as robots, that is, automatons which take over work formerly done by humans. Finally, some suggestions will be made as to how a library might be restructured in light of these developments.

Chapter 2
    Marshall Breeding, Library Technology Guides, USA

This chapter focuses on the changes in integrated library systems (ILS) over the past thirty years as the focus shifts from collecting physical items to electronic and digital materials. The relationship between the ILS and new specialized applications, including link resolvers, knowledge bases of e-content, electronic resource management systems, digital asset management systems, discovery services, and institutional repository platforms is discussed and placed in context. In addition to looking at workflows with these new systems, a general discussion of how academic libraries are likely to engage with these new systems, the time frames in which we can expect availability and widespread adoption, and any cautions or concerns to have in mind when selecting or implementing these systems.

The history of library automation can be traced to early printing methods of the 7th century A.D. The earliest collectors of books were usually religious scholars who amassed the religious texts of the day. Monks from East and West travelled great distances and often at great peril to gather meticulously hand-copied texts. Early inventions of woodblocks, and, later the printing press, enabled the mass-production of books that resulted in libraries' expansion into the secular world. Librarians have continued to bring technological advances into their work, combining web services, programming scripts, and commercial databases and software in innovative ways. The processes of selection, deselection, and assessment have been enhanced through these new products and services. The authors discuss a variety of technological applications for collection activities that have allowed collection managers to work more efficiently and better understand the use of their print and electronic collections. The effects of automation on the people involved in collection management are also explored.

Chapter 4

LibX is a platform that allows libraries to create customized web browser extensions that simplify direct access to library resources and services. LibX provides multiple user interfaces, including popups, context menus, and contextualized cues to direct the user's attention to these resources. LibX is supported by two toolbuilder applications - the Edition Builder and the LibApp Builder – which allow anyone to create, manage, and share LibX configurations and applications. These tools automate the process of software creation and distribution, allowing librarians to become software distributors. This chapter provides background and history of the LibX project, as well as in-depth analysis of the design and use of the LibX Edition Builder that has helped enable its success.

Chapter 5

Traditionally, migration from one integrated library system (ILS) to another has been an arduous, difficult task; so much so, that libraries may choose to stay with an unsatisfactory ILS longer than they would if a viable and easy alternative were available to them. However, this institutional inertia is not necessary, if the library (possibly in cooperation with a vendor) develops a method for avoiding unnecessary and problems. In this chapter, a process is described that, when implemented, maximizes results, while minimizing pain and stress on the library and its' staff.

Chapter 6

The authors discuss their experience with using artificial intelligence and chatbots to enhance their existing web sites and information services in public library settings. The chapter describes their budget driven motivations for embarking on this project and outlines the development and implementation of the bots in their library settings. They show how the bots are positioned to enhance existing services and describe the various reactions to the bots from their patron base, and staff. Different implementations of the bots are highlighted (text only, animated talking avatar, mobile site, desktop help icon) as well as the differing levels of complexity of these different implementations. They address the oft posed question "Does AI spell the end of Reference?" and describe the InfoTabby code sharing project.

Chapter 7

The Mathewson Automated Retrieval System (MARS) is the second largest automated library storage system in the world. Housed in the University of Nevada, Reno's spectacular Mathewson-IGT Knowledge Center, MARS provides storage for half of the print collection, and nearly all government documents, special collections materials, and multimedia equipment. This chapter will explore automated library storage management, including maintenance and care of the equipment, safety, stewardship of the collection, and how automated storage challenges our beliefs about the purpose and function of libraries.

Chapter 8

The automation of university libraries in Brazil underwent a restraint of trade on computers and software, which took place in the country between 1980 and 1990, restricting the initial use of automation systems. However, they were often developed in creative ways: systems and applications were created and used in various universities, some as free software, others based on the ISIS platform from Unesco, in addition to using modern foreign systems, which only occurred in the 1990's. This chapter provides a historical overview of the development of automation in the country's university libraries, from the moment in which Brazilian researchers began to disseminate information technology, creating an automation culture in higher education institutions. Many people and institutions have also contributed to promoting and implementing automation in university libraries. This paper is on future perspectives of academic library automation in Brazil with discovery tools, next generation cloud-based systems and library automation equipment. Some possible future developments are also presented.

Chapter 9

Lai Ying Hsiung, University of California, Santa Cruz, USA
Wei Wei, University of California, Santa Cruz, USA

The current economic downturn has resulted in constantly shrinking budgets and drastic staff reduction at the University of California, Santa Cruz (UCSC) Library. Meanwhile, rapid shifting to digital formats as well as dramatic growth in social networking, mobile applications and cloud computing continues. To face these challenges, the Technical Services (TS) at the university library at UCSC need a transformation. This chapter discusses how the authors have adopted the strategy of maximizing technology in utilizing "robot-like" batch processing tools in house to minimize the risk of becoming ineffective or irrelevant. In aligning human resources to apply those tools to achieve our goals in tandem with the mission of the library, the authors learn to work with the various issues and the barriers that we have encountered during the past decade. The authors are examining the changes brought to the department through the process, highlighting a plan of action, and providing guidance for those interested in bringing about a technological transformation that will continue into the future.

Chapter 10

Denise A. Garofalo, Mount Saint Mary College, USA

Exploring technology and academic libraries concerns more than just machines, functions or processes; the human factor is as important as the equipment. Implementing successful technology changes requires attention to the people involved, and academia is no exception. Technology can be divided into either disruptive or sustaining technologies, and these technological changes impact students, faculty, and staff. In higher education, technology changes are shifting knowledge transfer to a more participatory environment and a more synergistic experience. The academic library is in a transitive state of change, evolving from a warehouse of things to a collaborative learning destination for resources. Both the library and the academic environment must adapt to survive. Overcoming the challenge of changes to the delivery of instruction may lead to extensive restructuring of courses and curriculum. The academic library can serve as a collaborative partner with faculty, leading by example to incorporate technological changes.

Chapter 11

Regina H. Gong, Lansing Community College Library, USA
Dao Rong Gong, Michigan State University Libraries, USA

The first application of robotic technologies in libraries is in the area of storage and retrieval of library materials. This chapter discusses past, present, and future developments in robotic technologies in the area of library circulation. Issues and challenges libraries face in light of rapid developments in the electronic realm are discussed in relation to circulation. This chapter also highlights future trends and technologies for library lending, as well as possibilities for advancement with the increasing shift towards electronic content in libraries.

Preface

When deciding to write a book called "Robots in Academic Libraries" one of the first thoughts that occurred was how the cover would look. While this is a scholarly work doomed to a scholarly cover, two images leaped immediately to mind. One was of a cartoon robot chasing a cartoon librarian. The other, and the one that actually reflects the intention of this book, was a cartoon frog jumping out of a pan of water before it starts to boil.

This is how the library landscape seems to the author and this book is meant to sound warning bells. As these changes take place library workers must adapt or leave. What is needed is not some luddite equivalent of throwing wooden shoes into jacquard looms but an understanding by library workers at all levels that the time to acquire new skills is now. The jobs being done by nearly every facet of library workers can already be automated to some extent. Whether it is shelf ready cataloging or reference chatbots it is foolish to think that the skills that have been needed in the last twenty years will be at a premium in the next twenty. Technologies that look primitive and nascent now will not stay that way for long. Nor can humans really compete with Moore's Law.

Above all this book is a call to look at the places where humans do have advantages that cannot be easily replicated. Anything that cannot be taught as an algorithm fits in here so logical areas of exploration include community building, education and user experience. Hopefully the chapters in this book will inspire others to not only feel the water getting warmer but to jump. What are we jumping from? Library Automation has had a very established meaning for many years. Usually this phrase refers to Integrated Library Systems that handle Acquisitions, Cataloging, Serials, Circulation and the public interface. Slowly these siloed operations have been outsourced and automated to the point where minimal human interaction is needed. Cataloging departments are largely obsolete for the purposes of copy cataloging. Acquisitions is increasingly the realm of either Patron Driven Acquisitions or elaborately created profiles that are part of approval plans used by book vendors to simply send the books which your library most likely needs. Circulation is increasingly automated with the use of RFID technologies that allow for not only self-service checkouts but for automated workflows in re-shelving items. The back end of the library is becoming overrun with robots whether that means physical machines that do tasks formerly done by humans or software that does the same thing.

At the same time there is an increasing need for skills that move the focus of library talent from inventory control and description to improving user interface design, algorithmic control and management of personnel. Collection Development librarians must make sure that profiles are correct. Patron Driven Acquisitions needs human intervention to tweak parameters. As more items become electronic, discovery platforms must be improved in terms of usability and scale as more and more items the library does now own are accessed.

There is obviously a huge change happening in libraries and much focus has been placed on budgetary cuts, changing roles in reference and the resistance of library staff to change. What has not been covered as much is the tools of this change. This book looks at the library landscape not asking whether some things should change or even if they are changing but from the point of view that this change is happening and it is pointless to debate. The tools of automation have not undergone radical change but have simply evolved to their logical points. What is the goal of software that automates a task if it is not to eliminate the need to have a person doing that task?

The chapters in this book reflect a wide array of viewpoints from all over the United States and even Brazil. Starting with my own chapter on the inevitability of technological change there is a focus on how much is changing and what not to trust as steady. The following chapter focuses on what is now the state of the industry where libraries have to make difficult choices with limited resources. Breeding's chapter is as far as I know the most complete and up to date overview available.

Annette Bailey was lead author of the next two chapters although they are very different. While Breeding does for an overview of new technologies available Bailey goes in depth and looks at concrete examples of workflows in Technical Services and how they have changed because of recent trends. Her accompanying chapter on LibX is a good example of how a tool can be created in house that does what vendors cannot. LibX is a tool that "helps automate workflows such as finding known items, but the builder tools we created automate the process of creating library software."

Ruth Bavousett's chapter on LS migration is an unapologetically technical and brilliantly written piece on how to tackle one of the thorniest problems in library land: ILS migration. Many in the field, including the author, bemoan the antiquated state of our current ILS systems and the exorbitant prices demanded by vendors. Ruth is a leader in the field of Open Source systems which is one of the areas of great hope. Only by completely starting over and abandoning the business models of current vendors can libraries hope to really change. Ruth gives us one of the keys which is data migration.

The chapter on Chatbots by McNeal and Newyear is one of the few that really mention robots in the traditional sense of AIs. The tale of their reference chatbot doing virtual reference at Mentor Public library should sound alarm bells for reference librarians. After all, who cares if it is not as good, if it is good enough and does not need a salary or sleep? The technology is still new, but given a few years of development and patrons are unlikely to know or care if they are talking to a real human.

Carolyn Adams chapter on automated storage is describing what may become "the new normal." As libraries face the fact that employees are their most expensive resource anything that makes use of automation to replace employees makes sense. In the case of UNR there was a massive undertaking to automate the management and storage of a vast collection. This was a customized system built just for that library funded by "a $10 million joint donation from former Chief Executive Officer Charles N. Mathewson and International Game Technologies" but again this is a glimpse of the future.

Viana's chapter on library automation in Brazil is a little different but still within the scope of this book. He describes a country that went from a great resistance to automation to "creating an automation culture in higher education."

"Transforming Technical Services: Maximizing Technology to Minimize Risk" is a real story from the trenches. Hsiung and Wei describe in detail how the University of California Santa Cruz faced major budget cuts by automating everything they could to "maximizing technology in utilizing "robot-like" batch processing tools in house to minimize the risk of becoming ineffective or irrelevant." The grim statistics show quite eloquently the non stop trend towards replacing employees with technology.

Denise Garoalo's "Empires of the Future" looks at the human changes that need to happen to accommodate technological change. Looking at the "human factors" in going "from a warehouse of things to a collaborative learning destination for resources." This chapter really looks at how library spaces are changing and how staff and patrons are coping.

The final chapter by Regina and Dao Gong looks at the historical application of "robotic techniques" in libraries and gives an excellent historical introduction bringing us all the way up to the present day with technologies like RFID. Looking at how self-service check in and check out has altered the landscape not just in the US but around the world. Their description of a wide array of technologies is incredibly valuable for anyone wanting a quick update on the state of the art.

Overall, this book should be of interest to a wide variety of library professionals. It is hoped that management especially pays attention to this trend. They will have to deal with inevitable budget cuts and it is only by being prepared that they can make informed decisions about the future of their libraries and the employees that are their responsibility.

Chapter 1
The Inevitability of Library Automation

Edward Iglesias
Central Connecticut State University, USA

ABSTRACT

Library automation is considered in terms of technological directionality citing sources from various disciplines including the work of various theorists in the field. A brief history of library automation is followed by a look at library organizational structure and how it might be affected by technology in the future just as it has been by technology in the past. Finally, with a strong nod to pioneering economic theorists Brynjolfsson & McAfee there is a discussion on how Artificial Intelligence will affect library jobs and organization in the future. This chapter looks at the history of library automation within the context of technological directionality. Much has been written about the history and evolution of libraries, but less as to the eventual consequences of automation. The author seeks to correct this by looking at how current workflows and departments will be impacted by the use of Artificial Intelligence in automated processes to take over work formerly done by trained library professionals. For the purposes of this chapter, these AIs and automated processes are referred to as robots, that is, automatons which take over work formerly done by humans. Finally, some suggestions will be made as to how a library might be restructured in light of these developments.

A BRIEF HISTORY OF THE AUTOMATION IN LIBRARIES

The history of the automation of libraries is well documented and this section merely seeks to point the reader in the direction of some of the best sources. For a timeless look at library technology in general see Norman Steven's excellent chapter A

Popular History of Library Technology in Library Technology 1970-1990: Shaping the Library of the Future (Nelson, 1991, pp. 1–3). Starting at a similar point John Burke's Library Technology Companion illustrates "ten key developments in information technology" starting with writing and paper and making its way through to "A Techno-Savy Populace and a Society that Requires Technology" in eight pithy pages (Burke, 2009, pp. 5–13). Of greater relevance to this work is

DOI: 10.4018/978-1-4666-3938-6.ch001

Burke's observation of the "two main goals" of library technology adoption "better servicing the needs of the library's community and streamlining the workflow of the staff" (Burke, 2009, p. 4). It is this motivation of library technology that is the key in understanding the logical outcomes of such efficiencies in productivity. Finally, the recent book Parents of Invention: The Development of Library Automation Systems in the Late 20th Century by Christopher Brown-Syed is excellent especially for his intellectual bracketing of the "Era of Conceptualization" vs. the Era of Commercialization" (Brown-Syed, 2011, pp. 1–4). His tracing of the current developments of LAMP based servers, ILS systems and discovery layers to the days of microcomputers and well thought out decisions is truly inspiring.

Looking further abroad to over arching concepts in technology there are several books that are quite relevant. In What Technology Wants Kevin Kelly argues for a sort of evolutionary determinism that he refers to as the Techium. The Techium is defined as "the greater, global, massively interconnected system of technology vibrating around us" (Kelly, 2010, chap. 1). Kelly argues that certain inventions have an inevitability about them and even goes so far as to postulate technology as another Kingdom that in a sense evolves. What technology wants is, among other things, increasing complexity. This is true of all biological creatures that evolve so why not technology? From the view of a Librarian looking at technology this view is fascinating. Libraries have always been involved in technology. After all, all writing and books are a form of technology. Some of the earliest records extant are protocunieform clay tablets have written on them an accounting of agricultural production so the organization of information is also one of our earliest inventions(Powell, 2009, p. 70). A central storehouse of information or library was an early technology that librarians were the inheritors of. Whatever the Techium may be, we have a long relationship with it.

If we accept that there is an inevitability to technological progress then it stands to reason we should at least try to understand where that current of technology is going in our libraries. The question is not whether we will be swept along, but whether we will be drowned. Right now there are many in the library field struggling valiantly to keep certain esoteric arts from dying. These include humans doing copy cataloging, checking books in and out, Inter-Library Loan not to mention Authority control. All of these actions can be performed better and more cheaply by automata, robots if you will. It does not matter for purposes of this discussion if the robot is a piece of software that does Patron Driven Acquisition drastically changing what a Collection Development Librarian is and does or shelf ready books that eliminate backlogs and are orders of magnitude cheaper than a cataloging department. What matters is that machines or pieces of software are doing the jobs that used to be done by humans more cheaply and effectively. This is inevitable. The only choices left to those who work in libraries currently are to leave the field or acquire new skills. Undoubtedly there will be a few who resist change and just want to hang on until retirement but they will need to find new ways in which to contribute.

This issue needs to be addressed on a wide scale. Other chapters will focus on factors from the demise of the cherished ILS to the actions that will become necessary by library directors as they struggle to keep the library relevant in an Academy that itself is under attack from multiple fronts to case studies from pioneering libraries who have already taken the first few steps. Nothing here should be earth shaking or overly controversial. It has been spoken about at library conferences, written about in many articles, but to my knowledge this is the first systematic treatment of the effects of automation as a whole in the near future of academic libraries.

Currently Academic libraries are in a state of flux. Many have shifted from the highly siloed departments of the past but a few holdouts still look --organizationally at least-- much like they did in the 1970s. This is one typical organizational model of what a library might look like.

- A Library Director
  - Departments
    - Acquisitions
    - Access Services
    - Collection Development
    - Cataloging
    - Electronic Resources
    - Reference
    - Serials
    - Systems

Certain details should be readily apparent from looking at this structure. First is that it is built around modules in an Integrated Library System. Whether the departments caused the ILS vendors to build their systems that way or not our academic libraries are operationally ensconced as parts of a monolithic system that many agree is not relevant to the needs of those that use it. The other detail that should be readily apparent is that some of these categories overlap so much that their very names are nonsensical. Electronic Resources used to refer to databases the library acquired such as Lexis-Nexis or Academic Search Premiere. Now it can refer to e-books (Acquisitions), electronic journals (Serials) maintenance of an electronic repository (Systems) or editing the web site (Web Developer/ Information Architect/ Web Master).

The very structure of Libraries in the academy needs to change. The direction of this change will be what users (members, patrons etc...) want from their library. Assuming that they want what they have always wanted (hardly a certainty) then they will want access to research facilities that allow them to access information, primary and secondary, as well as assistance in that research. Much of our past structure has been focused on acquiring the resources that contain information and then organizing it. Now that information

comes (mostly) prepackaged and pre organized. This does not mean that one can simply buy a resource and put it up. Customization for user needs, adequate assistance and training as well as a physical space to allow for research and teaching must be of paramount importance in visualizing the academic library of the future.

Imagine a library where everything more or less happens automatically. What would this look like? The purchasing of materials would be largely driven by Patron Driven Acquisitions. Faculty and students would request a book through a web form or perhaps see a link to a book that is being considered for purchase on some sort of discovery layer. Immediately a computer algorithm would determine whether this is an appropriate item for purchase based on pre-existing parameters or whether it is better to simply order it from another library. All of this would happen automatically and the patron would be notified one way or the other. The book arrives shelf ready so that it merely has to be scanned and loaned out. Compare this workflow to the current method in many libraries where people have to decide whether an item is worth ordering, look to see where a copy may be found, place an order, receive the book, place a ridiculous paper band around it, and then give it to the user. While this workflow can be made more efficient, it will never match the efficiency of automatically ordering the item without staff intervention. There is simply no need for a separate ILL department. The circulation department would simply receive the book and check it out. Similarly if an order is received to borrow a book through ILL does the library really need a separate department to grab a book off the shelf and mail it. No decisions would need to be made. It would be a student workers job. Eventually it will be a robot's job.

Now think about what happens in Acquisitions and Collection Development. With the increase in automation a great deal of focus would shift to tweaking profiles to make sure approval plans are effective and accurate as well as on contract negotiation with vendors. Some of these tasks are familiar but others are not. Certainly it is already impossible to be an effective Acquisitions or Collection Development librarian without extensive computer expertise though this is mostly at a basic level where mastery of vendor's proprietary systems is the goal. Now Acquisitions and Collection Development must be concerned with the heuristics and algorithms involved in selection and purchase of materials. If they do not become expert in these techniques the library becomes dependent on what vendors tells us is happening. Just as we would not pay a bill without knowing we had received the item so we must not simply trust that vendors are doing what they say they are without understanding what it is they are doing. Those in charge of making decisions about what is purchased need to be able to follow the bit trail just like they currently follow the paper trail. This problem could be solved with third party open source tools that check on the effectiveness and audit programs. Much will depend on vendors' willingness to be audited and provide transparency about how they make choices.

Cataloging as it exists currently simply cannot go on. The funding is not there for precious records only viewed by the catalogers that create them. The copy cataloger will soon be as extinct as the buggy whip maker. These techniques are already in place today and there is simply no reason for the vast majority of academic libraries to employ copy catalogers. Shelf ready books cost less and there is no backlog. Original cataloging will still be needed but, depending on the library, for a small percentage of total items. Certainly none of the electronic items require cataloging. As for Serials with nearly 100% of serials online the real skill set will shift to finding statistics using standards like SUSHI and working them into an ERM (Electronic Resource Management system) to help develop the collection. This assumes that there even is a serials collection. Several libraries have already moved in the direction of "on-demand" articles

finding that it saves money on subscriptions that are unused as well as staff to maintain them. (Cheng, Bischof, & Nathanson, 2002)

Electronic Resources and Systems will bear the brunt of new tasks and will be where new staff are likely to find themselves. As libraries seek to differentiate themselves in terms of excellence of their collections, these collections will continue to drift online. Digital Archives and Repositories will need constant maintenance. Those former copy catalogers will likely end up doing original cataloging in Dublin Core for these archives. Depending on the institution this will either be a full time occupation or simply a student worker job. Either way this is very low level work compared to original MARC cataloging and smart libraries will soon figure out that an MLS is not needed.

Systems as a unit are likely to disappear altogether. As more solutions become cloud based there will be no need for a local server, local backups and local upgrades. There will still need to be individuals with technological savvy in these fields but a greater emphasis will be on Information Architecture, Website Development and of course vendor management. An exception to this may be the increasing role of Open Source systems. Really smart libraries will realize the advantage of not just using these systems but also of participating in the development process. Thus systems will transform from Systems Administration to User Experience. Whether the library is ready is another matter recently discussed in Library Journal (Schmidt, 2011).

Reference is likely to be the area that looks the most familiar. As long as there are students and faculty who have reference questions or need training in whatever Information Literacy morphs into there will be a need for physical humans to guide them through an increasingly complicated maze. The difference is likely to be how these services are delivered. Libraries are already making extensive use of virtual reference and as Colleges and Universities go further abroad to find students where they can, often offering late night classes or distance classes, then it makes sense that this trend will continue. There are some libraries investigating alternatives using Artificial Intelligence. The current generation are chatbots are rather primitive but good enough to handle directional questions and some basic ready reference. The question is not whether these technologies will be used but when. When will it be more likely that a student around the world is chatting to an AI and not a "real" librarian? For certain types of questions these bots are very likely to be the preferred method of communication as in the aforementioned easy categories (directional, ready reference) but also perhaps in areas where there is no equivalent human expertise. Imagine a situation where there is foreign language expertise needed but no one on staff is fluent. A chatbot fluent in that language would be very helpful. An increasing focus on teaching is likely to continue the call for Information Literacy. The idea of a reference librarian at a reference desk is likely to disappear. Between the focus on mobility as well as ubiquity on campus the virtual or scheduled reference interview will become the norm. Virtual Reference is a natural fit for AI chatbot systems that are at least as effective of some of the students hired to handle directional questions or low level ready reference. Academic libraries that choose to staff reference areas with highly trained and paid professionals to handle questions that can be answered with a quick Google search are simply making poor use of human resources. A current generation chatbot can answer who the 14th President of the United Sates was more quickly and cheaply than the average reference librarian and does not require a retirement plan.

A PROPOSED STRUCTURE

If all of these traditional departments are going away there needs to be something to replace it. This structure will not work in all academic libraries but should be easily adaptable to many.

Large research institutions will always need very specialized staff and very small libraries may be outsourced altogether leaving some kind of information commons staffed virtually with no physical items. Most medium size academic libraries will eventually need to adapt to something more streamlined and flexible.

Director

The director will have to transition from chief administrator to chief strategist. This does not mean the director will have to come up with all of the new ideas but simply that he or she will have to start viewing the library as an entrepreneurial enterprise that can be seen as a profit center within the university. These are unpalatable terms to many in the library field but unless we face the reality that the library must provide measurable value to the university there is no reason we should survive. The director must be keenly aware of this and be constantly searching for ways to show the library's usefulness to the larger community of constituents.

Strategic Planning Committee/ Steering Committee

This would be a group appointed by the director of individuals in the library who have a vision of the library's role in the university and how to achieve its mission. It is very important that buy in be sought for this group but ultimately this group must have clear authority to make change happen.

Coordinators

Rather than Department Heads centered around modules in an ILS there should be revolving Coordinators. Every full time professional librarian should be made to function as a coordinator periodically much like academic departments take turns having heads. Every librarian should have to deal with budgets and the consequences of decisions. Too often there is a refrain of powerlessness in our libraries. Unless everyone in the library is forced to lead they will never understand the importance of strategic planning and budgeting. All coordinators would have to argue for resources--physical and human--for their departments.

After this it becomes very individualized. No two libraries are likely to need the same people or departments. Certainly the departments formerly known as cataloging, acquisitions, serials, and ILL can all be eliminated in favor of automated solutions. Depending on the library and how much they invest in RFID technology and automated sorting circulation can probably go from a department to a part time duty. Reference is likely to be at least partially handled by AI chatbots leaving reference librarians to become either highly specialized or primarily teachers. The highly specialized would be needed for the type of reference question that cannot be easily outsourced to chatbots or the web at large. Questions dealing with theory, ambiguity, specialized knowledge or language are the least likely to be outsourced. This has much to do with the way that AIs are programmed. There are two relevant concepts that are traditional weak points in Artificial Intelligence that can also be seen as human strengths. One is the so called frame problem. According to the Stanford Encyclopedia of Philosophy

The first significant mention of the frame problem in the philosophical literature was made by Dennett The puzzle, according to Dennett, is how "a cognitive creature ... with many beliefs about the world" can update those beliefs when it performs an act so that they remain "roughly faithful to the world" (Shanahan, 2009)

This translates as machines have a hard time understanding the things about the world you "just know". St. Patrick's Day has to do with green beer, a catholic saint and being Irish. An AI can be programmed to know all of this but still have no clue what drunkenness involves, how to believe never mind have faith, and not know what a human

being is let alone nations or nationalities. As for vagaries like "everyone is Irish on St Patrick's Day," good luck. Local knowledge of weather is easily programmed but whether the weather is good enough to go to a parade is more difficult. Matters such as what constitutes a long commute vary so much by region that there can never be one answer programmed that will work for everyone.

These are areas where reference librarians will be invaluable, areas where it is very hard to anticipate questions that require significant other information to make sense. Human minds are very good at handling the frame problem and AI's are traditionally very bad. Thus it makes sense to hire and support reference librarians who understand things that are hard to program. Maps can be digitized but they do not convey information in the same way to an AI. They cannot, at present, discover something new. The big lesson is to learn subjects that machines have problems mastering, subjects that depend on a knowledge of the world at large.

Ambiguity is another area where humans have an advantage at least for now. Ever since Watson the IBM Supercomputer beat its human competitors on the game show Jeopardy, this problem has been starting to crumble. Still as the Stanford Encyclopedia of Philosophy states about "Lexical Ambiguity":

The lexicon contains entries that are homophonous, or even co-spelled, but differ in meanings and even syntactic categories. 'Duck' is both a verb and a noun as is 'cover'. 'Bat' is a noun with two different meanings and a verb with at least one meaning. 'kick the bucket' is arguably ambiguous between one meaning involving dying and one meaning involving application of foot to bucket. (Sennet, 2011)

This is where humans have a clear advantage in the form of the reference interview. When computers do it there is a feeling one is playing 20 questions. Reference librarians have the uncanny knack to help you find what you are looking for not merely what you say you are looking for. If a user asks for "that book on the Irish" an AI is of no help.

ACQUISITIONS AND COLLECTION DEVELOPMENT

These two areas go hand in hand as Patron Driven Acquisition becomes a more realistic option for libraries that want to have relevant not merely good collections. Because of this the Collection Development librarian must become an expert in profiling what the library will buy. Just because the system is patron driven does not mean that everything will automatically be purchased. Realistic budgetary limits will have to be set and all this must be programmed into the system that automatically makes on the fly determinations about purchases. Similarly there will be other factors that determine whether one subject area is more suited for growth than another. It is very hard to program AIs with knowledge such as "the robotics program just got through the curriculum committee and we need to start ordering materials now". Due to the extreme interdependence in area expertise it is very likely that acquisitions and collection development will be merged tasks.

ELECTRONIC RESOURCES AND SYSTEMS

These are two other areas that seem ripe for merging perhaps including serials as that becomes exclusively an electronic resource. Together this group will likely take on UX and Information Architecture duties. The real change will come as this group coordinates and distributes what used to be "techie" duties to other less technically skilled staff. This is already happening with the library website thanks to products like LibGuides,

WordPress and Drupal that let average users populate the content of the website. Currently there is some resistance to this but it should be obvious that the web is now the primary source of communication for the library and those who do not participate in that communication are already marginalized as invisible and non productive. With the pervasive use of content management systems the separation of content and design is now a given. The technically competent staff members will increasingly have to deal with vendor issues in the case of hosted solutions and troubleshooting in the case of locally managed ones. Skills in design, UX and communication will be paramount as technical skills such as programming are likely to be increasingly specialized.

THE OPEN SOURCE WILDCARD

The major factor that could affect this structure is the adoption of Open Source technology. This is appealing to librarians on several levels. First there is the guarantee that the data owned can continue to be accessed. Too often libraries give vendors data they have created and pay large amounts of money just to access it. Then there is the customizability of Open Source software. Many libraries have unique collections and needs that cannot be met by average off the shelf software but requires something bespoke. Think of a military library that needs discoverability but also limited visibility of items based on clearance level. The increasing prevalence of data sets in academic libraries will likely provide some challenges as new systems are developed to try and access highly heterogeneous data. Finally, even with full support and hosted solutions Open Source systems will often cost less and have better support than closed source systems. They do come at a cost however. Changes can be made to systems to customize them. This requires programming expertise either in house or outsourced on a fairly regular basis. The Open Source model also assumes that the code is released back to the main project. Some administrators may have a hard time with the University's intellectual property being shared in that fashion though as this model becomes more and more pervasive this is less of a problem than it used to be.

AIS AND SELF CORRECTION

It is tempting to think of libraries becoming a sort of vending machine for information so that one sticks in a query at one end and the answer drops out like a can of soda. No direct interactions just what you want immediately. There are several problems with this scenario. First there is the fact that libraries are more than just query engines returning answers. Much of the panic in our field has been over this notion that libraries are in competition with Google. If so we lost a long time ago. Google does what it does better than any library. The library however does a great deal that Google does not do. Librarians and libraries can help find answers to undefined questions:

I'm looking for a book that will change my life.

We can work on fuzzy problems:

Well if it is not available here, where could it be?

We can work in various dimensions of time, space and vagueness:

Can you teach my class Thursday? I'm not sure what we'll be covering.

In short, humans are self correcting. We are able to figure out when the question asked would not generate the answer wanted and figure out how to help researchers in the process of discovery. The difference between finding an answer that already exists as a fact and aiding someone in discovering something new is enormous. AIs can help with

what might be termed "ready reference" types of questions but not with deep multifaceted research. If all a librarian can offer is ready reference type of help they ought to be replaced by a computer. On the other hand if users need answers to difficult questions let them come to a library. Let the reference desk be spared from students asking what the capitol of Ohio is. The answer should be to look it up in Google. Now if that student wants to know where to find the original plans for the Ohio state Capitol and more about the architect let him or her come to the library.

ECONOMICS

From an economic perspective no one has looked at the problem of automation in the workforce in general quite as succinctly as the book Race Against the Machine (Brynjolfsson & McAfee, 2011). While the traditional model of economics goes something like "yes jobs are eliminated but better jobs are created" often called the "end of work" argument, Brynjolfsson & McAfee argue something a bit different.

The end-of-work argument is an intuitively appealing one; every time we get cash from an ATM instead of a teller or use an automated kiosk to check in at an airport for a flight, we see evidence that technology displaces human labor. But low unemployment levels in the United States throughout the 1980s, '90s, and first seven years of the new millennium did much to discredit fears of displacement, and it has not been featured in the mainstream discussion of today's jobless recovery. (Brynjolfsson & McAfee, 2011)

Brynjolfsson & McAfee place technology into the center of the discussion rather than at the fringes. With that reasoning they have many valuable observations about the role of humans in the workforce to come. Among them:

We don't believe in the coming obsolescence of all human workers. In fact, some human skills are more valuable than ever, even in an age of incredibly powerful and capable digital technologies. But other skills have become worthless, and people who hold the wrong ones now find that they have little to offer employers. They're losing the race against the machine, a fact reflected in today's employment statistics. (Brynjolfsson & McAfee, 2011)

These are the choices before us, to adapt and survive or grow extinct. A few years ago the Purchasing department at our university asked the Acquisitions librarian what was the difference between their jobs? After all, the purchasing department bought things and so did the Acquisitions librarian. The librarian patiently explained all the nuances in purchasing out of print books, the hassle of vendors who change contracts yearly not to mention the rather strange relationships we have with serial vendors ("what do you mean you sometimes get money back?") the purchasing department eventually concluded that acquisitions was indeed a dark art best left to the library. This may not always be the case.

As libraries continue to be merged with IT departments library values are likely to decrease in importance. IT departments in a University setting have traditionally handled a wide scope of tasks. If the library is increasingly automated and the library itself does not have the expertise to maintain its own systems IT will likely be called on to take over. There is a very real problem for our users in that. IT tends to focus on a customer service model based on giving the user a limited number of options which are very well understood and getting a user to the point where the service is working to an agreed on level. Software purchase decisions are usually top down and based on what industry standards are.

Compare this to the library. Often a librarian will have broad subject expertise and perhaps specialized knowledge in one or two areas but have only a generalized knowledge of the millions of items available for access. Buying decisions are often collaborative with faculty and based on accreditation as well as departmental preference. Often the reference librarian must work with a user to learn a system for the first time and only have a general knowledge of the underlying structure of information to guide them. This might be referred to as the "I know it's there" problem. Unless you deal with this information you don't know it exists and where it exists. In short, IT lives in a world that tries to assure mastery and stability over a limited universe while libraries deal with the vagueness associated with an unbounded universe of questions and answers.

If libraries and IT centers are to be combined they will require a greater understanding by both parties of the others culture and mission. There is a real possibility that most academic libraries will continue to collect resources and offer access to them requiring only a computer lab of some sort and a retrieval system. There would not be anyone "in the back" because all of those positions will have been automated. The replacements for those positions are likely to go to IT since currently libraries do not go out of their way to be seen as a center for technical competence. Eventually all of the reference folks could be replaced with chatbots or virtual librarians hired at a cheaper rate.

THE FUTURE IS UNCERTAIN

While the above scenario seems gloomy I doubt it will play out completely. Some parts will without a doubt, but there is much libraries can do to make themselves more valuable to the academy as a whole.

RE AFFIRM VALUE/VALUES

Not only do we need to get rid of waste and prove how we measurably contribute to the bottom line of the university but we also need to know what our values are and will continue to be. For a look at the value of academic libraries see Value of Academic Libraries: A Comprehensive Research Review and Report (Oakleaf, 2010). This report thoroughly covers areas where libraries make valuable contributions to the University and the community as a whole. This is usually where librarians get stuck and for good reason. Budgets are tight and there is always a need to justify our existence. That said, Universities have to justify their existence as well. It is imperative that academic libraries find common ground and this may be found by looking to more universal values of which libraries are a part.

CHANGE OUR ROLE IN THE UNIVERSITY SYSTEM

As libraries move increasingly from centers of physical information bearing entities to entry points to greater information resources there needs to be a shift from just research to content creation. This is already happening in many libraries as tighter integration with writing centers takes place. Similarly many libraries offer the tools to create audio video content and music. There are many opportunities for growth in this area including centralized data curation, maker spaces and high level computational resources.

DATA CURATION

Recent changes in the law regarding NIH and NSF grants (NIH Data Sharing Information - Main Page, n.d.) have led to a scramble to make publicly funded research viewable by the public. This means many researchers often having their data stored on one computer must now figure out how to make it accessible. For the library this is a golden opportunity to become a centralized repository of unique and valuable data. There will be a need for database experts who can migrate the data and place it within the larger realm of the discovery layer.

HIGH LEVEL COMPUTATIONAL RESOURCES

Along the same lines as Open Source software and cheap computers have decentralized and distributed computing from IT there are new problems creeping in that never occurred before. From professors who set up Moodle boxes to compete with the campus CMS and become overwhelmed with the administration overhead to the competing Beowulf clusters that all get hacked because the department had no security help setting them up libraries are in a unique position to help with these research and pedagogical tasks. While IT is great at mission critical infrastructure and software it may well be that the library and librarians are a more comfortable fit for many faculties.

MAKERSPACES

Increasingly creation of works is not limited to writing. As well as providing the tools for creating audio and visual media libraries can perform a vital role in distributing technology that may exist on campus but is limited by department. As areas like graphic arts embrace 3D printing and Chemistry uses this technology for rapid prototyping the library becomes an essential makerspace in which students and faculty can create things. Certainly with the current focus science technology and math this is an area likely to increase in usage and demand as well as funding.

While libraries will always need experts in information management, technology, and business practices, there is a need "for a new type of librarian" who has training in an academic discipline and an understanding of digital technology.

Responding to those new needs, the Council on Library and Information Resources, with a grant from the Andrew W. Mellon Foundation, has developed a new program of postdoctoral fellowships for recent Ph.D.'s in the humanities as "an alternate entry path into the profession of librarianship" (Johnson, 2003).

The emphasis here is on a "new type of librarian." This is becoming especially clear in the arena of "big data" and infomatics. The niche for Librarians is spelled out in a blog post by Lorcan Dempsy

The rise of webscale services which handle large amounts of users, transactions and data has made the management of big data a more visible issue. At the same time, as more material is digital, as more business processes are automated, and as more activities shed usage data, organizations are having to cope with greater volume and variety of relatively unstructured data. Analytics, the extraction of intelligence from usage data has become a major activity. Here is a helpful characterization by Edd Dumbill on O'Reilly Radar. (Dempsy, n.d.)

Here is the strange alchemical beast that librarians must morph into with the head if a researcher in bioinformatics the body of a librarian able to communicate to patrons and figure out what they need and perhaps the tail of a data management specialist. Add to this the previously mentioned UX expertise and you have librarians curating, organizing and helping users to discover new information they never had access to.

To conclude there are many different skills that librarians have and can acquire that differentiate them from the various automata that threaten to take over the library. A focus on continued education and flexibility will be key attributes of the librarian that rules over and is not ruled by

the various robots we will employ in the library. The human traits of intuition, imagination and creativity are not programmable. As long as users desire spaces where humans can help with vague questions that have no definite answer there will be a need for humans in the library.

REFERENCES

Brown-Syed, C. (2011). *Parents of invention the development of library automation systems in the late 20th century*. Santa Barbara: ABC-CLIO.

Brynjolfsson, E., & McAfee, A. (2011). *Race against the machine: How the digital revolution is accelerating innovation, driving productivity, and irreversibly transforming employment and the economy*. Digital Frontier Press.

Burke, J. J. (2009). *Neal-Schuman library technology companion: A basic guide for library staff*. Neal-Schuman Publishers.

Cheng, R., Bischof, S., & Nathanson, A. J. (2002). Data collection for user-oriented library services: Wesleyan University Library's experience. *OCLC Systems & Services*, *18*(4), 195–204. doi:10.1108/10650750210450130.

Dempsy, L. (n.d.). *Big data.big trend*. Lorcan Dempsey's Weblog. Retrieved April 4, 2012, from http://orweblog.oclc.org/archives/002196.html

Johnson, M. D. (2003). Turning Ph.D.'s into librarians. *The Chronicle of Higher Education*, *50*(8), C4.

Kelly, K. (2010). *What technology wants*. New York: Viking.

Nelson, N. (1991). *Library technology 1970-1990: Shaping the library of the future: General session entitled "Mainstreets, landmarks, and cross roads: Mapping library technology."* 5th Annual conference on computers. Westport, CT. London: Meckler.

NIH Data Sharing Information - Main Page. (n.d.). Retrieved August 29, 2012, from http://grants.nih. gov/grants/policy/data_sharing/

Oakleaf, M. (2010). *Value of academic libraries: A comprehensive research review and report*. Assoc. of College and Research Libraries.

Powell, B. B. (2009). *Writing: Theory and history of the technology of civilization*. Chichester, U.K. Malden, MA: Wiley-Blackwell.

Schmidt, A. (2011). Ready for a UX librarian? *Library Journal, 136*(18), 24.

Sennet, A. (2011). Ambiguity. In E. N. Zalta (Ed.), *The Stanford Encyclopedia of Philosophy*. Retrieved from http://plato.stanford.edu/archives/ sum2011/entries/ambiguity/

Shanahan, M. (2009). The frame problem. In E. N. Zalta (Ed.), *The Stanford Encyclopedia of Philosophy*. Retrieved from http://plato.stanford. edu/archives/win2009/entries/frame-problem/

Chapter 2
Beyond the ILS:
A New Generation of Library Services Platforms

Marshall Breeding
Library Technology Guides, USA

ABSTRACT

This chapter focuses on the changes in integrated library systems (ILS) over the past thirty years as the focus shifts from collecting physical items to electronic and digital materials. The relationship between the ILS and new specialized applications, including link resolvers, knowledge bases of e-content, electronic resource management systems, digital asset management systems, discovery services, and institutional repository platforms is discussed and placed in context. In addition to looking at workflows with these new systems, a general discussion of how academic libraries are likely to engage with these new systems, the time frames in which we can expect availability and widespread adoption, and any cautions or concerns to have in mind when selecting or implementing these systems.

INTRODUCTION

For the last thirty years or more, academic libraries have relied on integrated library systems to help them manage and provide access to their collections and services. The ILS was designed at a time when academic library collections consisted primarily of physical items and they provided automated support for a very broad range of the tasks related to the management and access of

these materials. The transition to ever increasing proportions of electronic and digital materials pressed the ILS beyond the limits of what it was designed to manage.

In order to deal with these new formats, the current phase of library automation leaves the ILS in place, but surrounded by other specialized applications, including link resolvers, knowledge bases of e-content, electronic resource management systems, digital asset management systems, discovery services, and institutional repository platforms. This approach, while filling in needed functionality, results in duplication of effort and

DOI: 10.4018/978-1-4666-3938-6.ch002

inefficiencies for library personnel in the way that they work and makes use of library resources difficult for patrons as they attempt to navigate through a complex matrix of interfaces and services.

This chapter will address the technologies emerging now that address the current and future needs of academic libraries. The chapter will describe and give perspective on some of the projects and products emerging in this context. An early section will describe some of the general concepts embraced by each, including unified workflows across all collection formats, highly shared data models, open API's and engagement with community developers, as well as general technology trends such as multi-tenant software-as-a-service. Sections will be devoted to some of the major products: Ex Libris Alma, OCLC WorldShare Platform, Serials Solutions Web-scale Management Solution, Innovative interfaces' Sierra, and Kuali OLE

The chapter will conclude with some analysis, perspective, and projections on how academic libraries are likely to engage with these new systems, the time frames in which we can expect availability and widespread adoption, and any cautions or concerns to have in mind when selecting or implementing these systems.

The State of the Integrated Library System

The model of the integrated library system has been the bedrock of library automation for around more than thirty years. The first automation systems introduced in the early to mid-1970s embraced a model of organization that grouped functional tasks into stand-alone systems specializing in a specific area of library operations, such as circulation, cataloging, acquisitions or serials management. Over time, these standalone modules consolidated into more comprehensive systems and have steadily gained more nuanced functionality, but many of the foundational prin-

ciples of the early automation efforts designed for a print world persist through the integrated library systems in use today.

The earliest days of library automation systems were created to address specific processes. Some offered cataloging and circulation capabilities, such as Libs 100 system introduced by CLSI in 1971 or the Gaylord Circulation 100 system in 1975. Others specialized in acquisitions, including Innovative's INNOVAQ. In these early days managing even a more narrow scope of functionality was a giant leap beyond the manual procedures previously in place. The ambition, from the beginning, was more comprehensive automation, and many of these early systems evolved accordingly, bringing together bundles of functionality.

The mold of the fully integrated library system germinated out of the previous generation of special-purpose applications, taking root by the 1980 when products began to be marketed that combined multiple modules that shared common databases and interfaces. The libraries of this era saw all their energies concentrated on managing collections of physical materials—books, serials, microforms, and manuscripts. Monographs existed only in printed form, and were purchased through approval plans, firm orders from publishers or book jobbers. Journals were published in print, and the functionality to perform check-ins to record issues received, make claims for missing issues, and to facilitate the binding into volumes was an enormous help over the manual procedures previously followed for tracking a library's holdings of newspapers, periodicals, and scholarly journals. This was also the time when researchers would find articles related to their areas of interest using printed indexes, such as the *Readers Guide to Periodical Literature* or one of the more specialized indexes.

It was in this print dominated time that the integrated library system entered the scene, with modules designed to work together to provide a comprehensive array of functionality. These integrated library systems might be configured for a single independent library, including those with multiple branches or for consortia of multiple independent libraries. The standard integrated library system modules included:

- **Cataloging:** To create, import, or update bibliographic records. Common functionality would include the ability to create new MARC records when needed and to connect to bibliographic services such as OCLC Cataloging Service, or to major libraries that offer a Z39.50 service to identify and acquire records previously created that match an item to be cataloged.
- **Authority Control Module:** May be an optional product or a feature of cataloging that allows the library to add consistency and linking among the records in their bibliographic database making use of standard authority terms, such as the Library of Congress Subject Headings, the Library of Congress Name Authorities, MeSH, or those from other national libraries or cataloging agencies.
- **Circulation:** To manage the loans, renewals, returns, and overdue processing for physical materials. Circulation might include capabilities to support direct consortial borrowing for libraries that allow patrons to make requests from their partner institutions.
- **Acquisitions:** To handle the business details involved in the procurement of library materials, including the management of funds allocated for departments or disciplines, management of a database of the vendors from which the library orders materials, issuing invoices for items ordered, creating vouchers to issue payments for items received, check-in of items received, annual fiscal year close to reconcile budgets and funds, and reports related to budgets, funds, and vendors.

- **Online Catalog:** To allow library patrons to search or browse for library materials
- **Serials Management:** For the specialized tasks related to newspapers, periodicals and scholarly journals, such as issue check-in, routing lists, claims for issues not received, and invoice processing for serials suppliers.
- **Binding Module:** Might be available as an option or provided through a third-party vendor to facilitate the process of creating bound volumes out of the single issues received.

The earliest versions of these integrated library systems ran on mainframe computers, but were also implemented on lower-cost midrange systems and servers as those became available.

The integrated library system has survived major transitions in computer architectures. Text-based interfaces, based on menus and transaction directives were employed during the reign of mainframe and midrange hardware. The basic deployment involved a central processor which did all the work with text-based terminals that provided staff and end-user access to the modules. A typical arrangement would involve physical terminals provided within the library at circulation desks, throughout technical services, and in public areas for patron access. Through terminal server controllers and other communications devices, access to the integrated library system could be provided outside the library on campus networks, to dial-in users, and eventually on the Internet. Terminal emulation software made it possible for those with personal computers to access these systems.

An age of client/server computing emerged in the 1990's which aimed to tap into the powerful capabilities of the desktop computers increasingly deployed in both office and home settings. These personal computers had ample processing power, ran multi-tasking operating systems, and offered graphical interfaces that could be operated more intuitively by persons without a technical background. Mainframes eventually were phased out due to their high cost and modest storage and computing capabilities compared to the server-class computers that come onto the market.

Driven by this transition in the broader information technology realm, a new line of integrated library systems were created that followed the client/server architecture. These new systems operated on servers under operating systems such as some version of Unix, OS/2 from IBM, or Windows NT from Microsoft. The client software for these new client/server integrated library systems would operate on desktop computers running Microsoft Windows or the Macintosh OS and offered graphical interfaces that used menus, selection, and editing techniques that could be operated using a mouse rather than the cryptic commands and directives of the text-based systems. These new client/server systems included: Horizon from epixtech, Taos from DRA, Millennium from Innovative Interfaces, Virtua from VTLS, Library.Solution from The Library Corporation, or Polaris from Gaylord Information Systems. Some of the incumbent text-based systems morphed into the client/server realm such as Unicorn from Sirsi Corporation with the introduction of the InfoVIEW and later WorkFlows clients.

Although a new cycle of technology drove the change to a new roster of integrated library systems, the basic operational approach and functionality did not change dramatically. By the mid-1990's libraries continued to manage collections of mostly physical materials. Some forms of electronic information were emerging, primarily in the area of indexing and abstracting products, but library collections themselves remained primarily in print during the onset of the client/server era. The new products consequently continued to offer the same basic modules with the same general sets of features as the previous era. The key challenge was to bring forward all the nuanced functionality present in the mature mainframe-based systems in to the nascent client/server products.

With these new client/server products mostly already in place, the age of electronic information came into libraries with great force. The first wave of changes came in the serials side, with a rapid, and by now nearly complete transformation to electronic publication. Academic libraries shifted ever larger proportions of their journal subscriptions to electronic form as well as to great number of indexing and abstracting services and aggregated databases of citations and full-text articles available on the Web. This growing body of subscriptions to electronic resources, unfortunately, found little help in the existing functionality of the integrated library systems of the time. To meet these needs, a variety of additional products and services emerged to support libraries with the management and access of electronic resources. These included: OpenURL-based link resolvers, federated search environments, and electronic resource management systems.

- **OpenURL Link Resolvers** to deal with the unsustainability of manually coding links to e-journal titles and to individual articles, providing an infrastructure based on OpenURLs embedded in content sources, parsed and interpreted by a library's link resolver, which relied on a knowledge base of detailed holdings data to create context-sensitive links to the target resource. Examples of these OpenURL-based link resolves include SFX from Ex Libris, 360 Link from Serials Solutions, LinkSource from EBSCO, or community-based projects such as CUFTS/GODOT. [For a detailed report on OpenURL-based link resolvers and their supporting knowledge bases, see Breeding, Marshall. "Knowledge Base and Link Resolver Study: General Findings."

- **Electronic Resource Management Systems:** Assist libraries with the procurement and management of their subscriptions to content products. Their features include specialized acquisitions capabilities for the procurement of new resources and subsequent renewals, administration of the terms of licenses, authorization details, vendor profiles, management of the detailed holdings within each content package, and many other features. While some of these features overlap with the acquisitions and cataloging modules of the integrated library system, they operate mostly independently. The points of integration with the organization's link resolver were stronger than with the integrated library system. Some of the major electronic resource management systems include Verde from Ex Libris, 360 Resource Manager from Serials Solutions, and Electronic Resource Management from Innovative Interfaces, Inc., and ERM Essentials from EBSCO Information.

- **Federated Search Tools:** Allowed users to search many different library resources simultaneously, providing a simplified research process compared to serially searching many content products. The federated search tools, such as those from MuseGlobal, MetaLib from Ex Libris, WebFeat, or 360 Search from Serials Solutions, were based on sending queries to multiple remote information resources in real time, intercepting responses, and presenting results to users. This methodology had the benefit of allowing users to more easily simultaneously search multiple related resources, but these tools were limited in the number of targets that could be addressed simultaneously.
- **Discovery Interfaces:** Have emerged in recent years to provide more modern and comprehensive ways for uses to gain access to library collections and relevant services than the online-catalogs packaged with integrated library systems. These discovery interfaces include relevancy-based search technologies, faceted browsing to help us-

ers narrow and navigate through search results, and offer many of the features that uses have come to expect in a Web-based search environment.

Libraries increasingly expect discovery services to provide access to broad representation of their collections. In addition to the traditional scope of the online catalog such as books, media materials, journal titles, it is increasingly desirable to include access to the individual articles contained within a library's subscriptions to electronic resources. The genre of index-based discovery services has emerged to address this expanded scope. While implementation details vary, the general approach involves the creators of a discovery service making arrangements with providers of information resource to provide content that can be loaded into a large index to support end-user discovery. Items selected for use from the discovery service would be fulfilled by the original publishing source. These index-based discovery services include Summon from Serials Solutions, EBSCO Discovery Service, Primo Central from Ex Libris, and OCLC's WorldCat Local.

A typical academic library would not only operate its integrated library system to manage its print collection, but will likely also depend on a cluster of these additional management and discovery applications to fill out missing functionality related to electronic and digital resources. This assemblage of independent systems imposes a high level of overhead for the library, both operationally and technically.

TRANSITION FROM CLIENT/ SERVER TO CLOUD COMPUTING

In the same way that client/server computing displaced an earlier generation of mainframe-based systems, the current wave of technology is based on a service-oriented architecture, Web-based interfaces, and designed to be deployed through multi-tenant software-as-a-service. The previous shift was driven by the presence of powerful desktop computers and the unsustainability of mainframes. One of the problems with the client/server architecture involved the client software that needed to be developed for multiple operating environments. Developers need to choose to create clients for different types of computers, such as Windows or the Mac OS. New versions of the application would require users to download and install new software. Today, Web browsers have become the preferred end-user client platform. The service-oriented architecture involves applications created out of small units of functionality that can be combined together to create more complex features. Lower-level services can be re-used in many different parts of an application, resulting not only in programming efficiency, but in the ability to compose new functionality through the recombination of existing services. The services can also be exposed to customers or to third parties, resulting in application programming interfaces that can be used to extend functionality or for interoperability.

The other key technology trend involves cloud computing. Rather than design applications to be installed on a server housed at the customer's site, they can be implemented on highly-scalable clusters of computing equipment capable of supporting large numbers of organizations on the same instance. Systems designed for multi-tenancy can be expanded incrementally as new organizations or individuals implement the service, with data segregated or aggregated as needed. For both personal use and business, cloud computing has become increasingly dominant.

Any new applications will be created with these characteristics and the legacy client/server architecture eventually diminishes. In the library automation arena, were in a time where products developed in recent years have been designed for multi-tenant software-as-a-service using Web-based interfaces with some of the existing client/server products evolving toward this architecture.

In addition to the technology shift underway, new applications developed today have the opportunity to take into consideration the fundamental changes that have transpired in the nature of library collections and operations. Instead of simply replicating the functionality of the traditional integrated library system focused on print materials into new technology platforms, a variety of products have emerged that address the current reality of library collections comprised of materials in many different media or formats.

THE EMERGENCE OF A NEW GENERATION OF LIBRARY SERVICES PLATFORMS

The major trend in the library automation industry in play since about 2008 involves the development of products to address both the failings of the traditional integrated library system and the advent of this new age of cloud computing. This new generation of library services platforms brings together a more comprehensive approach to providing support for current library realities and current technology architectures.

This new genre of software differs substantially from the incumbent integrated library system. These products diverge significantly from the functionality addressed and in technical architecture and constitute a new product category. Some of the organizations developing these products posit their own labels for this new category, tailored to their design and concepts. Ex Libris uses "unified resource management," OCLC and Serials Solutions both use the label "Web-scale." In order to have a vendor-neutral label for this group of products, the author has coined "library services platforms" to describe this slate of products that embody these characteristics. While each product offers a distinctive approach and may emphasize some characteristics more than others, they differ substantially from the scope, functionality, and architecture of integrated library systems.

Some of the key characteristics of library services platforms include:

- Comprehensive management of library materials including electronic, digital, and print.
- Designed to be deployed through multi-tenant software as a service.
- Reliance on shared knowledge bases for bibliographic and e-content holdings.
- Service-oriented architecture.
- Reliance on Web-based interfaces for staff functionality.

In general, these systems provide a more comprehensive approach that can displace a set of independent applications with a single unified framework. In addition to the core integrated library system, these new library services platforms also incorporate functionality that may have also been separately managed in electronic resource management systems, or digital asset management systems. The groupings of functionality within the products may also differ from the traditional integrated library system. Features related to circulation, resource sharing, document delivery, and interlibrary loan, may be organized into a broader set of workflows centered on resource fulfillment, for example.

MAJOR PRODUCTS

The transition toward these new products is just now underway. The inception of this new generation of library management products began in 2009, with product announcements coming out through 2011. The roll-out of the products lags behind a couple of years. We can anticipate the competition to be fully underway by about 2013, with larger proportions of research and academic libraries adopting these products over the course of subsequent years. The tipping point where library

services platforms dominate integrated library systems can't accurately be predicted, but may be anticipated to occur by about 2016.

OCLC WorldShare Platform

Organizational Background and Strategy

OCLC, a global library services organization was originally founded in 1967 to provide cataloging services to libraries in Ohio, and has steadily expanded geographically and in the breadth of its products and services. The organization today offers a wide range of products and services including resource sharing, interlibrary loan, collection analysis, virtual reference, among dozens of others, all generally oriented toward amplifying the efforts of individual libraries through cooperation.

Organized as a non-profit in the state of Ohio, OCLC has an executive management team overseen through a system of governance that includes a Board of Trustees as well as set of advisory global and regional councils. OCLC operates throughout the world, with almost 26,000 member libraries in 170 countries. OCLC has also grown through the acquisition of other organizations and for-profit companies. In the bibliographic services arena, organizations acquired include UTLAS (1991), Bibliocentre (1997), authority control services from Blackwell North America (1997), WLN (1999), Library Technical Services (1999), and the Research Libraries Group (2006). OCLC has also acquired a number of for-profit companies involved in the library automation industry, including PICA (1999-2007), Sisis Informations systeme (2005), Fretwell-Downing Informatics (2005), Openly Informatics (2006), Amlib (2008), EzProxy (2008), CONTENTdm (2006), and BOND (2011). Through these acquisitions, OCLC has gained ownership of relevant technology products, responsibility for the support of a number of traditional integrated library systems, and a pool of experienced developers, product managers, and executives. In recent years, OCLC has channeled its attention not only to the ongoing maintenance and support of these acquired products, but it has been especially focused on developing a new generation products for resource discovery and library management.

WorldCat Local Discovery Services

Leveraging the massive WorldCat bibliographic resource, OCLC developed WorldCat Local as a discovery services that could work with a library's existing integrated library system. Based on the holdings set in WorldCat, patrons are able to find items owned by the library, supplemented by a vast number of additional materials from other OCLC member institutions. WorldCat Local is able to show the status of items held in the library through behind-the-scenes interrogations to the local ILS. Consistent with the trend toward extending the scope of discovery services to also include electronic resources at the article level, OCLC has made arrangements with publishers and other content providers to index an ever expanding array of these materials in WorldCat.

WorldCat Local was launched as a pilot product in April 2007. OCLC reported over 1,500 libraries using WorldCat Local in 2011.

WorldShare Management Services Design and Concepts

In a move that takes OCLC even deeper into the library automation industry, OCLC has embarked on a strategy to provide products and services for the management of library operations. With products and services already in place that provide cataloging, resource discovery, and interlibrary-loan, providing additional capabilities including circulation, acquisitions, and electronic resource management could be accomplished with incremental, though still substantial development. But with these services in place, a library would no

longer need to maintain much of its local infrastructure, including its integrated library system.

OCLC articulates a vision that amplifies the impact that libraries can have on their clientele and that the broader library community can have on society through cooperation on a global scale. Instead of individual libraries operating individual and isolated automation systems and catalogs, they can join together, facilitated by OCLC's services, to not only operate with much more efficiency, but to provide powerful resources that compete on a par with major commercial enterprises such as Amazon.com. OCLC characterizes this global cooperation as Web-scale, or operating "at the network level."

In order to support these new library management services, OCLC has created a new technology infrastructure capable of supporting a massive volume of transactions with global redundancy. This new infrastructure, now called the WorldShare Platform, not only was designed to handle a transaction load equivalent to global library activity, but also provides access to key data resources, such as the WorldCat bibliographic database, the WorldCat knowledge base of e-journal holdings data, and other collaboratively managed information resources. The WorldShare Platform follows a service-oriented architecture and exposes a full suite of application programming interfaces and Web services. OCLC has used this infrastructure to create its own applications such as the WorldShare Management Services and WorldShare License Manager, which offer functionality for the management of a library's print and electronic resources.

Beyond these and other applications that OCLC might develop, OCLC member libraries and authorized third-party organizations can also develop additional services or applications on the WorldShare Platform. The same API's that OCLC uses to create its own major applications, such as WorldShare Management Services will be made available to support applications created by programmers external to OCLC. OCLC provides an "App Gallery" as a repository for libraries to share the applications and services they create using the WorldShare Platform. Once an application has been completed, tested, and certified by OCLC, it can be made available to other libraries. Some applications may use content resources freely available, others may tap into content or services made available only through specific subscriptions. The applications that libraries use will need to align with their own subscriptions to OCLC products and services. It would not make sense, for example, for a library to make use of an app that extends some aspect of WorldShare Management Services if they did subscribe to that product.

The BIBSYS consortium in Norway, for example, has a long history of developing its own library management systems and subscribed to WorldShare Management Services with the intention of extending its capabilities through their own development efforts using the APIs. The 105 member BIBSYS consortium, which includes the National Library of Norway and all the major academic libraries in the country, selected OCLC WorldShare in November 2010, following a competitive procurement process. In August 2012 BIBSYS discontinued its project to implement WorldShare Management Services.

Consistent with its ever expanding global strategy, OCLC has established data centers in multiple global regions to support the WorldShare Platform. OCLC's original data center at its headquarters in Dublin, OH has been protected through a redundant site in Westerville, Ohio for several years. OCLC has since activated data centers outside the United States including the United Kingdom (December 2011) and Sydney (March 2012).

Functional Design

OCLC WorldShare Management Services provides functionality traditionally been made available to through an integrated library system. Rather than rely on an integrated operated either for an individual library or through a limited number of libraries through a consortium, municipal, county, or regional system, WorldShare Management Services provides a platform designed to be shared at a global level to provide support for library operations. Extending OCLC's existing services of cataloging, resource sharing, and discovery, WorldShare Management Services addresses functionality related to circulation, acquisitions, and serials management. An additional set of services, the WorldShare License Manager, provides tools to help libraries manage their subscriptions to electronic resources.

The functionality in WorldShare Management Services builds on the massive WorldCat bibliographic database, which held around 240 million records in May 2012 and continues to grow rapidly, both through new cataloging performed directly on the system by OCLC members and through records loaded from national catalogs and other sources. The basic data model underlying OCLC's new services is based on moving libraries away from individually managed bibliographic and other supporting databases toward basing management tasks on linkages made to shared records in WorldCat.

The cataloging functions of WorldShare Management Services are based on the existing OCLC Cataloging capabilities. Bibliographic records in MARC format can be created and enhanced according to well-established procedures and authorizations, with support for new metadata formats such as Dublin Core.

WorldShare Management Services provides acquisitions functionality with capabilities to handle both print and electronic materials. A global database of vendors is provided, with the capability for each library to add local data as needed. Workflows are included to order materials, receive items, to process and manage invoices, to manage budgets, and to manage vendor records. The system automatically creates holdings records associated with the bibliographic record in WorldCat for new orders, making the items available for circulation upon receipt.

WorldShare License Manager is available as a separate optional subscription and provides additional tools for managing electronic resources. It relies on the WorldShare knowledge base of detailed holdings within electronic resource packages to support tasks related to their acquisition, management, linking, and access.

Development and Deployment Roadmap

The strategy to develop the WorldShare Platform has been underway since before 2009 when OCLC made its original announcement for Web-scale Management Services, as the product was called prior to the launch of WorldShare as the brand for this suite of services to support library management. A group of early adopter libraries began implementing WorldShare Management Services in July 2010, and the service has been available in general release since July 2011. OCLC launched WorldShare License Manager in January 2012 bringing in additional capabilities to manage electronic resources.

Target Library Segments

The OCLC WorldShare Platform has been designed for all types of libraries, consistent with the broad scope of the organization's membership and commercial customer base. Early adopters represent a diverse set of libraries, both in size and type.

Early marketing efforts were focused in the United States, attracting a variety of libraries as early adopters of WorldShare Management Services. Examples of the early libraries to place the product into production include:

- Craven-Pamlico-Carteret Regional Library System, with 10 public library members in North Carolina (Jan 2011)
- Boundary County District Library (Dec 2010)
- Lawrence Technological University (Feb 2012)

- Ogeechee Technical College, in Statesboro, Georgia (Oct 2011)
- Bucknell University
- Spring Hill College

By February 2012, OCLC reported that 35 libraries had completed migration to WorldShare Management Services to replace their previous integrated library system.

In January 2012, the University of Delaware announced its intent to implement WorldShare Management, the first member of the Association of Research Libraries to select the product.

Alma from Ex Libris

Organizational Background and Strategy

Ex Libris, based in Israel and with a broad international presence, specializes in developing products and services for academic and research libraries. Today the company ranks as one of the largest companies in the global library automation industry and offers a broad slate of products related to the discovery and management of library resources.

Ex Libris traces its roots to Hebrew Union University in Jerusalem, where the original Aleph software was created for the needs of this library, beginning around 1980. Its success in its home university lead to interest in other universities, leading to the establishment of Aleph Yissum founded through the technology transfer unit of the university to commercialize the system, initially with Israel and beginning in 1986 through a new company called Ex Libris, Ltd., among other countries. In 1995 the two separate companies were merged together into a single corporate structure, Ex Libris Group. Ex Libris has grown steadily through the adoption of Aleph in many regions of the world, both through distributors and in corporate offices established for specific regions. The company has also grown through the acquisition of other companies, beginning with Dabis in 1997 and with Endeavor Information Systems in 2006.

The ownership of Ex Libris has gone through several transitions through its organizational history. Initially Hebrew Union University and Ex Libris founder Azrael Morag shared ownership. In 1999 venture capital firms Walden Israel and Tamar Ventures gained ownership stakes. Francisco Partners, a large private equity firm, gain exclusive ownership in 2006 through a deal that also resulted in the merger of Endeavor Information Systems into the company. In August 2008 the company was sold to Leeds Equity Partners.

The business strategy of Ex Libris has consistently been based on investing in the development of new products for the academic and research library market. The company has continually developed its own Aleph integrated library system, taking the product through multiple generations of technology, beginning with its original COBOL-based mainframe versions, to midrange systems, to the current version using Unix and Oracle-based servers and Windows-based clients. Following the acquisition of Endeavor, Ex Libris also channeled development resources into Voyager. In addition to its core ILS products, Ex Libris has created other products, mostly related to the management and access of electronic materials in library collections. Ex Libris acquired the SFX technology from the University of Ghent in 2000, performed additional development and commercialized it to establish the genre of OpenURL-based link resolvers. Other link resolver products have also entered this market and Ex Libris has continued to maintain its lead in terms of the total of installations. The company has also developed MetaLib, announced in July 2000, as its offering in the federated search arena. In 2004 Ex Libris introduced Verde to help libraries manage their electronic resources.

Primo for Discovery and Delivery of Library Resources

With a slate of well-established, through aging products, Ex Libris began its initiative to develop a new generation of technologies initially with a discovery layer for end-user access to library resources, to be followed later by a new platform for internal library operations and resource management. The company described this strategy as "unified resource and delivery" responding to the increasing strategic priority of electronic and digital resources in academic libraries, involved bringing all types of resources into access and management platforms instead of the previously established model that treated them separately.

Ex Libris announced Primo as its end-user discovery and delivery platform in mid-2006. The early versions of Primo aimed to provide access to all types of library resources using a locally maintained index that a library would populate with metadata from its integrated library system, local digital repositories, as well as other collections, supplemented by the use of MetaLib to bring in results from the library's subscribed electronic resources. Primo was designed to be highly customizable for each library, allowing the library to set its own indexing and presentation rules, relevancy weightings, as well as many other configuration details.

In July 2009, Ex Libris announced Primo Central, a hosted index of scholarly resources that could be seamless integrated into Primo or other discovery interfaces. Primo Central brought Primo into the ranks of the new Web-scale or index-based discovery services initially established by Serials Solutions with the introduction of Summon in January 2009. Ex Libris has continually expanded Primo Central with the addition of collections of e-journal aggregations, e-book collections, and other scholarly resources.

Alma Design and Concepts

Ex Libris began its early design of Alma beginning in 2009 based on the concept of unified resource management. The company's existing product line included Verde for the management of electronic resources as well as Aleph and Voyager that followed the traditional integrated library system model generally optimized for print resources. One of the foundational concepts of Alma involves bringing the management of all types of library resources under a single platform rather than providing separate products for print, electronic, and digital resources. Primo brought together end-user access of the multiple components of a library's collection; Alma aims to bring the same unification to staff operations relative to different types of resources.

Alma has been designed for a multi-tenant software-as-a-service deployment, though implementations on local infrastructure will be possible as well. Ex Libris has established data centers to host Alma and its other products, including on in the Chicago area of the United States and another in The Netherlands (activated in July 2011), with additional sites planned in other geographic regions.

Though offered through a proprietary software license, Alma will include APIs that expose all areas of functionality so that libraries can access the data and functionality of the system through external programs and applications. These APIs, for example, make it possible to access Alma through other discovery layers other than Primo. Libraries will also be able to use the APIs to enable Alma to dynamically communicate with other components of campus infrastructure, such as enterprise resource planning systems (ERP) or learning management systems.

Alma will include knowledge base components in support of various aspects of resource management. The management of electronic resources requires knowledge bases that describe such details as the specific titles and issues associated with any given aggregated database or other resource to which a library might subscribe. Profiles of the library's subscriptions can then be applied to the knowledge base to determine the specific materials available within the library's collection. Products such as the SFX link resolver and the Verde electronic resource management system relied on a knowledge base maintained by Ex Libris. An extended version of this knowledge base will also drive the management of resources in Alma and will be made available as a component of the Community Zone, available to all the institutions that sue the product.

Alma follows a hybrid data storage model, with some databases shared across all the users of the product in a set of data stores called the Community Zone and other data specific to individual institutions stored in segregated areas called the Local Zone. Content in the Local Zone can still be hosted in Ex Libris data centers, but each library's Local Zone cannot be accessed by other institutions.

Ex Libris will provide a bibliographic database in the Community Zone to which any library can attach its holdings to shared records. Libraries may also choose to implement their own separate bibliographic database in ways similar to current ILS implementations. Initial deployments, for example, may take place initially with bibliographic data held in a Local Zone, with subsequent transition to a more shared bibliographic data model. This hybrid approach for the management of bibliographic records contrasts with OCLC's WorldShare that requires all libraries to share records and does not support the concept of individual bibliographic databases per institution.

Functional Design

The staff interface to Alma is entirely Web-based, requiring no installation of software on local workstations. Since the product is deployed through software-as-a-service, the library also does not operate local servers.

Once a library staff member signs into Alma, he or she will be presented with a customized page, called a dashboard, which provides access to the specific aspects of functionality appropriate to that job position or role. System administrators have the ability to selectively activate the tasks available to any given staff member, not only simplifying the menu of options, but also securely limiting access to only those features authorized for use.

Alma offers functionality to support all the standard operations expected in an integrated library system, though in many cases workflows may be somewhat different. Libraries can expect full support for their routine operational tasks such as circulation, acquisitions, cataloging, and the management of print and electronic serials. Alma includes Metadata Management System that supports multiple metadata formats, initially including MARC, Dublin Core, and MODS. Implementation strategies may also impact workflows, such as whether the library chooses to place its bibliographic records in the open Community Zone, or to keep them within their Local Zone.

Development and Deployment Roadmap

Ex Libris has been working with a group of development partners since the initial announcement of the product in 2009. The libraries engaged as development partners include Boston College, Princeton University, Purdue University, Katholieke Universiteit Leuven Libraries in Belgium which includes the LIBIS network of 30 independent libraries.

Alma has been developed incrementally, with multiple releases provided to development partners for testing and feedback. The following partner releases were completed:

1. **June 2010:** Basic functionality from a selection of modules, including basic circulation features, management of staff accounts, staff search capabilities, basic acquisitions, and system repository.

2. **December 2010:** Including unified acquisitions for print and electronic resources, cross format search through staff client, a metadata editing workbench with MARC support,

3. **May 2011:** Including advanced integration capabilities with discovery layer interfaces, support for multiple metadata formats for resource description, complete acquisitions workflows for ordering, processing invoices, and receiving, and the initial set of functionality for electronic resource management, and workflows for digitization of materials upon patron request, and role-based user dashboards, providing access to features required for library personnel based on their functional position.

4. **September 2011:** Advanced integration with Primo, including complete end-user account functionality and other standard online catalog features, improvement to the user interface, further refinements to the Alma resource editor for streamlined management of inventory and cataloging features, and enhanced fulfillment capabilities

5. **November 2011:** A version delivering all remaining functionality planned for the initial release of Alma, subject to final testing and quality assurance.

Alma was first put into production at Boston College on July 2, 2012.

In addition to the initial group of development partners for Alma, a number of libraries signed commitments to become early adopters, including additional universities in Australia, Europe, and the United States. The planned implementation of Alma in its development partner and early adopter sites from 2012-13 mark the beginning of the deployment phase of the product.

Target Library Segments

Alma, like all Ex Libris products, has been designed for academic and research libraries as well as national or state libraries, and consortia.

It will be marketed across all geographic regions. While many of the initial development partners will be migrating from the company's own Aleph or Voyager integrated library systems, the group also includes libraries previously using products from competing companies.

Intota from Serials Solutions

Organizational Background and Strategy

Serials Solutions operates as a wholly owned business unit of ProQuest, a major provider of content products to libraries. ProQuest is owned by Cambridge Information Group.

Serials Solutions was founded in March 2000 by brothers Peter McCracken, Steve McCracken, and Mike McCracken along with Chris Pierard. The initial product of the company was a database that described the e-journal titles held within various content packages to which a library subscribes. The effort involved in assembling and organizing this data was beyond the resources of most libraries, making it difficult to maintain lists that described the individual e-journals available to users. Serials Solutions worked with e-journal publishers and aggregators to acquire detailed data regarding these content products and incorporated them into a knowledge base that could then be used to dynamically generate e-journal lists and other tools for access and management of these resources. The effectiveness of these products depended on improving the data received by publishers, resolving inconsistencies, omissions, and errors. This emphasis on high quality data as the basis for creating products for libraries has been an ongoing strategy for Serials Solutions from its founding through its latest products.

ProQuest Information and Learning acquired Serials Solutions from its founders in March 2005. In December 2006 ProQuest became part of Cambridge Information Group. Serials Solutions as a technology company specializing in products to help libraries manage and provide access to their electronic resources operates as an independent business unit of a large corporate entity that offers a wide range of content products to libraries.

Serials Solutions has developed a growing suite of products, built on its knowledge base of e-resource holdings to facilitate the management and access of a library's electronic resources. The company has developed a set of products, including the 360 Link OpenURL link resolver, a federated search environment called 360 Search, and a full electronic resource management product called 360 Resource Manager. The knowledge base that drives each of these products has been branded KnowledgeWorks.

Summon an Index-Based Web-Scale Discovery Service

Serials Solutions launched Summon in 2009, the first major commercial discovery service based on an index that includes the print materials owned by a library and the individual articles held within its subscriptions. Summon also ties into KnowledgeWorks for OpenURL link resolution and into 360 Core for profiling the content to be made available in the discovery service according to the library's subscriptions. Consistent with the broad strategies of Serials Solutions, Summon relies on aggregating and managing data to drive the functionality of a service. In this case, the data component takes the form of an index created from the metadata or content contributed by publishers which facilitates the discovery of that content.

All the technology products created by Serials Solutions have all been deployed as multi-tenant software as a service rather than as software that would be installed locally in a library. These products include the entire 360 product suite, KnowledgeWorks, and the Summon discovery service. All access to the products, both for staff management tasks and services accessed by patrons is delivered through Web-based interfaces, with no software installed on local servers or workstations. The products can be configured for each library through a Web-based customer portal for branding and cosmetic appearance, to profile content subscriptions, and to select functional options. These are not products where the company hosts individual installations for each customer site, but rather follow the multi-tenant software as a service architecture where the same instance supports all sites using the product.

Serials Solution Expands into Print Management through Intota

From its earliest days, Serials Solutions focused exclusively on products related to the management and access of electronic resources. In June 2011 at the American Library Association Annual Conference, the company announced that it intended to expand the scope of its products to also encompass a library's print materials and other resources managed within the traditional integrated library system. Serials Solutions strategy did not involve offering a traditional integrated library system, but rather to apply the techniques it employs for the management of electronic resources to the library's print inventory.

When initially announced, Serials Solutions gave the product the provisional descriptive name Web Scale Management Service. In January, 2012, the product was branded as Intota.

Serials Solutions positions Intota as designed to support the complete lifecycle of library materials, beginning with selection, procurement, description, discovery, through end-user fulfillment. In expanding the scope addressed in its management products, Serials Solutions aims to enable libraries to decommission their integrated library system, relying instead on functionality delivered through their software as a service offering.

Intota will follow the knowledge base driven approach to resource management as previously established in the 360 suite of products on electronic resources. The knowledge base will be ex-

panded to include bibliographic records in MARC format as the basis for tasks such as acquisitions and circulation of physical materials.

Intota will be designed to make use of Summon as its patron interface. Summon was conceived to provide access to both print and electronic materials, though relying on an existing ILS. With Intota, Summon will take on all the functionality needed for patron services related to print materials.

The design of Intota, as described by Serials Solutions, will include a full set of documented APIs that will allow libraries to take advantage of all aspects of its functionality and data. Although Summon was designed to use Summon as its patron interface, libraries interested in providing an alternate discovery service would be able to do so through the APIs. Likewise, libraries can make use the Summon APIs to allow other discovery interfaces, such as Blacklight or VuFind to tap into its index or to allow Summon to serve as the front end for other library services platforms or integrated library systems.

Development Roadmap

As of August 2012, six libraries have opted to work with Serials Solutions as development partners for Intota, including Oklahoma State University, Ball State University, Johnson County Community College in Kansas, Marist College, State University of New York in Genesco, and the University of Buffalo.

Target Library Segments

Serials Solutions from its beginning has focused primarily on providing products for academic libraries. Intota likewise has been designed for this segment of the library industry. All of the development partner sites are academic libraries, ranging large research institutions that belong to the Association of Research Libraries to smaller community college libraries. Initial adopters of

Intota will likely come from the ranks of those already using Serials Solutions products for electronic resource management or discovery. Intota will attempt to draw from the installed base of academic libraries using traditional integrated library systems, such as Voyager and Aleph from Ex Libris, Symphony and Horizon from SirsiDynix, or Millennium from Innovative Interfaces.

Serials Solutions comes into the library services arena as much as two years later than some of its direct competitors. While being first to market comes with significant sales advantages, given the long cycle of transition from integrated library systems to library services platforms there will be ample opportunities for the product once it is completed and enters its marketing and implementation phase.

Kuali OLE

Kuali OLE aims to create a new generation library management platform for all formats of materials designed for academic and research libraries managed as a community project to be made available as open source software. The Andrew W. Mellon foundation has provided significant support for Kuali OLE, funding three grant proposals, including a one-year planning process, a two-year software development project, and collaborative endeavor with JISC to develop a knowledge base for electronic resource management. Participating institutions provide matching funds and contribute in-kind resources.

Organizational Background and Strategy

The initial phase of the project was supported through a one-year planning grant awarded to Duke University to work toward an initial design of a new generation library management environment, consistent with current library needs and modern technology architectures. A grant award of $475,700 from the Mellon Foundation funded the Open Library Environment project that commenced in August 2008 and was concluded in September 2009. Led by Duke University, other institutional participants included University of Kansas, the University of Florida, Lehigh University, Columbia University, the University of Pennsylvania, Vanderbilt University, Library and Archives Canada, the National Library of Australia, the University of Maryland, and Whittier College. The work of the project included refining the general concepts surrounding a new system, gathering information on desired capabilities from libraries through a series of regional workshops, developing workflow charts for functionality through a business process modeling process.

Follow-up activities from the initial project included identifying institutions that would participate in a follow-up project to build the software and joining with the Kuali Foundation as its governance body. The Kuali Foundation oversees a variety of other projects that create enterprise software for higher educational institutions, including the Kuali Student System, Kuali Financial System, Kuali Coeus research management system, and several others. The Kuali Rice project developed the enterprise class middleware that provides support for each of the other Kuali projects. Rice provides a strong Service Oriented Architecture foundation for other enterprise applications. By relying on Rice, each Kuali project avoids the need to develop low-level support services but rather focus on higher-level functionality.

The Kuali OLE build project is a partnership among several major educational institutions, including Indiana University, the University of Pennsylvania, a consortium of members from the University of Florida system (representing: Florida State University, Florida International University, New College of Florida, University of Miami, the University of South Florida, and the Florida

Virtual Campus),, the University of Michigan, the University of Maryland, the University of Chicago, Lehigh University, Duke University, and North Carolina State University.

Support for the build phase of the project comes from $2.38 million awarded by the Mellon Foundation, with each partner institution making a significant financial investment and contributing personnel allocations. Indiana University serves as the lead institution. Brad Wheeler, the Chief Information Officer of Indiana University also serves as the board chairman of the Kuali Foundation. Much of the programming will be performed by a commercial software development firm. The Kuali OLE project engaged HTC Global Services as a strategic partner. HTC makes use of programmers in India in addition to its offices in Troy, Michigan.

An additional project was defined to focus on creating components required for electronic resource management. A collaborative project between Kuali OLE and JISC was announced in June 2012 to develop an open access repository of e-journal holdings and tools based on open source software for housing and managing the data. The Mellon Foundation awarded $499,000 to support this project, known as the Global Open KnowledgeBase, led by North Carolina State University and scheduled for completion in November 2013.

Discovery Services

The software created by Kuali OLE will focus on library management processes and will not include a patron discovery interface. One of the early decisions made in the initial planning grant was to consider discovery out of scope since there are several open source and proprietary products already available. The Mellon Foundation, for example, had previously funded the eXtensible Catalog project, which provides a variety of tools and utilities to support the creation and integration of discovery services. Other open source discovery

projects include Blacklight and VuFind that both rely on the Apache SOLR search technology. The key strategy for Kuali OLE is to create all the APIs needed to operate well with any of the discovery products.

Kuali OLE Functional Design

The Kuali OLE project embraces a vision of comprehensive resource management and will create a platform with the capability to manage all the many different types of materials that comprise library collections and their corresponding metadata. There will be a family of metadata editors that will be used as needed for MARC, Dublin Core, EAD, and other formats. Each type of metadata will be stored and edited in its native format.

The scope of Kuali OLE does not correspond directly to a traditional integrated library system. Its functionality is broader than an ILS in that it delivers support for the management of electronic materials, and accommodates multiple metadata formats other than MARC, and aims to accommodate other workflows prevalent in academic libraries. It also shifts some functions traditionally bound within the ILS to external systems to avoid redundancies with other business systems in the broader academic enterprise environment, such as a centralized authentication service. Kuali OLE is designed to be integrated with student management systems and enterprise resource planning environments.

One of the important strategies in the management of electronic resources involves the use of a knowledge base that tracks the different titles and holdings bundled in the many content products to which libraries subscribe. These knowledge bases such as KnowledgeWorks from Serials Solutions, the Global SFX Knowledge Base from Ex Libris, and the OCLC Knowledge Base play an important role in the management of electronic resources by allowing a library to track the individual titles and issues of e-journal titles available by creating a profile of the subscriptions to aggregated databases, e-journals, and other content products to which it subscribes. Kuali OLE launched the Global Open Knowledgebase project to develop this essential capability with the need to rely on proprietary knowledge bases and to provide capabilities beyond that delivered in the commercial products such as use of unique identifiers for organizations involved in the supply chain, data regarding the publication history if titles, and increased opportunities to repurpose the data.

Kuali OLE has been designed to be implemented in an enterprise environment rather than in as multi-tenant software as a service. At least initially, each of the development partner libraries will install the software locally, though in some cases the installation will serve a fairly large number of libraries. This approach differs from those such as OCLC's WorldShare Management Services or Serials Solutions Intota where a single instance of the software is shared by all the institutions subscribing to the service. Kuali OLE emphasizes the integration among systems across the campus rather than pooling resources among separate implementations.

Development and Deployment Roadmap

Kuali OLE will rely on software from other Kuali projects and other open-source components. The Kuali Rice Version 2 middleware provides the lower-level service-oriented architecture layer, including such support functions as workflow management, messaging, action triggers, and connections to identity management services. Kuali Ole incorporates the Kuali Financial System to support business transactions, such as those in collection acquisitions processes. The system will also use Apache Jackrabbit as its document store and Apache SOLR for to deliver APIs for faceted discovery. The use of these major components will allow the project's developers to focus on programming unique library-oriented functionality.

The project organized the development of the software into four incremental releases. (See http://www.kuali.org/OLE/Timeline for details).

- **November 2011:** Version 0.3. Implementation of Core Technologies
- **May 2012:** Version 0.6 Cooperative Infrastructure and Services
- **October 2012 (Planned):** Version 0.8 Describe and Deliver
- **First Quarter 2013 (Planned):** Version 1.0 Ready for Implementation

The initial release of the full Kuali OLE Version 1.0 is expected to be completed in the first quarter of 2013. This release will include data migration and implementation tools. Some of the development partner libraries are positioned as early adopters, such as Lehigh University and the University of Chicago, with implementations beginning in 2013. Others will migrate on a later cycle in 2014-2016.

It is early to anticipate how broadly Kuali OLE will be implemented beyond the initial development institutions.

Target Library Segments

The Kuali OLE project focuses directly on libraries associated with higher educational institutions. Although the planning phase included two national libraries, the current group of partner libraries consists entirely of academic libraries, with some organized in consortia.

Sierra from Innovative Interfaces, Inc.

Organizational Background and Strategy

Innovative Interfaces, Inc., based in Emeryville, CA, has been developing library automation products for libraries since its founding in 1978 by Jerry Kline and Steve Silberstein. The company's initial product allowed libraries using the Libs 100 circulation system to connect to OCLC for catalog records. The company created the INNOVAQ product to automate acquisitions in 1982, adding serials control in 1985, an online catalog in 1987, and circulation in 1989. This course of incremental development produced one of the early fully integrated library systems, named INNOPAC. Innovative launched a Web-based online catalog in 1995. INNOPAC was based on a proprietary database management system that Innovative created and it was implemented on minicomputers with text-based terminal interfaces.

In 1997, Innovative began the transition to Millennium, redeploying the functionality of INNOPAC through a client/server architecture to meet the growing expectations for graphical user interfaces and distributed computing. The Millennium clients were developed in a Java environment, allowing them to work with both personal computers and Macintosh computers. Innovative was one of the earliest adopters of Java for staff clients of an integrated library system. The transition from INNOPAC to Millennium followed an evolutionary approach, preserving the business logic that underlies the functionality of the system while morphing the product into a technology environment consistent with then-current expectations.

Innovative interfaces was owned by its two founders through 2001, when Jerry Kline acquired the interests of Steve Silberstein. The company recently went through an ownership transition when in March 2012 two private equity companies, Huntsman Gay Capital and JMI Equity made strategic investments in the company. Co-founder Jerry Kline retained partial ownership of Innovative and serves as Chairman of the newly-constituted board. In August 2012 Kim Massana was named as the company's new Chief Executive Officer, with Neil Block continuing as President.

Throughout its corporate history, Innovative has followed a technology strategy of incremental, evolutionary development and a business strategy of steady organic growth in library customers. The company has stayed mostly out of the mergers and acquisitions arena, with its 1997 acquisition of UK-based SLS as the only exception.

Discovery Services

Innovative Interfaces developed Encore as its strategic discovery product, initially released in May 2006, offering the features expected, such as relevancy-ranked results, faceted navigation, enriched display of results, including cover images, tables of contents, and a simple interface with a single search box. Innovative has developed a proprietary search technology, called Right Results, which aims to retrieve results and sort them according to library-specific criteria.

Encore Synergy, launched in April 2010, brings article-level content into the discovery process through real-time connections to remote resources using Web services. Encore does not follow the index-based search model in the same way as seen in Summon, EBSCO Discovery Service, or Primo Central. In June 2012 Innovative formed a partnership with EBSCO to develop the capability to allow libraries using Encore to incorporate results from the EBSCO Discovery Service index.

Functional Design

Millennium, through one of the integrated library systems generally regarded as having the most sophisticated functionality, especially for technical services support, also had a reputation as a relatively closed and proprietary application. Sierra, in contrast, follows a service-oriented architecture, exposing a full set of documented RESTful APIs for programmatic access to the functionality and data within the system, and a SQL-compliant database.

Sierra, though offered as proprietary licensed software, makes use of many open source components. Innovative selected PostgreSQL as the primary database technology for the transactional database for Sierra instead of its own proprietary technology or Oracle which were the options for Millennium. Libraries will be able to access data within Sierra through SQL, either through APIs or through third party reporting products such as SAP Crystal Reports. Apache Lucene supports the search and retrieval operations of Sierra needed for staff functions and for interactions with discovery services.

In contrast to Millennium, which divided functionality into separate modules for circulation, cataloging, acquisitions, and serials, Sierra offers a unified, non-modular staff client with access to all the capabilities of the system, activated selectively according to the security profile of the staff member.

One of the key design elements of Sierra is a services layer that resides above the business logic that presents all the functionality of the system through services delivered though SOAP wrappers. Higher-level applications can then be composed from these services, including the Sierra Desktop App and the Sierra Web App. Another module at this layer will organize lower-level services into bundles of RESTful Web services or APIs that can be accessed by library programmers and third-party applications.

Unlike the other competitors in the library services platform arena, Innovative has not created knowledge bases as part of its product strategies for the management of electronic resources. Ex Libris, Serials Solutions, and OCLC created knowledge bases of e-content holdings in support of their OpenURL-based link resolvers. These three organizations also leverage their knowledge bases to support their electronic resource management products and are extending them further in support of their comprehensive resource management platforms. Innovative offers the Web-bridge link resolver but does not populate it with its own knowledge base content. Rather, customers can import data from 360 Link or SFX, or can license a product called CASE (Content Access Service) from Innovative which is based on data provided through OCLC.

Sierra is also not designed for deployment through multi-tenant software as a service, such as with WorldShare Management Services and Intota where all libraries using the product share the same instance. Rather, deployment options will be similar as for Millennium where the software can be installed locally in a library or consortium or as separate instances hosted by Innovative.

Development and Deployment Roadmap

By taking carrying forward business logic and coding, Innovative was able to provide all the existing functionality in Millennium and Electronic Resource Management in the initial release of Sierra. Consistent with the company's long-standing evolutionary process, the transition from Millennium makes some wholesale changes in technology underpinnings while preserving the steady advancement of features and functionality. Innovative will also continue to enhance and support Millennium for many years, as it did for IN-NOPAC once the transition to Millennium began.

Innovative began its marketing efforts for Sierra in 2011. According to data provided for the annual *Library Journal* "Automation Marketplace" report, the company signed 206 contracts for Sierra, representing about 700 libraries and 1,616 individual facilities, setting an industry sales record. Almost all were to libraries already using Millennium, reflecting a very positive response from within the Innovative customer base.

Implementations of Sierra began in April 2012, with Hillsdale College as the first library to place the software into production use.

Sierra has also attracted libraries from outside Innovative's existing customer base. The PrairieCat Consortium of 155 academic, public, and school libraries in Illinois placed Sierra into production in July 2012, migrating from Sirsi-Dynix Symphony.

Target Library Segments

Sierra targets the same types of libraries as Millennium, which includes a mix of academic, public, and special libraries. Innovative has not developed its strategic automation products for any given sector of libraries, but rather has worked to include the functionality needed by a very broad range of sizes and types. Early implementers of Sierra are accordingly diverse, ranging from law, academic, and public libraries as well as a large multi-type consortium.

EVOLUTIONARY PATH

We've described the main products that were developed anew in this new genre of library services platforms. These new products will likely gain adoption through the next phase of library automation, especially in those libraries heavily invested in electronic resources. We've talked about these new products taking a fairly dramatic departure from the course set by the incumbent generation of integrated library systems.

In parallel to the development of library services platforms, the existing integrated library systems will evolve in ways to extend their viability into the future. The traditional integrated library system will continue to flourish long into the future, especially in support of public libraries.

Not all types of libraries face the same challenges. While research and academic libraries, for example, face enormous pressures relative to management and access of their electronic and digital resources, many public libraries deal with a pattern of continued, or even increased, circulation of their physical materials, coupled with dramatic interest in e-book lending. The kind of technical infrastructure needed for these use patterns differs considerably from that of academic libraries in dealing with large bodies of electronic scholarly resources.

The patterns of activity seen in public libraries have proven amenable to the evolution of existing integrated library systems. While the issue of integrating e-book management and access stands as an enormous challenge for public library automation, the responses so far have been at the level of the discovery layer rather than a more complete reinvention of the core system.

The library automation arena has generally favored the evolutionary approach over the revolutionary. Many established products have survived for multiple decades and through multiple technology cycles:

- **Innovative Interfaces:** INNOVAQ, INNO-PAC, Millennium, Sierra
- **Sirsi Corporation / SirsiDynix:** Unicorn to Symphony. Original text-based terminal clients, InfoVIEW graphical client to WorkFlows client, originally developed in C and later in Java.
- **Spydus:** Offered by Civica, represents a long line of evolution from the Urica automation system originally developed in South Africa.

In the case of Innovative, its latest generation product steps into the realm of library services platforms while the products of SirsiDynix and Civica remain more within the realm of integrated library systems. All continue to be viable products for their target library customers.

RE-CONVERGENCE OF DISCOVERY WITH MANAGEMENT SERVICES

One of the trends in place since about 2005 has been the use of discovery products in addition to, or instead of, the online catalog module supplied with the integrated library system. Consistent with the reality that library collections extend beyond what is directly managed within the integrated library systems, these discovery products not only provide a more modern interface, but also address electronic resources and digital collections. Most of these discovery services are based on indexes that incorporate a broad representation of the body of electronic resources to which libraries subscribe, which can be profiled according to a given library's specific subscriptions.

These discovery services have been designed to work with a library's existing technology infrastructure, generally supporting all of the major integrated library systems. These systems do, however, have affinities and dependencies. Discovery services make use of the knowledge bases that also support OpenURL link resolution. While it is possible to operate an OpenURL link resolver from another vendor, it may involve an additional layer of overhead in synchronizing knowledge bases. There is a general trend toward libraries coalescing toward discovery services and link resolvers from the same providers.

In the next phase, the coalescence will extend to bundle library services platforms and discovery services. While it will remain possible to use products provided by different vendors, it will require the library to implement and maintain a process of integration inherent when acquiring them together as a suite from the same provider.

LOOKING FORWARD

This survey of the offerings available within this new genre of library services platforms reveals that within this slate of new-generation library services platforms, several distinctive options are available as libraries consider their automation strategies. While these products have some commonalities, each brings its own strategic vision, technical architecture, and level of completeness and functional maturity.

The products vary in the extent to which they embrace cloud technologies. OCLC's WorldShare Management Services and Serials Solutions Intota fully embody the multi-tenant software as a service model, with all libraries using the service sharing a common infrastructure, including bibliographic and e-content databases. OCLC starts from an advantage on the traditional bibliographic database side with its massive WorldCat database; Serials Solutions comes in with deeper resources in the management of electronic resources. Ex Libris' Alma also follows the multi-tenant software as a service model, though with a hybrid option for common and locally scoped data stores. Kuali OLE and Sierra do not press as much into cloud computing, with dedicated installations for each organization deploying the software.

Three of these products have been developed specifically for academic and research libraries: Alma, Intota, and Kuali OLE. Sierra and WorldShare Management Services were designed for a wider range of library types.

We also see a range of development and deployment timelines and corresponding risk factors. OCLC's WorldShare Management Services has been out in its initial production release the longest. Sierra likewise has seen its general release and has the largest number of libraries using it in production environments. Ex Libris has completed the initial version of Alma with one library now using it in production. Kuali OLE and Serials Solutions are on a later development schedule and currently do not have libraries using them in production.

These varying implementation schedules give some of the candidate products a short term advantage in the ability to demonstrate real-world use. All the systems fall toward the beginning of their development cycles. Academic libraries that wish to wait until the products reach full maturity may incur less risk, but staying with legacy tools during those years may have strategic implications on their ability to fully manage their collections of electronic resources.

While it is not possible to project how well these products will compete among each other going forward, it does seem clear that over the next phase of library automation in academic libraries, playing out over a period of 5-7 years, we will see a near complete turnover from the current legacy of integrated library systems to the new slate of library services platforms. The need for more appropriate functionality, especially related to the management of electronic resources, and for more open technical architectures will drive a cycle of migrations. Just as the transition from mainframe computing to the client/server area brought about the demise of one set of library automation systems in favor of new ones in an earlier phase, we can expect a similar changing of the guard as cloud computing, service-oriented architectures become the mainstream technical platforms and as comprehensive resource management becomes the standard expectation in academic libraries.

ACKNOWLEDGMENT

The majority of the information about the products and companies comes from the author's own research gathered through interviews and e-mail exchanges with company representatives and libraries using their products, from product literature, or from technical information on company Web sites. This ongoing research is also informed previously published articles and reports that may provide more in-depth information.

REFERENCES

Breeding, M. (2010, June). Access articles through encore synergy. *Smart Libraries Newsletter*, *30*(6), 2–4.

Breeding, M. (2011, August). The beginning of the end of the ILS in academic libraries. *Smart Libraries Newsletter, 31*(8).

Breeding, M. (2011, November). Progress in the new generation of library service platforms. *Smart Libraries Newsletter*, *31*(11), 3–4.

Breeding, M. (2011, January). Ex Libris marks progress in developing URM. *Smart Libraries Newsletter, 31*(1).

Breeding, M. (2011, December 5). OCLC World-Share platform: OCLC brands and strengthens its Webscale strategy. *Information Today*. Retrieved from http://newsbreaks.infotoday.com/NewsBreaks/OCLC-WorldShare-Platform-OCLC-Brands-and-Strengthens-its-Webscale-Strategy-79208.asp

Breeding, M. (2011, May). Innovative interfaces to launch Sierra: A new generation automation platform. *Smart Libraries Newsletter, 31*(5).

Breeding, M. (2011, August). Serials solutions to build web-scale management solution. *Smart Libraries Newsletter, 31*(8).

Breeding, M. (2012, April 1). Automation marketplace 2012: Agents of change. *Library Journal, 137*(5).

Breeding, M. (2012, March). Looking forward to the next generation of Discovery Services. *Computers in Libraries, 32*(2), 28–32.

Breeding, M. (2012, May 1). *Knowledge Base and Link Resolver Study: General Findings*. Prepared for the National Library of Sweden. Retrieved from http://www.kb.se/dokument/Knowledgebase_linkresolver_study.pdf

Breeding, M. (2012, September). The Orbis Cascade Alliance selects Alma for consolidated automation. *Smart Libraries Newsletter*, *32*(8), 6.

Breeding, M. (2012, August). Crossing the threshold: Boston College places Alma into production. *Smart Libraries Newsletter, 32*(8), 3–5.

Breeding, M. (2012, March). Ex Libris: Alma, Aleph, and Primo. *Smart Libraries Newsletter, 32*(3).

Breeding, M. (2012, June). OCLC navigates the realm of open linked data. *Smart Libraries Newsletter, 32*(6).

Breeding, M. (2012, March). OCLC Launches the WorldShare brand and wins an ARL member. *Smart Libraries Newsletter*, *32*(3), 6–7.

Breeding, M. (2012, April). Innovative interfaces joins the private equity club. *Smart Libraries Newsletter, 32*(4).

Chapter 3
Automation and Collection Management:
A Short History and Recent Trends

Annette Bailey
Virginia Tech, USA

Leslie O'Brien
Virginia Tech, USA

Edward Lener
Virginia Tech, USA

Connie Stovall
Virginia Tech, USA

ABSTRACT

The history of library automation can be traced to early printing methods of the 7th century A.D. The earliest collectors of books were usually religious scholars who amassed the religious texts of the day. Monks from East and West travelled great distances and often at great peril to gather meticulously hand-copied texts. Early inventions of woodblocks, and, later the printing press, enabled the mass-production of books that resulted in libraries' expansion into the secular world. Librarians have continued to bring technological advances into their work, combining web services, programming scripts, and commercial databases and software in innovative ways. The processes of selection, deselection, and assessment have been enhanced through these new products and services. The authors discuss a variety of technological applications for collection activities that have allowed collection managers to work more efficiently and better understand the use of their print and electronic collections. The effects of automation on the people involved in collection management are also explored.

INTRODUCTION

This chapter discusses the automation of collection management. Historically, the availability and diversification of materials along with new developments in technology have ushered in improvements in library collection maintenance. Libraries have successfully transitioned from tending small collections of manuscripts to managing the multi-format, multi-location, electronically available, and even occasionally preserved only for the purposes of being accessed at the point of catastrophe with our technological infrastructure collection. The decisions collection managers must make today have never been more complex.

DOI: 10.4018/978-1-4666-3938-6.ch003

The rise of the electronic resource has introduced new challenges for librarians working on collection management tasks with all formats of resources. These challenges are met with open source and vendor systems that can bring together data across sources. New metrics have been created to analyze the value of publications. Patron-driven acquisitions introduces the concept of "just in time" as opposed to the traditional "just in case" method of selection. The preservation of millions of print materials in digital format by entities such as HathiTrust can potentially change the answer to fundamental questions about our collections.

Even given everything that technology can do for assisting with collection management, the decisions that must be made daily regarding our libraries are still made by librarians. Conversations with faculty, input from our users, online chat interactions with students, and the specialized subject knowledge required of our profession are still an integral part of how the business of collection management is done in the 21st century.

This chapter discusses how advances in technology have changed the selection, deselection, and assessment of library materials. It describes existing technologies for each of these areas of collection management. Finally, it examines the continued need for human involvement in this traditional area of librarianship.

BACKGROUND

When it comes to library collections, practitioners refer to associated professional duties and methodologies as collection management or collection development. Hazen (1991) takes up the difference between collection management and collection development, stating that the newer, preferred term of collection management "subsumes collection development…but it also encompasses preservation" (p. 291). Collection development, on the other hand, consists of policy formation, selection, and acquisition. Roughly a decade later in a classic

library and information science textbook, Evans (2000) prefers the term collection development and postulates its six components, all of which operate in a "constant cycle": selection policies, selection, acquisition, deselection, evaluation, and community analysis (p. 16). More recently, Gregory (2011) says that collection development represents a "subpart of collection management that has primarily to do with decisions that will ultimately result in the acquisitions of materials" (p. xiv). Conversely, the umbrella term collection management focuses on "information gathering, communication, coordination, policy formation, evaluation, and planning that results in decisions about acquisitions, retention, and provision of access to information sources" (p. xiv).

All three authors referred to earlier explicitly mention how technology drives change, both within libraries and within the practice of collection management (Gregory, 2011; Evans, 2000; Hazen, 2011). In the *Springer Handbook of Automation*, Kaplan concludes that libraries are "witnessing the opening wedge in the dissolution of the ILS into a series of independent modules that communicate with each other by means of Web 2.0 services," and that the "future will be one of distinct functional modules that communicate with one another by exploiting the concept of unified resource management" (p. 1296). While a great deal of the literature focuses on the impact of automation on collection management since the 1960s until the present day, Broadus (1991) points out that "the spread of printing" was the catalyst behind collection developments' shift toward "selectivity" after centuries when the "major challenge to those in charge of libraries was to find things to collect" (p.5).

Collection assessment, powered by automation and standardization, figures prominently in the last five years of collection management literature, particularly for making decisions regarding selection and deselection of monographs, databases, and journal subscriptions. Wilde & Level (2011), in naming resources used to make these decisions, cite "statistics from link resolvers (Ex Libris SFX), the OCLC WorldCat Analysis® tool, Journal Citation Reports, and more" (p. 222). Additionally, their survey of collection managers revealed that the "most common statistics that are collected by libraries are circulation statistics (51.97%) and interlibrary loan statistics (45.67%)" (p. 223). To store and manipulate usage data, many have opted to create databases in-house (Carroll & Cummings, 2010; Metz & Cosgriff, 2000; Tucker, 2009; Harrington & Stovall, 2010) Types of calculations used to aid in journal selection and deselection include cost per use, cost per title, return on investment; other data consulted include interlibrary loan data, number of searches, citation analysis, and number of downloads.

Regarding deselection more specifically, a process by which resources are stored, discarded, or discontinued, Gregory (2011) speaks briefly on what he terms weeding or deselection, conveying that school and public librarians must pay the most attention to this activity, and that for any library, "the collection development policy lays down the general guidelines" (p. 50). Mosher (1991) prefers the term "pruning," which he says is driven by the "same factors as initial purchase" (p. 376). Evans (2000) quotes McGraw's criteria for deselection: duplicates, unsolicited and unwanted gifts, obsolete books (especially science), superseded editions, damaged resources, unused, unneeded volumes of sets, and periodicals with no indexes (p. 416.) Mosher (1991) points out that deselection "remains one of the most sensitive functions of the librarian" (p. 374). Similarly, Ward & Aagard (2008) underscore the sensitivity of the deselection in describing deselection as a necessary evil, or "the dark side" of collection management (p.273); all of the above authors emphasize the importance of people, alongside hard data, in deselection assessment.

While Gregory (1991) does not explicitly name selection as an essential component of collection management in his introductory definitions, he speaks at great length about selection as a part

of the overall process and deems it "truly at the heart of library profession" (p.1). Further, selection should be based on, among other things, "anticipated users' needs" (p.61), and he provides a lengthy list of review resources to aid in selection. However, looking through centuries of library history regarding selection criteria, it becomes clear that who drives selection of library resources has vacillated through years (Leedham-Green & Webber, 2006; Drueke, 2001; Lerner, 1999). Currently, collections management places a great deal of emphasis on patron-driven acquisitions, also sometimes referred to as demand-driven acquisitions, or purchase on demand, and signifies yet another shift in who selects library materials. Enabling this shift of selection to the patron is progressively advancing technology, specifically the "popularization of e-book platforms" (Shen, p. 204). Further, as financial resources continue to shrink and as collection managers continue with data-driven collection assessment and learn of weak circulation statistics, many believe "just in time" selection rather than "just in case" selection to be a wiser use of funding (De Fino & Lo, 2011; Jones, 2011; Lugg, 2011).

Early Automation Impact on Collection Management

The radical transformation of automation factors significantly in the long history of collection management practices. Most recently, patron-driven acquisitions holds the potential to disrupt the current model of collection management, although we have yet to witness long term consequences or benefits. Given a closer look at this history, it is reasonable to argue that in some respects demand-driven acquisitions signals just another shift in who is driving book selection. It has, after all, fluctuated throughout the years. What is different about this newest model of demand-driven collection building rests squarely on the instantaneous, seamless acquisition of resources made possible by more technologically advanced automation.

Further, it could be argued that automation made its first impact on library collection management with the advent of early forms of printing presses. To strengthen this claim, the nature of library collection management practices or models prior to the invention requires some history.

The Impact of the Printing Press on Collection Management

Some confusion exists over whom first put to use automated printing—Gutenberg's invention circa 1455 or the Chinese roughly 1200 years earlier. While we will not confront that argument here, it is safe to say that the beginning of automated printing changed how and why libraries collected specific texts. Centuries prior to Gutenberg's invention, for example, Broadus (1991) says that 'librarie keepers' found themselves hard-pressed to obtain texts to add to collections, and acquisitions of materials even involved dangerous travel at times (p.8). Most of these librarie keepers were persons of religious distinction who built collections of papyri, parchment, tablets, woodblock, plates, seals, and stamps, with most selected primarily for the purpose of religious education. It should be noted, too, that these librarie keepers contributed to and used the collections they created, in addition to selecting any reading materials which were not received by gift or through inheritance—the predominant means of acquisitions of the day.

Buddhist library keepers managed collections similarly. According to Ranasinghe (2008), in the 1st century B.C., well after Buddhism had been introduced into Sri Lanka, Buddhist monks operated libraries, called Mahavihara centers, where the Buddhist manuscripts, or Tripitaka, and commentaries were written and kept. For centuries, those in Sri Lanka depended upon visiting missionaries for oral storytelling, but after great famine, political upheaval, and realignment of regional clans, the missionaries stopped traveling. At that time, Buddhist monks committed to the preservation and collection of the Tripitaka and

commentaries by putting the teachings in writing. Monks deposited the originals, thought to be gold plates, in the temple rock called Aloka Lena, but not before making copies on palm leaves. Those copies were then distributed to the Mahavihara and Abhayagiri centers, where monks collected the texts both because they were considered sacred but also because copying the scriptures was considered meritorious. Though copies of Buddhist charms were created first using a process where stamps were used to impress images on paper, McMurtie (1967) argues that it was not until about mid 7th century A.D. that the same type of technology was used but with larger wooden blocks and stone steles to impress copies of text and images. These copies made their ways to new temples, all of which eventually served, too, as educational centers, and monks operating the libraries within the centers throughout the East Asia began collecting secular works as well. Further, Lerner (1998) asserts that by the 10th century in China, printers using woodblock "could produce well over a thousand copies in a day...thus it was possible to distribute a 130-volume edition of the Classics widely across China" (p. 59).

New technology enabling greater mass production and larger collections of texts saw a second surge in growth in China and Korea in the 10th and 11th centuries A.D., when movable type first came into existence—at least in the East Asia. Western European monastic libraries of the same time period generally consisted of small, manually copied manuscript collections, most of which included Christian scriptures and corresponding commentary. Due to centuries of warring and raiding, Michael Harris (1995) says:

The 10th and 11th centuries marked another low period in the development of libraries and literature in Western Europe...there was a decline in interest in the classics and many monks were even illiterate and allowed the books in the libraries to go unattended. (p. 97)

Libraries managing to survive through this period still depended on manually copied texts, with Rome as the main source of original texts.

Drueke (2001), author of "St. Osmund's New Legacy," maintains that monks like St. Osmund who established the library at the cathedral of Salisbury operated the libraries and scriptoria, and acquired religious and scholarly texts, mainly for the purpose of "establish[ing] the Norman Church and thus Norman rule in England" (p. 508). Acquisitioning texts, in stark contrast to today's heavily automated process, often involved receiving gifts of or inheriting manuscripts, leaving collections quite uneven rather than comprehensive. Using lists of books available in other libraries throughout Western Europe and establishing connections with those holding other sought-after manuscripts, monks often set out on dangerous roads to meet with others to obtain an exemplar text—when someone was willing to lend one in exchange for another equally valuable item. In essence, the labor-intensive process of acquisitions and collection management in medieval Western Europe included locating scarce, suitable exemplar texts, and manual copying of the texts to add to the collection. Given such a laborious process, it seems unlikely that monks chose to deselect texts from their collections, although Sharpe (2006) suggests that some evidence does exist to convey how hundreds of books were removed from one library to make room for new materials (p. 238).

Gutenberg's Impact on Collection Management

As Gutenberg's invention of movable type in 1455 decreased the labor involved in producing books, at the same time it increased not only the number of texts available and general readership, but it also helped to increase the number of and types of libraries, particularly university libraries and national libraries. Collections diversified, and the main drivers behind book selection saw a shift as well, although most scholars would be quick to point out that dramatic change did not occur overnight but in the course of a few decades. It would be close to another 500 years before automation impacts library operations so significantly with the use of machine readable text. (For clarification, see Table 1 for a timeline regarding automation and other events affecting collection management practices.)

While new automation beyond Gutenberg's printing press appears largely absent for those 500 years, it cannot be said that automation and advances in that automation, did not continue to impact libraries and their collection and acquisition processes. Instead, it could be said that Gutenberg's printing press instigated the slow birth of the modern library during those centuries. Beforehand, virtually all libraries were small and few, Alexandria's great library notwithstanding. With the exception of small private libraries, most collections consisted of 100 or fewer religious manuscripts, perhaps supplemented with some educational material, normally grammar, Greek and Latin literature, and basic mathematics and sciences. Paternalistic in their roles as manuscript selectors, monks or friars in charge paid little attention to patron demand when it came to selecting exemplar texts. By and large, these libraries did not exist to provide leisure or entertainment; instead, most of the libraries existed as storage places for the manuscripts and as devotional areas for spiritual tasks, chiefly copying religious texts, and, perhaps, writing commentaries on those texts.

Automation Trends Continue to Shape Collection Management

As Harris (1995) points out in *History of Libraries in the Western World,* subscription, circulating, and mechanical libraries (or trade libraries) represented new models, as a result of increased book availability, and the new models arrived with new collection purposes and acquisitions processes. Circulating libraries—small businesses more or less run by booksellers—allowed the

Table 1. Automation Technologies and Correlating Events Affecting Library Collections Management

Period	Event
1st century BC	Rise of Buddhist centers with libraries
Late Classical to Early Medieval	Growth in Christian monasteries and their libraries
7th century AD	Mass production of Buddhist charms by stamping
11th century AD	Wood block printing with movable type in Asia
12th century AD	Advent of the European university
14th century	Birth of the Renaissance
15th century AD	Gutenberg' printing press with movable, alphabetical type
16th century AD	Reformation
	Dissolution of monasteries under rule of Henry VIII
18th century AD	Age of Enlightenment
	Beginning of the Industrial Age
19th century AD	Birth of Romanticism
1877	Standardization of catalogue by Dewey
20th century AD	
1930s	First punched cards by libraries
1940s	Blanket orders by libraries
1950s	Production of mainframe computers
	Libraries first utilize computers
	Most universities turn over book selection to librarians
1960s	Library automation associated with computer
	Early OPACs
	Use of MARC
1970s	Personal computers and ARPANET invented
	Rise of the integrated library system
	Digitization of bibliographic and citation indexes
	Dawn of Information Age and Post-Modern period
1990s	Tim Berners Lee invents WWW
	Shelf-ready books
21st century AD	Robotic retrieval for collections
	Advent of self-checkout machines in libraries
2009	Widespread use of dedicated e-book reader devices
2005-2012	Implementation of patron-driven acquisitions

general public to pay a fee to rent books, most of which were popular reading items, particularly the newly popular genre—the novel (p.152). The changes created by automated printing created at least three new concepts here: the prominence of the bookseller, the advent of popular reading items, and book lending for a fee. Perhaps two others are not as obvious: an increased readership among the masses and greater choice in reading material. Library collections changed accordingly. Subscription libraries found success with similar concepts, yet instituting fees that gave one a membership to a group, somewhat resembling book clubs, and while reading choices were also greater in number as well, most books spoke to high-browed readers. Conversely, Harris (1995) states that mechanical libraries built collections upon the need for cheaper, practical trade-driven books, and by the end of the 19th century, 700 mechanical libraries thrived on the British Isles. Of the subscription libraries, the London library and the Leeds Library still exist today (p.153).

In addition to shifts in types of subjects and genres collected, means of selection and acquisition changed (as did space and services). Merchants operating these libraries did not rely on gifts as much as monastic libraries; instead, as Forster & Bell claim (2006) it was not uncommon for circulating libraries, for example, to coexist with another shop of various descriptions. Fees and profits from the business typically aided acquisition of new and used books, and, ultimately this helped with creating a much more varied collection. Perhaps the most dramatic divergence from traditional monastic libraries, however, occurred in the actual selection of books; the paternalistic selection methods of the monastics gave way to a more democratic process: subscription members, for instance, voted for members of a selection committee who chose reading material for the members (p. 148). Even as these libraries grew larger and the need to employ a full-time librarian evidenced itself, "members were anxious to keep control of book selection" (Forster & Bell, p.152).

Compared to the relative speed with which subscription, circulating, and mechanical libraries grew and diversified as a consequence of automated printing, university and college library collections meandered at a snail's pace until the Enlightenment. Just like monastic libraries dissolved in the 1540s, college library collections holding similar texts suffered as well. Jensen (2006) asserts that in 1535, Henry the VIII ordered the removal of "a number of scholastic texts—Duns Scotus, Walter Burley, Antonius Trombetta, Thomas Brico and Stephanus Brulefer—...from the Cambridge curriculum" in his quest to drive religious reform (p. 346). In addition to the impact of religious reform, for quite some time scholars held fast to orthodox values and teaching methods, and as a result there was little change to what was considered the canon. Eventually, however, the proliferation of new texts, cheaper prices of those texts, and the commodification of current scientific findings and debate in print became hard to ignore by universities and colleges.

Like most other libraries, university and college libraries depended almost entirely on donations, and such dependency left these collections, too, out of date and spotty. Even while these institutional libraries still relied on donations for some time after the advent of Gutenberg's printing press, the content and scope of those donation collections began to change and show signs of the increase in print. Jensen (2006) points out that those owning private libraries increased dramatically. Starting in the late 16th century, as printed copies of texts became more affordable and portable, scholars passed down their collections to colleges or universities to which they were affiliated, collections began to change shape both in the type of books acquired but also in dramatically increasing numbers. Keep in mind that books held in monastic and university library collections of the late middle ages were too large to carry room to room, much less outdoors, and books were most often chained to the lectern, forcing scholars to engage with the text in its permanent place. Because of these

changes in text availability and because universities finally recognized the need to change curricula to keep scholars abreast of current findings, library keepers, often faculty members who had another position in the university, implemented new and deliberate acquisition practices, like asking donors for fees rather than or in addition to the usual book donations. To acquire new texts for university and college libraries, library keepers worked directly with publishers, worked with scholars to create shared collections, and evaluated bibliographies and catalogue lists of the best libraries. Lists from publishers were also consulted, and it became possible to import texts in a wide variety of languages; periodicals, a new type of publication containing current scholarship and commentary, provided book reviews for librarians and scholars alike.

As texts and libraries multiplied and as collections grew and diversified, the need for more employees working in libraries to manage resources became apparent—in other words, as text proliferated and labor costs associated with retention and management increased. Local catalogues needed to be created and maintained. New closed stacks, created as a result of smaller books, needed a person to retrieve books for patrons, and, of course, fees and circulation required recording and managing. As employment grew within libraries and work became more complex, especially in the late 19th century, faculty opted to hire a person dedicated to the library rather than manage the library themselves, though the position required little or no training. Despite hiring a librarian, it was still quite some time before the decisions regarding book selection was turned over to librarians. In fact, Broadus (1991) points out that "librarians actually considered such responsibility dangerous" (p. 8). For the most part, libraries provided faculty with funds and review resources to select books, and librarians typically facilitated acquisition and maintenance thereafter.

As precarious as selection and collection building might have seemed for librarians in the 19th century, by the 1950s a great deal of universities and their faculty handed over that very responsibility to librarians. Why the change? One can speculate that in the post-WWII world, academia saw a combination of surging technological advances, an economic boom, growth in universities, and, in the United States, at the very least, because of the G.I. Bill, a large increase in student population; conceivably, faculty found themselves spending more time in the classroom and in the labs, with little or no time to select books. Perhaps, too, it is no coincidence that 20 years after Ranganathan's influential 1931 work, *Five Laws of Library Science,* librarianship as a profession found itself maturing. Regardless of the precise impetus, we can assert that from the 1950s onward, librarians and the means by which they managed collections saw—and continue to see—dramatic change. Of all automation technologies affecting collection management and acquisitions in the 20th century, perhaps only the radical change caused by the rise of the personal computer coupled with the invention of the world wide web compares to the impact of movable type.

RECENT AUTOMATION TRENDS IN COLLECTION MANAGEMENT

One area where recent computing technologies and the use of web have impacted collection management the most centers on data collection and assessment, the latter of which collection managers perform to make important selection and deselection decisions. The use of quantitative and qualitative metrics of collection assessment is nothing new in collection management, but the ease with which data is collected, manipulated, and managed has certainly improved, and it has enhanced the decision making processes compared to the early days of manual recording of data. Today, collection managers use these data for a wide variety of purposes. For instance, selection of new materials might be based on data from patron recommendations, as well as data from cir-

culation patterns, citation lists, and core literature lists. Using these data along with assessing the physical condition of materials, the duplication of content, and outdated or superseding content helps facilitate making decisions during the deselection process. Before the use of computers and the web, performing collections and acquisitions functions and compiling data about collections necessary for informed decision-making consisted of tedious and costly labor-intensive processes. With the advent of online catalogs and circulation systems, desktop computing and office software products, gathering collections data has been automated. In the 21st century, a myriad of tools and programs have emerged to gather, manipulate, and analyze the data needed to acquire, maintain, and deaccession library materials. As Franklin (2005) states, "(a)bout a decade into the digital information environment, we already know considerably more about digital content use than we ever did about print journal and book use." (p. 242)

Selection

The process of selecting library materials and the expectations of skills necessary for librarians working in collection management has changed with technological advances. A review of position descriptions for selection managers in the mid-1990s describes how subject specialists typically selected new titles for their collections (Rowley, 1995). Librarians reviewed vendor selection slips, Library of Congress proof slips, book reviews, and bibliographies in scholarly journals, and printed publisher catalogs. While approval and standing order plans relieved librarians of some aspects of this labor-intensive process, much selection was, until recently, still being done in this manner. For decades before the 1990s, the process was painfully slower. Those responsible for book selection tracked down books to purchase, title by title, either through publisher catalogs received in the mail or from libraries with comprehensive collections, reviews in journals, and patron recommendations.

Considering the time lapse between a monograph's publication date and the time necessary to then send one by mail to a reviewer (who then had to manually write or type the review and then send the review to reviewing journal to be published), it now seems unfathomable that this represents just one segment of an achingly slow process. Generally speaking, any titles considered for purchase had to be verified, which usually involved searching a printed copy of the National Union Catalog. After choosing a book to order, librarians typed out the orders in carbon copy triplicate or quadruplicate and sent them through the mail to be processed and shipped. Print files were maintained for these order forms.

The selection process was revolutionized in the latter half of the 1990s by the introduction of vendor websites and electronic notification slips that can be emailed or viewed within a vendor selection database. In the 21st century, subject selectors are as likely to use blogs, RSS feeds, and social media to guide their new title selections. Libraries also use social media such as Facebook to facilitate patron suggestions for new purchases (Petit, 2012).

The introduction of approval plans by bookjobbers was another step forward towards automating the selection of library materials. Warzala believes that "book approval plans achieved their predominant character in the early 1960s" by predicting the "book purchasing habits of academic libraries" (p. 514). More specifically, rather than spend all the time necessary to select books title by title, librarians and bookjobbers worked together to create a profile specific to institutional needs. This profile, based on criteria like types of degrees awarded in specific disciplines, LC classification, book award winners, and best-seller lists, drove the automatic, routine shipment of books to the institution. To supplement the collections built on the institutional profile, "firm order" funds were used to purchase items title by title, when necessary.

Web services and the growing opportunities to learn programming skills among non-systems librarians have created the potential to automate many more routine selection tasks. In 2004, Cornell University Library implemented an automated program to replace the manual book selection process. Subject specialists used printed cards from the Library of Congress Alerts Service to make their selections, marking them with fund information and other notes before sending them to the ordering unit. When the Library of Congress replaced the printed service with using ftp to transmit MARC records, Scott Wicks, Head of Acquisitions, Bibliographic Control and Documents for the library saw an opportunity to automate the process, so he worked with collection managers and library programmers to design a streamlined selection process. Wicks envisioned using the MARC record data throughout the awareness, selection, ordering and cataloging processes to save manual input and rekeying data. The resulting program, called Integrated Tool for Selection and Ordering at Cornell University Library (ITSO CUL) was written with PERL, C, and PHP. In 2006, Cornell partnered with OCLC to distribute ITSO CUL as WorldCat Selection™.

JTacq is another purchasing application designed to streamline the selection and ordering workflow. The system imports lists of titles for potential selection from text files, MARC files, OCLC WorldCat®, and citation software such as Zotero into an interface for acquisitions staff or subject selectors. Like ITSO CUL, JTacq retrieves pricing and availability from multiple vendors, as well as any local library holdings. Library staff can compare vendor data across a wide selection, apply order data and submit orders in batch mode. MARC records are retrieved via the Z39.50 search protocol, and the system sends email notification to the requestor. Both JTacq and ITSO CUL are compatible with most major integrated library systems.

Using interlibrary loan (ILL) request data presents yet another way to drive new book selection, although compiling the necessary data has not always been easily captured or transmitted.

ILL systems are rarely connected to a vendor's ordering database, and sometimes they are not even integrated with the library catalog or acquisitions system. In response to this lack of interoperability, libraries developed systems that seamlessly pass patron-supplied bibliographic data through to the selector or acquisitions staff. Atlas Systems, Inc., a software development firm, developed "addons" for its ILLiad™ system that search vendor databases, like Amazon for requests initiated from an ILL form. University of Delaware Library enabled the ILLiad™ addons so that patron requests matching specific criteria for purchase are processed through a custom queue. The addons work by searching the local catalog, OCLC WorldCat®, Amazon and other book vendors to locate the requested title. If available for purchase, the library buys a copy rather than borrowing it from another library (Gaffney, 2011).

Similarly, the Getting It System Toolkit (GIST) is free, open source software that channels ILL requests into collection purchases. Using GIST, the library can set parameters within their ILLiad™ system for price, publication date, academic level, and more. When a patron request satisfies one or more of these parameters, GIST sends patron requests from ILLiad™ directly to the selected book vendor. The search returns price comparisons and availability information, which is then passed through a message queue to ordering staff. The goal of the system concentrates on facilitating demand-driven selection of items, at the same time as streamlining the acquisitions process. GIST was developed in the library at State University of New York, Geneseo with support from Atlas Systems, OCLC®, the IDS Project, and the Rochester Regional Library Council (Pitcher, 2011).

In addition to using ILLiad™ patron data to drive collection purchases, the rapid adoption of e-readers in combination with increased e-book production touched off another patron-driven purchasing model: demand-driven acquisitions (DDA), also referred to as patron-driven acquisitions (PDA).

Patron-driven acquisitions signify an approach to the selection of library materials that enables patrons to directly make purchasing decisions. While the librarian still provides some professional intermediation by determining which titles are made available through the PDA service, patrons have never had such a direct impact on the selection and purchasing of library materials. A confluence of technological innovations had to be developed and widely adopted by both librarians and patrons to make the development of PDA possible. Even though the search for literature on PDA reveals that 89% of the results were published during the years 2010-2012, many would argue that the concept of PDA is not new, that through patron suggestions librarians have been purchasing on demand for decades. As stated earlier, however, it is the seamless automation in which these patron-driven purchases are conducted that make this model new.

For many libraries ready to engage in PDA, the first step is to work with their bookjobber to create a profile for potential demand-driven purchases. This profile is typically designed with criteria similar to an approval plan. Once a profile has been created, each week the bookjobber will automatically FTP to the library short MARC records with a URL in the 856 field linking to the e-book aggregator's platform. The library will then upload the batch of records into the catalog, making them accessible to users through an OPAC or discovery layer so the e-book are discoverable by patrons as they perform routine information seeking tasks. That a title is part of the PDA plan is typically not made apparent to users. However, for management purposes, one or more local notes are created in the MARC record to track these titles.

From this point, what happens in terms of purchasing once the patron discovers a desired e-book depends upon parameters agreed upon between collection managers and the bookjobber. Prior to starting PDA, both parties design a specific plan where the library must choose which e-book provider is preferred. In addition to choosing an e-book provider, the library must also stipulate how many short term loan (STL) uses by a patron will trigger an outright purchase of an e-book. STLs consist of e-book downloads, wherein a patron looks through it briefly, perhaps four to five or up to 10 minutes, but moves on to another citation. The cost of the STL generally represents about 10% to 20% of the total e-book purchase price. In many cases, libraries have the option to bypass short-term loan purchases; instead, each time a patron downloads a title, the bookjobber charges the library for the e-book. At the point of purchase, the bookjobber automatically sends invoices and cataloging records; in some cases, paying an additional fee will entitle access to enrichment data for better cataloging records.

There are several challenges to the ongoing maintenance of PDA for libraries. The discovery of duplication with already purchased e-books is complex but necessary as budgets tighten. Libraries are scripting solutions to compare lists of titles from the library catalog with lists from vendors (Stahl, Correa-Morris, & Hwang, 2012). This matching can be accurate for many, but not for all titles. Another challenge rests in the discovery and removal of PDA records from the library system. A variety of approaches are being used, including the addition of notes or the creation of series in the MARC records.

Assessment

The responsibility for evaluating print collection usage typically falls to collection managers, but assessment methods vary greatly. As stated above, collection managers have used a variety of qualitative and quantitative metrics to evaluate print collections, some of which include comparing holdings to peer institutions and to core bibliographies (Rowley, 1995). Conspectus, an early example of an automated assessment tool, enabled libraries to create and maintain a local collection assessment database, and was widely adopted by

research libraries. It was originally available from the RLIN interface in the early 1980s and later became a PC-based tool before it was discontinued by OCLC® in 1997. Other commercially produced automated tools include OCLC WorldCat Collection Analysis™, Ulrich's™ Serials Analysis Service, and Serials Solutions® Overlap Analysis, among others. These tools help with evaluating the library's collection against quality or quantity standards. For example, WorldCat Collection Analysis™ provides benchmarking against peer library holdings and local holdings. Through that comparison, the analysis report reveals areas where titles should be acquired to strengthen holdings to that of a peer library, or to build support for a new program area.

Thanks to the development of the NISO COUNTER Code of Practice, the SUSHI protocol, and third party statistics gathering tools, analysis of electronic resources usage has exploded in a relatively short timeframe. Since the first release of the COUNTER Code of Practice for journals and databases in 2003, dozens, if not hundreds, of libraries have documented efforts to develop a "decision database," or matrix for deciding which e-resources to cancel or renew. COUNTER provides simple, standardized usage data that collection managers can build upon by adding a cost per use metric. Some electronic resource management systems (ERMS) provide a means of incorporating COUNTER reports and matching a title's use with what the library paid for the subscription. If a library does not have an ERMS due to cost or other restrictions, assessment requires more tools to pull together data from disparate sources. Acquisitions data, vendor knowledge bases, catalog holdings, consortia or big deal titles lists must be aggregated and stored in a database using software like Microsoft Access.

Once the data has been harvested, the values must be normalized and match points identified in order to efficiently create the database. If a publisher uses a print ISSN while the library uses an E-ISSN to describe the same title, matching can be difficult. Harrington and Stovall describe the use of the ISSN-L or Linking ISSN table to match Virginia Tech's library catalog holdings with COUNTER reports to correctly link a title with its usage data (Harrington & Stovall, 2011). Similarly, the x-ISSN tool helps librarians track journal publishing history. The need for such services as ISSN-L helps to illustrate the complexities of managing metadata for electronic resources. Assessing electronic resources will be greatly facilitated by the development and implementation of programming tools that enable metadata linking between systems and knowledge bases.

Alongside quantitative data, sound assessment practices require additional qualitative metrics. Libraries are using best practices to incorporate data such as impact factors in their analysis of their electronic resources. To acquire additional qualitative data, librarians consult Thomson Reuter's Journal Impact Factor and Local Journal Utilization Report or Eigenfactor™. Franklin (2005) discusses a variety of web-based assessment tools that can be used to pull other data into the assessment process. He and Plum developed MINES for Libraries™, an online survey tool that gathers demographic data on patrons' research habits. Specifically, it captures information on where and how patrons are using the libraries resources. This data, combined with COUNTER usage data and cost, can help libraries assess how resources are marketed, and inform collection decisions. Many libraries have incorporated Google analytics into their assessment toolbox to track users as they move through the catalog, databases, and web pages. New emerging indices that may also be included in future assessment of collections consists of the most recent release of Google Scholar Metrics. Google Scholar Metrics use a five-year h-index and h-median metrics. While it is limited in many ways, it presents collection managers with many approaches using data to generate metrics that can be used for assessing the "scholarliness" of a publication.

Deselection

In truth, librarians never completely understood how print collections were used. (Franklin, 2008, p.4).

The flip side of collection building is deselection, or withdrawing items from the library's collection. Because deselection is not always met with enthusiasm, Ward and Aagard (2008) note that librarians view deselection as the "dark side" of collection management. While collection managers craft and adhere to policies regarding the retention, storage, or withdrawal of materials these policies vary by library, given that each organization has unique circumstances, philosophies, or policies governing deselection. Regardless of the criteria used for deselection, all libraries are faced with the challenge of how to apply useful technology to what is typically a highly labor-intensive process. In his seminal work for libraries on weeding, Slote (1997) recommended programming the library's computer system to use the "Slote Method" of weeding (p. 35). During this period of second generation library systems, he envisioned:

...having the computer perform all of the functions required, including recording and summarizing the shelf-time periods at the circulation stations, recording the date of the in-library use of each volume, recording and summarizing the shelf time periods experienced by the books not currently in circulation, predetermining cut-points, and identifying weeding candidates. (Slote, p.35)

As the ILS becomes more flexible and better tools emerge to manipulate the data from those systems, librarians are better positioned to make well-informed decisions about their physical collections.

As academic libraries have largely embraced subscribing to electronic journals, a major focus for collection managers at academic libraries in the

21st century has been the archiving of print journal collections. Whereas academic library rankings and stature have been predicated on the number of serial subscriptions and print holdings, libraries are increasingly relying on electronic journal archives to replace those bound volumes. Regional print repositories and electronic "dark" archives abound. Libraries must pull together data about their journal holdings from a variety of sources in order to decide what to store or withdraw. Serials Solutions® and other "knowledge base" services have an overlap analysis feature to compare journal coverage at the title and volume level.

To assist libraries with weeding print journals, Ithaka S+R has developed a decision support tool. The tool uses the arguments in the "What to Withdraw" report as a basis for analyzing local library holdings against large print journal repositories. The interactive spreadsheet is used by many JSTOR member libraries to make responsible decisions about which printed journals can be safely deaccessioned. The tool uses simple filtering in a pre-populated Excel spreadsheet to arrive at recommendations. Local or regional consortia members use the tool to determine what journal titles to preserve in a shared repository. The Print Archive Preservation Registry (PAPR) from Center for Research Libraries is one of the newest online databases of journal repositories. It provides an interactive tool that displays which regional or state repositories are housing specific journal titles. Databases such as these help collection managers feel secure in their deselection decisions, knowing that print preservation is being addressed by the library community.

GIST (Getting It System Toolkit) Deselection Manager, the companion software to the acquisitions module from SUNY Geneseo mentioned earlier, uses similar features of the acquisitions module to aid libraries in deselection. The library loads a file of ISBNs from the library catalog into the software, for example, a list of books with no circulation counts. Using a series of APIs, the program searches WorldCat®, Google Books,

and HathiTrust holdings, then checks the price of the item and availability in Amazon. GIST lets the library know if the discarded item can be sold to a second hand book vendor. There are options for batch processing, even batch deletion of WorldCat® holdings.

Automated Retrieval Systems & Deselection

Automated Retrieval Systems (ARS) offer unique benefits to academic libraries. However, there are emerging complexities to performing traditional collection management tasks using the ARS. The efficiency of retrieving items from the system can be hampered by the system also being used for other functions such as circulation. Staff time is dedicated to placing the materials on a shelf for further review. Multi-volume sets are difficult to evaluate as they cannot simply be pulled from a single or set of bins in the ARS. Decisions of what to evaluate rely on the quality and quantity of information in the catalog record. The ARS can be a barrier to ongoing weeding and maintenance of the library's print collection.

PEOPLE IN THE PROFESSION

Nothing will ever be attempted if all possible objections must first be overcome – Samuel Johnson

To this point, much of the discussion involving both human resources and automation has centered on labor costs. However, the increasing emphasis on automating collection management does not negate the continuing importance of people in the process. For many, however, the nature of their roles will change, sometimes significantly. As certain functions become more centralized, bibliographers will have time to devote to other needs. These can include higher-level tasks relating to the collection or may involve other aspects of their job. Such changes in work assignments and responsibilities can be stressful, thus good communication is critical to a successful transition.

Selectors have built a substantial fraction of many academic library collections one book at a time. They have searched publisher catalogs, read book reviews, checked lists of core titles and award winners, and tried to keep up with the changing needs and research interests of faculty in their assigned disciplines. As Horava (2010) notes, "Traditionally, pride and prestige were imbued in the hundreds of daily actions of building a permanent collection that would serve our community's present and future needs" (p.143). As a result, subject bibliographers can develop strong attachments to their respective collections. Especially in those disciplines where the bibliographer has particular expertise or a strong interest, they may be reluctant to cede authority over collection-related decisions affecting their areas.

For many still in the profession today, this traditional model of collection management, with subject bibliographers carefully weighing decisions about making individual items may have been the only approach they were taught, with little consideration of other methods. But while it still has its place, this traditional model has become increasingly untenable in many libraries. Facing challenges such as a shrinking workforce and increased user expectations, librarians who have collection management responsibilities may be asked to cover more disciplines or to take on more active roles with instruction and outreach. Library and information science programs are adapting their curriculum to reflect some of these new realities but even those who teach collection development course may struggle to find the right approach in a rapidly changing environment (Holley, 2012).

On the selection side, approval plans were among the first steps in many libraries towards implementing a more centralized approach to collection development. By setting parameters for publisher, format, material type and the like, an approval plan offers a way to streamline the decision making process considerably. In addition to saving time, libraries can negotiate favorable pricing terms. Given their benefits, they have become an integral part of the library landscape and many institutions have closely integrated an approval plan into their materials workflow as a tool to help build the collection. To remain effective, such plans need to evolve over time to match changing needs rather than be used in a "set it and forget it" manner. It is important for libraries to assess and monitor their approval plans and the use of materials acquired in this way to ensure that they spend their money wisely. An in-depth assessment of the approval plan at Penn State found that 31% of titles had not circulated during the study period while a similar analysis at UIUC found a 40% non-circulation rate (Alan, Chrzastowski, German, & Wiley, 2010).

Whether it is collection managers, subject bibliographers, or support staff, those involved with an approval plan should have an understanding of how it works. In particular, the criteria used to determine what qualifies for shipment in print or digital form along with the rationale behind those decisions. Close monitoring is especially important at the outset of a new approval plan or when any significant changes are made. A good customer service representative can also help suggest ways to fine tune a plan to meet local needs. They can also work with collection managers to run reports that allow exploration of hypothetical "what if?" scenarios. For bibliographers, continuing to review receipts of titles and slips from approval plans on a regular basis still provides a way to stay involved in the selection process. Of course, some bibliographers are far more scrupulous in reviewing their titles than others, so central oversight is important.

When it comes to automating the deselection or storage of materials there can be more resistance among librarians. This can range from fairly minor concerns to outright opposition. Among certain individuals there can be an unwillingness to consider the entire idea of weeding the collection in the first place. But even when there is agreement that something needs to be done, there may be differing opinions as to the best approach to reducing onsite collections. As with an approval plan, one key factor in building acceptance is that those involved should have a role in shaping the criteria and parameters that will be used to identify titles. In both cases the underlying goal is the same—to build a pool of titles based on distinguishing characteristics that set them apart in a systematic way. There can be significant differences across disciplines in how the corresponding collections are used and subject specialists are often the most attuned to those variations. Just as journal impact factors should not be compared across widely divergent subjects, so too it can be hard to implement a single set of deselection criteria across the board. What is low use in one field might be commonplace in another. A monograph on mathematics may have a long useful life while one in a fast-changing discipline like genomics may be out of date almost as soon as published. Being sensitive to these differences and incorporating input from bibliographers can help determine the most appropriate criteria to use.

The library's mission statement, collection management policies, and strategic plan can all help provide guidance and direction. Another means of helping to build consensus is to focus on the desired outcome. If a library reduced on-site collections, how will it then utilize that space? Consider also the broader institutional needs, policies, and politics. The days of collection size being the primary metric in judging the value and standing of a library have passed. While the sheer physical presence of row upon row of stacks may still represent to some a powerful symbol of scholarship, campus administrators today are more likely to focus on outcomes such as improvements in student learning.

This focus on external needs is also important for communications outside the library. While a small-scale deselection project may go unno-

ticed, anything big enough to make a difference will likely generate questions among faculty and students. Having comprehensive, up-to-date collection policies in place can help address such concerns when they arise (Demas & Miller, 2012). The periodic surveys of academic faculty across the United States conducted by Ithaka S+R have shown a continuing drop in the percentages who believe that "regardless of how reliable and safe electronic collections of journals are, it will always be crucial for my library to maintain hard copy collections of journals" (Schonfeld & Housewright, p.19). Being able to show what similar steps other libraries at peer institutions have already done can also help.

Coordinating the work of multiple librarians on a deselection project can prove difficult and there is a need for clear guidelines and criteria. While exact percentages of material withdrawn will vary from one area to another, there must be a perception of fairness across different subject areas if the process is to succeed. Compiling the initial title lists through a central process can reduce inconsistencies and help ensure that everyone takes part in reviewing their materials. Monitoring decisions at a higher level can also help prevent any gross inequities across disciplines with regard to determining what titles are appropriate to retain. Ward and Aagard (2008) reviewed the challenges of past attempts to review items housed in the repository at Purdue University and emphasized the benefits of providing bibliographers with relevant data on likely candidates for withdrawal rather than asking them to review everything in their subject areas. Having an intern do much of the preliminary work also enabled the librarians involved there to focus more time and effort on making decisions based on their respective subject expertise.

A well-trained library staff is also essential to keep digital resources running smoothly. This includes handling the initial acquisition, ensuring links are in place, publicizing resources to users, and troubleshooting problems. This latter issue is particularly important and requires ongoing attention. With numerous platforms and a wide range of technical expertise, library users may often require assistance. If someone reports difficulty accessing an online resource, library staff must first determine if the issue is specific to a particular user or method of access or if it is in fact a broader problem. In the event of an interruption of service procedures should exist for reporting the issue and flagging any affected resources on the system. Fortunately, most modern integrated library systems will allow posting of a resource advisory for multiple resources from a single parent record. Keeping an up-to-date listing of contact information for vendors and publishers can also help expedite the process.

FUTURE RESEARCH DIRECTIONS

The future of automating collection management lies in the development of technologies, and the use of existing technologies, that allow librarians to work efficiently with large, varying sets of data. At North Carolina State University, collection managers are using SAS software to make key decisions about millions of print volumes. They found that using office software and their ILS, they were limited in the amount of data that could be processed and thus analyzed. They presented on this approach and other "techie tools" that they are using for timely decision making at the 2011 Charleston Conference (Norberg, 2011).

Systems will be developed to manage information about growing and evolving electronic resource collections. Kent State gave a presentation at the Innovative Users Group in 2012 on a "Pre-ILS" system they have developed to manage electronic resources prior to their inclusion in their ILS. This web-based system offers a variety of key features missing from ERMs. This demand for rapid development of electronic resource system will only continue to increase as electronic resources multiply and diversify.

Discovery systems give collection managers new insights into what their library users need and how they want to use information. Libraries are still in the early stages of deciding how best to use data from these systems to manage their collections. Over the next few years we will learn what impact discovery systems will have on selection, deselection, and assessment of library materials.

New web services are emerging from publishers and aggregators that enable libraries to place information in library web pages. It remains to be seen how widely adopted these services will be by academic libraries. They are already having an impact on how usage data is reported by some vendors such as Institute of Scientific Information who now separates out in their COUNTER report the queries that are from web services. Collection managers must decide how to interpret this data and communicate it to electronic resource managers, subject selectors, and administration.

The practice of weeding print collections has been better documented than weeding electronic collections. Deselection of electronic journals and databases occurs as a routine process—libraries cancel electronic journals if cost per use or subscription prices are deemed too high, if article downloads don't justify maintaining a subscription, or if titles fall out of packages or big deal licenses. Electronic books are another matter. Librarians have raised the issue of deselecting e-books on a title basis, but in 2012 it does not appear that anyone is putting this into practice. It is more often the case that collection managers will cancel or subscribe to e-book packages based on usage or an assumption that demand will grow. In the near future we should begin hearing more about libraries' decisions to withdraw electronic book titles from their catalogs, and we look forward to learning more about the rationale to do so.

CONCLUSION

The history of technological advances in printing and collection management provides a roadmap to understanding current issues. The need to collect and process large quantities of data across collections of millions of materials drives the ongoing developments in automating collection management. However, these systems and data are not only processing information about how our collections are used, they are also giving us valuable information about our users so we can best tailor what our libraries offer. Librarians and staff are acquiring new skills, such as programming, enabling them to perform better collection analysis and management.

Case Study

Background

Virginia Tech is a land grant institution serving more than 30,000 students in 150 graduate programs and 65 bachelor's programs. The University Libraries are comprised of the main library, Carol M. Newman, and three branches: Veterinary Medicine Library, Art & Architecture Library and the Northern Virginia Resource Center. The Libraries are members of ARL, OCLC, ASERL, CRL, LOCKSS, and SPARC, as well as VIVA, an electronic library consortium for higher education in Virginia.

This case study describes the recent projects to further automate collection management at Virginia Tech University Libraries. The motivation to automate practices has come from a change in leadership and a change in focus on how physical space is used at the Libraries. Virginia Tech hired in a new dean in 2010. Since that time, in response to student requests, new services have been implemented, such as the main library building being open more hours. Increased traffic into the library building and the desire to collaborate with other campus services necessitated the Collection Management department to automate in order to respond accurately and efficiently to changes in the collections.

Patron-Driven Acquisitions

In spring 2012 University Libraries launched a patron driven acquisitions program with book vendor Yankee Book Peddler (YBP). Less than 25% percent of the total book budget will be dedicated to patron-driven acquisitions (including short term loans and purchases) for this pilot. With the expressed interest of several of our college librarians, specific disciplines were chosen for the initial focus, including the social sciences and business. The goal of implementing patron driven acquisitions at Virginia Tech is to transition the approval plan to the patron-driven model for e-books. Only e-books are being selected for inclusion in the project.

The YBP GOBI system is designed to help prevent some of the duplication that is potentially a problem with e-resources acquisitions. We have selected to sign contracts with the three aggregators providing patron-driven acquisitions titles through YBP: EBSCO, EBL, and, Ebrary. From YBP we will receive MARC records via FTP to ingest into our III Millennium catalog and then into Virginia Tech's discovery product Summon.

Several key decisions still need to be made before moving forward with our implementation of patron driven acquisitions. We need to set a threshold of clicks for short term loans and final purchase of the e-book. These thresholds can be set differently for each aggregator. The plan can also be modified based on data about how users are interacting with these titles.

An evaluation is planned for winter 2012 to reevaluate, and, given success of the project, to expand the percentage of the collections budget used.

Deselection

University Libraries is shifting its mission to be more in line with current trends in academic libraries. We are creating an information commons space in Newman Library that will allow for more student study areas. This space and others throughout the library are being evaluated for redesign as new classrooms and other opportunities for services on campus to move into the library. For example, the print government documents area is being consolidated and we are transitioning to the Documents without Shelves service from Marcive. This space is under review for being transformed into a Student-Centered Active Learning Environment for Undergraduate Programs (SCALE-UP) classroom.

The Libraries launched its very first physical inventory of print collections in 2008 beginning with our Art and Architecture branch. With that project completed, we are now taking an inventory of all print materials in the main library. Staff are walking through the stacks with laptops, scanning barcodes, and using the III Circa system to evaluate what we actually have in the building. This project has raised questions about whether we should invest in RFID and realize the benefits recently mentioned in the NISO Recommended Practice "RFID in U.S. Libraries" (Walsh, 2011). This option is currently being explored.

In 2011, University Libraries began participation in the ASERL Collaborative Journal Retention Program. Program members agree to store selected titles from their print journals in secure facilities for a minimum of 25 years, ensuring that a copy of last resort is available to ASERL members if needed. Through this project, we aim to commit 500 print serial titles from our collection to ASERL's Collaborative Journal Retention Program. Meeting this goal will result in more space for group study, teaching spaces, and collaboration between students and faculty. By placing the collection into storage with an express agreement to preserve these materials, the University Libraries is able to meet both its mission to serve the campus community and the library community at large.

Deciding which collections to move to storage relies on accurate information about the items. The efficient and effective shifting and deselection of collections demands software tools to track the location of each item and the decisions made. We are using external tools such as Ithaka S+R's interactive decision support tool, a pre-populated Excel spreadsheet. It is also imperative for our participation in projects like ASERL.

To track and store our usage statistics, we rely on our III Millennium ILS and a Microsoft Access database. The Access database allows us to include additional data that we need when making decisions regarding the collection such as database renewals and new e-book acquisitions. We are analyzing what collections we have in both print and electronic formats, and which print collections would be more useful to our patrons as readily available electronic resources. Based on this data, we are purchasing backfiles of serials collections.

Another area we are currently evaluating for deselection is our monographic print collection. Using III Millennium's create list functionality, we generated a list of all single part monograph titles that were published before 1990 with zero or one circulations within the call number ranges physically located on the second and third floors of the Newman Library. Taking into account key data not limited to in house use, circulation counts, and data from other sources such as HathiTrust holdings and WorldCat® holdings, we have generated reports for our subject selectors. They can use this data in conjunction with feedback from the faculty in their disciplines to make careful decisions about what materials should be sent to storage or weeded from the collection entirely.

The WorldCat® holdings data was harvested using a script that called the WorldCat® holdings API; further, we requested the three letter code for each library that has indicated that they have at least one copy of that monograph. We ran this script over four days on a limited number of requests, parallelizing the requests for efficiency since we needed the holdings for more than 48,000 unique items. The data has been further processed to represent holdings limited to Virginia libraries and holdings outside of Virginia. This project was inspired by the work of Sustainable Collections Services™, a library consulting company that has worked with other academic libraries on large scale deleselection projects.

Standards and Electronic Resources Management

A suite of standards and recommended best practices is currently available from NISO that support the management of the data necessary to manage electronic resource collections. The COUNTER Code of Practice sets the standard for the international reporting of electronic resource statistics. The COUNTER code is periodically revised to reflect changes in electronic resources. We are harvesting COUNTER usage statistics and ingesting them into an Access database. While this standard defines how usage should be counted and reported, getting accurate title and ISSN lists from publishers and aggregators has ranged from straightforward to notoriously complex. Normalization of titles and ISSNs has been a hurdle to being able to successfully manage this information. By extracting the linking ISSN from the ISSN-L table, we have been able to normalize more than 70% of our journals. Being able to effectively combine the title list and the COUNTER report list of titles has been critical to our being able to rely on our usage data to make decisions.

Another related standard, the SUSHI Protocol, defines the communication of usage statistics between systems. We tested using the SUSHI protocol with our III Millennium system's ERM module and with the Serials Solutions®' MISO SUSHI client in 2010. While both options worked to automatically retrieve some electronic resource statistics we had several concerns that led to us tabling the use of SUSHI. The first concern was that upon closer evaluation, the statistics that we had received via SUSHI were not the same as the statistics we received by downloading the data directly from the vendor. Another concern was that if there was any interruption in the network or other problem, then the usage would not be harvested. A final concern was that many so-called SUSHI compliant publishers could not respond to SOAP requests, or had other data issues such as errors in their XML elements within the namespace which would prevent the successful transmission of data. We found that we spent more time troubleshooting the attempts to retrieve the data and the data delivered by SUSHI than we would have spent simply visiting the vendor's website and downloading the data. Automated harvesting of our statistics using the SUSHI protocol worked best when we briefly subscribed to the Scholarly Stats service and only had one request to send to that service. Our current practice to collect usage statistics is manual, biannual harvesting. We are planning to reevaluate the use of SUSHI, considering new developments such as the recently released Python library SUSHI Py.

WorldCat Collection Analysis™ Tool

The WorldCat Collection Analysis™ (WCA) Tool was used by Director of Collection Management Paul Metz in 2006 to evaluate Virginia Tech's print collection compared to peer institutions. Based on the subject reports he was able to generate using the WCA tool, he distributed overlap data. Title-by-title data was used to make purchasing decisions to ensure that Virginia Tech had access to key monographs for subjects that were not as widely held by peer institutions.

REFERENCES

Alan, R., Chrzastowski, T. E., German, L., & Wiley, L. (2010). Approval Plan Profile Assessment in Two Large ARL Libraries. *Library Resources & Technical Services*, *54*(4), 179–179.

Bowersox, T., Oberlander, C., Sullivan, M., & Black, M. (2011, March). GIST: Getting it System Toolkit: a Remix of Acquisitions, Collection Development, Discovery, Interlibrary Loan, and Technical Services. Paper presented at the OCLC ILLiad International Conference, Virginia Beach, VA. Retrieved April 14, 2012 from http://www.atlas-sys.com/2011-illiad-international-conference-session-archive/

Broadus, R. N. (1991). The History of Collection Development. In C. Osburn & R. Atkinson (Eds.). Collection Management: A New Treatise. (Vol. 26. Part A, pp. 3-28). Greenwich, CT: JAI Press.

Carroll, D., & Cummings, J. (2010a). Data Driven Collection Assessment Using a Serial Decision Database. *Serials Review*, *36*(4), 227–239. doi:10.1016/j.serrev.2010.09.001.

De Fino, M., & Lo, M. L. (2011). New Roads for Patron-Driven E-Books: Collection Development and Technical Services Implications of a Patron-Driven Acquisitions Pilot at Rutgers. *Journal of Electronic Resources Librarianship*, *23*(4), 327–338. doi:10.1080/1941126X.2011.627043.

Demas, S., & Miller, M. (2012). Curating Collective Collections — What's Your Plan? Writing Collection Management Plans. *Against the Grain*, *24*(1), 65-68.

Drueke, J. (2001). St. Osmund's New Legacy: The Scriptorium Informs Electronic Text. *Libraries & Culture*, *36*(4), 506–517. doi:10.1353/lac.2001.0066.

Evans, G. E. (2000). *Developing Library and Information Center Collections*. Englewood, CO: Libraries Unlimited.

Forster, G., & Bell, A. (2006). The Subscription Libraries and Their Members. In Leedham-Green, E., & Webber, T. (Eds.), *The Cambridge History of Libraries in Britain and Ireland* (*Vol. III*, pp. 147–168). Cambridge, England: Cambridge University Press. doi:10.1017/CHOL9780521780971.014.

Franklin, B. (2005). Managing the Electronic Collection With Cost Per Use Data. *IFLA Journal*, *31*(3), 241–248. doi:10.1177/0340035205058809.

Franklin, B., & Plum, T. (2008). Assessing the Value and Impact of Digital Content. *Journal of Library Administration*, *48*(1), 41–57. doi:10.1080/01930820802029334.

Gaffney, M. (2011, March). *Item Shipped!: Purchase on Demand and ILLiad 8 Addons*. Paper presented at the ILLiad conference, Virginia Beach, VA. Retrieved April 14, 2012 from http://www.atlas-sys.com/ILLiadConf/Presentations/ItemShipped.pdf

Gregory, V. L. (2011). *Collection Development and Management for 21st Century Library Collections: An Introduction*. New York, NY: Neal-Schuman Publishers.

Harrington, M., & Stovall, C. (2011, November). Contextualizing and Interpreting Cost Per Use for Electronic Journals. Paper presented at Charleston Conference, Charleston, SC. Retrieved April 23, 2011 from http://www.slideshare.net/group/2011-charleston-conference/slideshows/3

Harris, M. H. (1995). *History of Libraries in the Western World*. Metuchen, NJ: Scarecrow Press.

Hazen, D. (1991). Selection: Function, Models, Theory. In Osburn, C., & Atkinson, R. (Eds.), *Collection Management: a New Treatise* (pp. 273–300). Greenwich, CT: JAI Press.

Holley, B. (2012). Demise of Traditional Collection Development. *Against the Grain, 24*(1), 30-31.

Horava, T. (2010). Challenges and Possibilities for Collection Management in a Digital Age. *Library Resources & Technical Services, 54*(3), 142–152.

Hot new tool? ITSO CUL. (2004). *Backstory, 1*(1). Retrieved April 10, 2012 from http://www.library. cornell.edu/backstory/v1n1/itsofeature.htm

Jensen, K. (2006). Universities and Colleges. In *The Cambridge History of Libraries in Britain and Ireland* (*Vol. 1*, pp. 345–362). Cambridge, England: Cambridge University Press. doi:10.1017/CHOL9780521781947.016.

Jones, D. (2011). On-Demand Information Delivery: Integration of Patron-Driven Acquisition into a Comprehensive Information Delivery System. *Journal of Library Administration, 51*(7/8), 764–776. doi:10.1080/01930826.2011.601275.

Kaplan, M. (2009). Library Automation. In Nof, S. J. (Ed.), *Springer Handbook of Automation* (pp. 1285–1298). Berlin: Springer. doi:10.1007/978-3-540-78831-7_72.

Leedham-Green, E. S., & Webber, T. (Eds.). (2006). *The Cambridge History of Libraries in Britain and Ireland*. Cambridge, England: Cambridge University Press. doi:10.1017/CHOL9780521781947.

Lerner, F. A. (1998). *The Story of Libraries: From the Invention of Writing to the Computer Age*. New York, NY: Continuum.

Lerner, F. A. (1999). *Libraries Through the Ages*. New York, NY: Continuum.

Lugg, R. (2011). Collection for the Moment: Patron-Driven Acquisitions as Disruptive Technology. In Swords, D. A. (Ed.), *Patron-Driven Acquisitions: History and Best Practices* (pp. 7–22). Berlin, Germany: De Gruyter Saur. doi:10.1515/9783110253030.7.

McMurtrie, D. C. (1967). *The Book: The Story of Printing & Bookmaking*. New York, NY: Oxford University Press.

Metz, P., & Cosgriff, J. (2000). Building a Comprehensive Serials Decision Database at Virginia Tech. *College & Research Libraries, 61*(4), 324.

Mosher, P. H. (1991). Reviewing for Preservation, Storage, and Weeding. In Osburn, C., & Atkinson, R. (Eds.), *Collection Management: a New Treatise* (pp. 373–391). Greenwich, CT: JAI Press.

Norberg, B., Orcutt, D., & Vickery, J. (2011, November). New Tricks For Old Data Sources: Mashups, Visualizations and Questions Your ILS Has Been Afraid to Answer. Paper presented at the Charleston Conference, Charleston, SC. OCLC. *Creating the Conspectus*. Retrieved April 15, 2012, from http://www.oclc.org/research/activities/past/rlg/conspectus.htm

Petit, J. (2011). Twitter and Facebook for User Collection Requests. *Collection Management, 36*(4), 253–258. doi:10.1080/01462679.2011.605830.

Pitcher, K., Bowersox, T., Oberlander, C., & Sullivan, M. (2010). Point-of-Need Collection Development: The Getting It System Toolkit (GIST) and a New System for Acquisitions and Interlibrary Loan Integrated Workflow and Collection Development. *Collection Management, 35*(3/4), 222–236. doi:10.1080/01462679.2010.486977.

Ranasinghe, R. H. I. S. (2008). How Buddhism Influenced the Origin and Development of Libraries in Sri Lanka (Ceylon): From the Third Century BC to the Fifth Century AD. *Library History, 24*(4), 307–312. doi:10.1179/174581608X381602.

Rowley, G. S., & Association of Research Libraries/Office of Management, S. (1995). *Organization of Collection Development*. United States of America: Association of Research Libraries. Office of Management Services.

Schonfeld, R. C., & Housewright, R. (2010) *Faculty Survey 2009: Strategic Insights for Libraries, Publishers, and Societies*. Retrieved April 21, 2012 from http://www.ithaka.org/ithaka-s-r/research/faculty-surveys-2000-2009/faculty-survey-2009

Sharpe, R. (2006). The Medieval Librarian. In Leedham-Green, E.S. & Webber. T. (Eds.). The Cambridge History of Libraries in Britain and Ireland (Vol. I, pp. 218-241). Cambridge, England: Cambridge University Press.

Shen, L., Cassidy, E. D., Elmore, E., Griffin, G., Manolovitz, T., & Martinez, M. et al. (2011). Head First into the Patron-Driven Acquisition Pool: A Comparison of Librarian Selections Versus Patron Purchases. *Journal of Electronic Resources Librarianship*, *23*(3), 203–218. doi:10.1080/1941126X.2011.601224.

Slote, S. J. (1997). *Weeding Library Collections: Library Weeding Methods*. Englewood, CO: Libraries Unlimited.

Stahl, R., & Correa-Morris, C. Hwang. S. (2012, April). Solving the Complexities of Ebook Record Management in Millennium. Paper presented at the Innovative Users Group Meeting, Chicago, IL.

Tucker, C. (2009). Benchmarking Usage Statistics in Collection Management Decisions for Serials. *Journal of Electronic Resources Librarianship*, *21*(1), 48–61. doi:10.1080/19411260902858581.

Ward, S. M., & Aagard, M. C. (2008). The Dark Side of Collection Management: Deselecting Serials from a Research Library's Storage Facility Using WorldCat Collection Analysis. *Collection Management*, *33*(4), 272–287. doi:10.1080/01462670802368638.

Warzala, M. (1994). The Evolution of Approval Services. *Library Trends*, *42*(3), 514–514.

Wilde, M., & Level, A. (2011). How to Drink From a Fire Hose Without Drowning: Collection Assessment in a Numbers-Driven Environment. *Collection Management*, *36*(4), 217–236. doi:10.1080/01462679.2011.604771.

ADDITIONAL READING

Anderson, K. J., Freeman, R. S., Herubel, J.-P. V. M., Mykytiuk, L. J., Nixon, J. M., & Ward, S. M. (2010). Liberal arts books on demand: A decade of patron-driven collection development, Part 1. *Collection Management*, *35*(3/4), 125–141. doi: 10.1080/01462679.2010.486959.

Andrade, J. L., D'Amour, H., Dingwall, G., Su, J. T., & Dyas-Correia, S. (2011). 'Shelf-ready' print serials acquisitions. *Serials Review*, *37*(1), 29–34. doi:10.1016/j.serrev.2010.12.008.

Baker, G., & Read, E. J. (2008). Vendor-supplied usage data for electronic resources: A survey of academic libraries. *Learned Publishing*, *21*(1), 48–57. doi:10.1087/095315108X247276.

Barnhart, A. C. (2010). Want buy-in? Let your students do the buying! A case study of course-integrated collection development. *Collection Management*, *35*(3/4), 237–243. doi:10.1080/01462679.2010.486986.

Booth, H. A., & O'Brien, K. (2011). Demand-driven cooperative collection development: Three case studies from the USA. *Interlending & Document Supply*, *39*(3), 148–155. doi:10.1108/02641611111164636.

Bracke, M. S. (2010). Science and technology books on demand: A decade of patron-driven collection development, Part 2. *Collection Management*, *35*(3/4), 142–150. doi:10.1080/01462679.2010.486742.

Bravender, P., & Long, V. (2011). Weeding an outdated collection in an automated retrieval system. *Collection Management*, *36*(4), 237–245. doi:10.1080/01462679.2011.605290.

Breeding, M. (2011). The systems librarian. Library tech strategies: Efficiency of innovation? *Computers in Libraries*, *31*(8), 28–30.

Breitbach, W., & Lambert, J. E. (2011). Patron-driven ebook acquisition. *Computers in Libraries*, *31*(6), 17–20.

Carpenter, M., Graybill, J., Offord, J. J., & Piorun, M. (2011). Envisioning the library's role in scholarly communication in the year 2025. *Portal: Libraries & the Academy*, *11*(2), 659–681. doi:10.1353/pla.2011.0014.

Carpenter, T. A. (2010). ONIX for publications licenses: Getting an electronic grip on license information. *The Serials Librarian*, *58*(1-4), 79–86. doi:10.1080/03615261003623054.

Carroll, D., & Cummings, J. (2010a). Data driven collection assessment using a serial decision database. *Serials Review*, *36*(4), 227–239. doi:10.1016/j.serrev.2010.09.001.

Chadwell, F. A. (2011a). What's next for collection management and managers? *Collection Management*, *36*(4), 198–202. doi:10.1080/01462679.2011.607555.

Chadwell, F. A. (2011b). What's next for collection management and managers? Sustainability dilemmas. *Collection Management*, *37*(1), 3–8. doi:10.1080/01462679.2012.633322.

De Fino, M., & Lo, M. L. (2011). New roads for patron-driven e-books: Collection development and technical services implications of a patron-driven acquisitions pilot at Rutgers. *Journal of Electronic Resources Librarianship*, *23*(4), 327–338. doi:10.1080/1941126X.2011.627043.

Dominguez, D. V., & Ovadia, S. (2011). What's next for collection management and managers? *Collection Management*, *36*(3), 145–153. doi:10.1080/01462679.2011.580427.

Fei, X. (2010). Implementation of an electronic resource assessment system in an academic library. *Program: Electronic Library & Information Systems*, *44*(4), 374–392. doi:10.1108/00330331011083257.

Gregory, V. L. (2011). *Collection development and management for 21st century library collections: An introduction*. New York, NY: Neal-Schuman Publishers.

Hazen, D. (2011). Lost in the cloud: Research library collections and community in the digital age. *Library Resources & Technical Services*, *53*(4), 195–204.

Henderson, K. S., & Bosch, S. (2010). Seeking the new normal. *Library Journal*, *135*(7), 36–40.

Herrera, G., & Greenwood, J. (2011). Patron-initiated purchasing: Evaluating criteria and workflows. *Journal of Interlibrary Loan. Document Delivery & Electronic Reserves*, *21*(1/2), 9–24. doi:10.1080/1072303X.2011.544602.

Hulbert, L., Roach, D., & Julian, G. (2011). Integrating usage statistics into collection development decisions. *The Serials Librarian*, *60*(1-4), 158–163. doi:10.1080/0361526X.2011.556027.

Johnston, M., & Weckert, J. (1990). Selection advisor: An expert system for collection development. *Information Technology and Libraries*, *9*(3), 219–225.

Jones, D. (2011). On-demand information delivery: Integration of patron-driven acquisition into a comprehensive information delivery system. *Journal of Library Administration*, *51*(7/8), 764–776. doi:10.1080/01930826.2011.601275.

King, D. (2009). What is the next trend in usage statistics in libraries? *Journal of Electronic Resources Librarianship*, *21*(1), 4–14. doi:10.1080/19411260902858276.

Kinman, V. (2009). E-metrics and library assessment in action. *Journal of Electronic Resources Librarianship*, *21*(1), 15–36. doi:10.1080/19411260902858318.

Leffler, J. J., & Zuniga, H. A. (2010). Development and use of license forms for libraries with and without electronic resource management systems. *Technical Services Quarterly*, *27*(3), 279–288. doi:10.1080/07317131003765977.

Matthews, T. E. (2009). Improving usage statistics processing for a library consortium: The virtual library of Virginia's experience. *Journal of Electronic Resources Librarianship*, *21*(1), 37–47. doi:10.1080/19411260902858573.

Nabe, J. (2011). What's next for collection management and managers? *Collection Management*, *36*(1), 3–16. doi:10.1080/01462679.2011.529399.

Nixon, J. M., & Saunders, E. S. (2010). A study of circulation statistics of books on demand: A decade of patron-driven collection development, Part 3. *Collection Management*, *35*(3/4), 151–161. doi:10.1080/01462679.2010.486963.

Paynter, R. A. (2009). Commercial library decision support systems: An analysis based on collection managers' needs. *Collection Management*, *34*(1), 31–47. doi:10.1080/01462670802548585.

Pesch, O. (2009). ISSN-L: A new standard means better links. *The Serials Librarian*, *57*(1/2), 40–47. doi:10.1080/03615260802669052.

Pesch, O. (2011a). COUNTER code of practice: A preview of release 4. *The Serials Librarian*, *61*(2), 140–148. doi:10.1080/0361526X.2011.591578.

Pesch, O. (2011b). Standards that impact the gathering and analysis of usage. *The Serials Librarian*, *61*(1), 23–32. doi:10.1080/036152 6X.2011.580044.

Pitcher, K., Bowersox, T., Oberlander, C., & Sullivan, M. (2010). Point-of-need collection development: The getting it system toolkit (GIST) and a new system for acquisitions and interlibrary loan integrated workflow and collection development. *Collection Management*, *35*(3/4), 222–236. doi:10.1080/01462679.2010.486977.

Shen, L., Cassidy, E. D., Elmore, E., Griffin, G., Manolovitz, T., & Martinez, M. et al. (2011). Head first into the patron-driven acquisition pool: A comparison of librarian selections versus patron purchases. *Journal of Electronic Resources Librarianship*, *23*(3), 203–218. doi:10.1080/194 1126X.2011.601224.

Street, C. (2010). Getting the most from a database trial. *Legal Information Management*, *10*(2), 147–148. doi:10.1017/S1472669610000551.

Swords, D. A. (2011). *Patron-driven acquisitions: History and best practices*. Berlin, Germany: De Gruyter Saur. doi:10.1515/9783110253030.

Tucker, C. (2009). Benchmarking usage statistics in collection management decisions for serials. *Journal of Electronic Resources Librarianship*, *21*(1), 48–61. doi:10.1080/19411260902858581.

Tyler, D. C., Melvin, J. C., Yang, X., Epp, M., & Kreps, A. M. (2011). Effective selectors? interlibrary loan patrons as monograph purchasers: A comparative examination of price and circulation-related performance. *Journal of Interlibrary Loan. Document Delivery & Electronic Reserves*, *21*(1/2), 57–90. doi:10.1080/107230 3X.2011.557322.

Tyler, D. C., Xu, Y., Melvin, J. C., Epp, M., & Kreps, A. M. (2010). Just how right are the customers? An analysis of the relative performance of patron-initiated interlibrary loan monograph purchases. *Collection Management*, *35*(3/4), 162–179. doi:10.1080/01462679.2010.487030.

van Dyk, G. (2011). Interlibrary loan purchase-on-demand: A misleading literature. *Library Collections, Acquisitions & Technical Services*, *35*(2/3), 83–89. doi:10.1016/j.lcats.2011.04.001.

Walsh, A. (2011). Blurring the boundaries between our physical and electronic libraries: Location-Aware technologies, QR codes and RFID tags. *The Electronic Library*, *29*(4), 429–437. doi:10.1108/02640471111156713.

Wilde, M. (2010). Local journal utilization report: Supporting data for collection decisions. *Collection Management*, *35*(2), 102–107. doi:10.1080/01462671003615058.

Wilde, M., & Level, A. (2011). How to drink from a fire hose without drowning: Collection assessment in a numbers-driven environment. *Collection Management*, *36*(4), 217–236. doi:10.1080/01462679.2011.604771.

Wilson, K. (2011). Beyond library software: New tools for electronic resources management. *Serials Review*, *37*(4), 294–304. doi:10.1016/j.serrev.2011.09.010.

KEY TERMS AND DEFINITIONS

Assessment: The use of qualitative and quantitative methods to analyze library collections.

Collection Management: "The process of information gathering, communication, coordination, policy formulation, evaluation, and planning that results in decisions about the acquisition, retention, and provision of access to information sources in support of the intellectual needs of a given library community." (Gregory 2011).

Deselection: Deselection of library materials is the process of removing items from the collection and is essential for the maintenance of an active, academically useful library collection. Subject specialists are responsible for conducting an ongoing de-selection effort in their designated disciplines and for maintaining the quality of the collection.

Selection: The process of choosing materials to be acquired for inclusion in a library's collection.

Chapter 4
Streamlining Access to Library Resources with LibX

Annette Bailey
Virginia Tech, USA

Godmar Back
Virginia Tech, USA

ABSTRACT

LibX is a platform that allows libraries to create customized web browser extensions that simplify direct access to library resources and services. LibX provides multiple user interfaces, including popups, context menus, and contextualized cues to direct the user's attention to these resources. LibX is supported by two toolbuilder applications - the Edition Builder and the LibApp Builder – which allow anyone to create, manage, and share LibX configurations and applications. These tools automate the process of software creation and distribution, allowing librarians to become software distributors. This chapter provides background and history of the LibX project, as well as in-depth analysis of the design and use of the LibX Edition Builder that has helped enable its success.

INTRODUCTION

Web-based resources and applications have taken on a dominant role in the daily workflow of researchers, students, and librarians alike. LibX (Bailey & Back, 2006) is a platform that allows libraries to provide services for these users at the point of need, by creating a presence for a library inside a user's browser. A plug-in, installed by the user, provides users with quick access to a library's catalogs, databases, and resources such as web services. Such access is provided through search interfaces, including popups and context menus, as well as through specially crafted "cues" embedded into pages a user visits which enrich those pages with metadata such as bibliographic record, holdings, and/or real-time circulation information.

LibX's widespread deployment and adaptation was enabled through the use of two accompanying tool builder applications: the LibX Edition Builder and the LibX Libapp Builder, both of which are web-based. The LibX Edition Builder enables librarians and interested members of a local

DOI: 10.4018/978-1-4666-3938-6.ch004

community to create a LibX Edition, which is a customized configuration reflecting the resources and services available to that particular community. The Libapp Builder allows the creation of Libapps (library applications) that interact with web pages a user is visiting. Both design tools that allow community-based sharing and distribution of editions and Libapps, respectively.

This chapter will present the history of the LibX project, from its early conception to its present state, explain the motivation, design, and functionality of the Edition Builder and Libapp Builder tools, and discuss the user experiences we collected. LibX meets the theme of this book, "robots in the library," in two significant ways: not only does the LibX plug-in itself help automate workflows such as finding known items, but the builder tools we created automate the process of creating library software.

BACKGROUND

In 2005, OCLC (OCLC, 2005) conducted a study on Perceptions of Libraries and Information Resources, which included a companion report on college students' perception in particular. This report found that search engines, rather than physical or online libraries, were most students' first choice when seeking a source or place for information. 72% of college students and 80% of overall respondents preferred search engines, compared to 24% and 17% who preferred physical or online libraries. These respondents valued search engines for their reliability, cost-effectiveness, ease of use, convenience, and speed. On the other hand, libraries outranked search engines in the survey participants' perception of trustworthiness and accuracy by a similarly wide margin. The conclusions we drew from the survey, as well as from our own and others (Fast & Campbell, 2004) concurrent observations, were twofold. First, we concluded that library search interfaces needed to be improved to compete better with search

engines. Most existing interfaces were simple databases that indexed only metadata and required users to use arcane search syntaxes. Since then, this deficiency has been recognized by library system vendors and publishers of abstracting and indexing databases, resulting in the proliferation of discovery systems such as ProQuest's Summon service. These systems adopt the innovations and the resulting user experience introduced by search engines, while providing access to full text indices of many licensed academic resources.

Second, we realized that like search engines, which are often directly accessible through a browser's user interface (also called the browser's "chrome"), library interfaces should be accessible in a similarly direct and easy-to-use way, integrated with the user's "webflow." Moreover, by integrating access to library resources into the browser, we expected to be able to provide integrated services than went beyond mere user-initiated searching.

Others had similar realizations. In 2005, Jon Udell started his Library Lookup Project, which provides a way to generate bookmarklets for searches in popular library OPAC implementations. Bookmarklets are canned snippets of JavaScript code that can be stored in a browser's bookmark file. Selecting the bookmark activates the code, which then performs a search in a given library's OPAC.

At around the same time, a shift in browser development took place. An offshoot of the open-source Mozilla project supported by Netscape, the Firefox browser started to gain popularity as an alternative to the then dominant Internet Explorer by Microsoft. Firefox uses a dramatically different design than existing browsers by allowing software extensions that had direct access and control over the browser's user interface and operation, allowing for user-specific functionality that had not been foreseen by the browser's designers. These "Firefox extensions" (since renamed to "Add-Ons") rapidly gained in popularity as a large community of developers and adapters developed many for purposes ranging from blocking advertisements to supporting custom buttons, links, or other user interface enhancements. They were relatively simple to develop as they relied on lightweight programming languages such as JavaScript and high-level user interfaces languages such as XUL, rather than C++ or Java like existing desktop applications. A small number of libraries started developing toolbars which provided an alternative search interface for their OPAC and offered them to users.

During the summer of 2005, we developed such a toolbar for Virginia Tech's University Libraries. In addition to providing the ability to search the University Libraries' then new III Millennium OPAC via a toolbar, it provided quick links for frequently used resources as well as a context menu that allow a user to select text on a page that could then be used in a search. Most exciting to many who saw this feature for the first time was the ability to place Virginia Tech's logo into other pages, including the search results page of Google and Yahoo. When the users clicked on the logo that appeared, the search they had just performed in Google was repeated against the library's OPAC. Thus, the library became present and available even to users who used search engines for their research.

We soon realized that other libraries were interested in offering such toolbars to their users in order to make their OPACs and resource more easily accessible. Consequently, we restructured the code such that all references to Virginia Tech resources and branding were factored out into a configuration file. We created a script that combined the source code of our extension, now called LibX, and the edition information in the configuration file and created a downloadable package users could install.

Even though the LibX source code was provided under an open source license, initially, only we had the expertise to create such editions. Interested librarians sent us the information about their library's resources, and we inserted it into the build process, adapting and correcting it as

necessary. During that time, LibX was adapted by several dozens of libraries. We added a fair number of catalog types to provide support for many of the most widely used OPACs.

This manual edition configuration process was very time consuming. We did not have time to accommodate all requests, and our adaptors wanted to be able to make changes to their configuration, or create new ones, without waiting for us. To address this issue, we conceived the idea of an Edition Builder interface that could allow anyone, without user interaction, to configure and build a LibX Edition for their user community. We applied for and were awarded a National Leadership Grant by the Institute of Museum and Library Services to pursue this idea. After 1.5 years of development, the Edition Builder went online in 2007. Shortly after its introduction, the adoption of LibX increased rapidly, at about a pace of 20 per month.

We also devoted efforts to creating a version of LibX for Internet Explorer (IE), which at that time still had the dominant market share among browsers. Unlike Firefox extensions, extensions for IE were much more difficult to develop. They required the use of a compiled language (C++ or C#) along with integration into the Windows COM component model, which was not designed to be easily extended by its vendor. Nevertheless, we succeeded in providing a version of LibX for IE, implemented in .Net (Baker, 2007).

In 2009, Google shook up the browser world with the release of the Google Chrome browser. Google Chrome introduced a number of innovations, both from the user interface perspective, as well as, from the perspective of how to design extensions. Its user interface no longer supported the use of toolbars, which were deemed too much of a burden in terms how they restricted the available screen real-estate. It instead encouraged instead the use of small browser buttons which users could access when needed to open a popup window that provides extension functionality. From the extension developers' perspective, the

Figure 1. The LibX 2.0 User Interface. Instead of a toolbar, a small button next to the browser's address bar opens a popup window, which provides access to LibX's services. LibX supports a preview of results from resources that provide a web services API, such as ProQuest's Summon service.

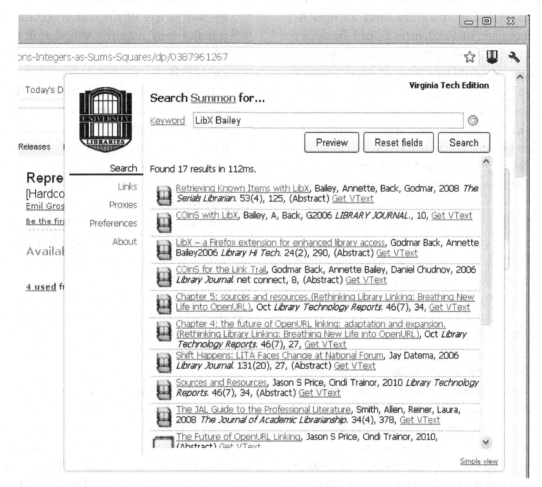

key novelty introduced by Chrome was that extensions could now be developed entirely in the web standard languages HTML, CSS, and JavaScript, whereas Firefox extensions still required some amount of a browser-specific XUL language coding. Chrome also introduced a security model for extensions, which as a practical implication made it simpler to distribute updates to an extension without user confirmation as long as these updates did not ask for more privileges than those the user agreed to when initially installing the extension.

We liked Chrome's design and anticipated its success, thus decided to base the next generation of LibX, which we called LibX 2.0, on Chrome

(Nicholson, 2011). We developed LibX 2.0 for Chrome, but structured the JavaScript code such that those portions that were specific to Chrome were separated from those that were not. As a result, we were then able to create a version for Firefox by replacing only the platform-specific portions with code specific to Firefox.

We also changed the way in which we distributed updates to our code. Whereas it previously was the responsibility of edition maintainers who had adopted LibX for their library to create a new distribution package for their users (after configuring and testing it), we now provide a single build of the source code, which then configures itself for a specific edition only after a user installs it.

Just like early LibX adaptors had voiced the desire to be able to customize LibX for their library, more and more adapters voiced the desire to be able to customize the cues placed by LibX into pages, or to add new types of interactions of their users with existing web content. Because the community release and adoption of the Edition Builder had been such a success and contributed to the wider use and adoption of LibX, we embarked on the development of the Libapp builder, which is now being released to the community.

LIBX

In this section, we will consider LibX from multiple perspectives: end users, edition maintainers, and Libapp builders.

End User Perspective

LibX end users are those who have the LibX plugin installed. Although we targeted LibX mostly at students and researchers, much of our user base consists of librarians who use LibX to automate their workflow. These users include reference librarians who use LibX to look up information during reference service interactions, or technical services librarians during selection and acquisition workflows. Anecdotal evidence we received after conference presentations suggests that many of those librarians have become to view LibX as an indispensable tool.

For end users, LibX provides the following facilities:

- **Catalogs and Databases:** Accessing OPACs and Discovery Systems. LibX supports more than 17 different OPAC types. Beyond that, LibX supports the use of URL templates to access any resources that support form-based access. Whether accessing a particular resource requires

direct support as a distinct OPAC type, or can be expressed using the URL templating mechanism, depends on the complexity and variety of the search parameters a resource may require. Although in theory all web-based systems can be accessed using HTML forms, and thus using the standard GET/PUT methods that are part of the HTTP standard, many existing OPACs required so much fine-tuning that we opted to use JavaScript code, rather than a template, to provide access.

While for most such resources LibX simply leads the user to the results page they would have obtained had they first navigated to the resource's search interface and performed their search there, LibX has recently started supporting a 'Preview' option that displays search results directly in the LibX popup. Such previews are supported by discovery systems that provide an open API that allows their programmatic use.

- **Context Menu Integration:** Oftentimes, a search term appears on a page a user is visiting. After selecting search terms, LibX users can initiate searches of via a right-click context menu against the configured catalogs and databases. This context menu is both configurable and context-sensitive. Configurability allows users to display only the most frequently used resources. Context-sensitivity implies that LibX senses when a user selects identifiers such as ISBNs or ISSNs, and display only the appropriate search options.
- **Branding and Quick Links:** A LibX edition can display a library's branding in the form of logos and icons. LibX can also display to users a preconfigured list of links, such as the pages displaying a library's news items or opening hours. We included these features so that users would feel at home with their edition of LibX, recogniz-

Figure 2. The configurable LibX context menu allows direct searches from selected text.

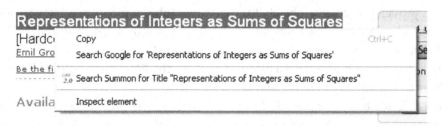

ing familiar symbols, and so that libraries could include them in their service offerings more easily. Many libraries have created customized pages where they offer LibX for download, often complemented with tutorials created by librarians for how to use it.

- **OpenURL Support:** The OpenURL framework described in NISO Z39.88 provides a way to avoid direct links to resources by replacing them with a context object, which is a metadata description of the resource being referred to. OpenURL resolvers (Sompel & Beit-Arie, 2001) are web services that process OpenURL metadata descriptions and identify how to access the resource. Most major academic libraries operate such resolvers, which are driven by knowledge bases that record the specific provider or aggregator through which this library's users have access to the object. LibX edition can be configured to record OpenURL resolver information and use it whenever they encounter such context objects on web sites. For instance, Wikipedia citations include COinS (Context Object in Span), which hide these context objects in otherwise inert HTML markup in a page. LibX processes these COinS by linking the user to their library's OpenURL resolver (Chudnov, 2006).

Recently, some OpenURL resolvers, notably Link/360, have started to provide programmatic APIs to query the OpenURL resolver. LibX uses those services to retrieve direct links to an item's fulltext, when available, and indicates the availability of full text to the user.

- **Proxy Support:** Many online publishers authenticate users based on the Internet Protocol (IP) address assigned to the computer from which the access occurs. For academic libraries in particular, this IP address is mapped to the network of a subscribing institution. When academic users attempt to access these resources after-hours from a computer located off-campus at their homes, they are using an IP address assigned by a commercial Internet service provider (ISP) and are denied access. A proxy is a program that acts as an intermediary between users and resources. To the resource, it represents an on-campus user because it operates from an on-campus machine. Off-campus users must authenticate with their proxy before accessing resource through it. Without LibX, a user must sign in (i.e., start proxying) their library homepage, and then navigate through the rewritten links until they reach the resource, a process which is burdensome. With LibX, users can invoke the proxy at the point of need – for instance, if a search engine led them to a publisher's page that refuses access.

- **Google Scholar Integration:** A frequent task for researchers is the retrieval of known items, such as articles, based on citations. Google Scholar is a full-text search engine that supports the OpenURL standard in that it acts as a provider of context objects for its search results. We integrated this ability into LibX in a sophisticated manner. A user invokes the search by dragging a piece of text from a web page or PDF document onto the LibX toolbar or button. Before displaying results to the user, LibX invokes Scholar to search for the item, and then analyzes the results using a similarity analysis. If the result is likely to represent a hit for a known-item search, we lead the user directly to the OpenURL resolver based on the context object provided by Scholar. In a study conducted in 2007 (Bailey & Back, 2007), we found that this "magic button" could lead the user to the resource in 81% of cases, with a precision of 94%, for 400 randomly selected citations occurring in the 4 most circulating journals at Virginia Tech from 4 distinct research areas.

- **AutoLinking:** When a user visits a webpage, LibX analyzes the text of that page for patterns that correspond to known identifiers, which include standard numbers such as ISBNs, ISSNs, Digital Object Identifiers (DOIs), and PubMed IDs. This recognition is based on a combination of regular expressions and checksum digit checks. For instance, ISBNs consist of either 9 or 12 digits, followed by a checksum digit or by an X. If LibX finds 10 or 13 digits, possibly with interspersed separation characters such as hyphens or spaces, it applies the ISBN checksum algorithm to see whether this combination of digits could be a valid ISBN. If so, LibX will replace it with a hyperlink that links to a search of the edition's first configured catalog (usually the primary OPAC of a community). LibX cannot know whether a 10 or 13 digit identifier is indeed an ISBN; the checksum heuristics has a 1 in 11 chance

of leading to false positives. Conversely, LibX sometimes misses standard numbers when they span multiple HTML elements, such as when the page's markup contains PMID: 17671256.

When the user hovers with the mouse over a hyperlinked identifier, LibX will contact a metadata service to retrieve metadata associated with that standard number and display this information in a tooltip. We use OCLC's xISBN (OCLC, 2003) and xISSN (OCLC, 2007) services to find information about ISBNs and ISSNs, respectively. We use NCBI's Entrez Utilities to retrieve information for PubMed numbers (Sayers, 2010-). For DOIs, CrossRef's web services provide the necessary metadata information. This service has two immediate uses. If a standard number appears out of context on a page (without metadata information such as title, author), a user can learn immediately which items this number refers to. If it appears in the context of metadata, such as in the list of references in an article published online, it allows users to immediately verify the correctness of the identifier with respect to the metadata.

- **Libapps:** LibX's auto-linking features are implemented using a more general facility, referred to as Libapps. These Libapps represent a generalization of the "cue" facility we included in early LibX prototypes. From the end user perspective, a Libapp does something meaningful when a user visits a page. For instance, it may sense that the user visited a book vendor's page where a book is offered, retrieve a standard number identifying the book being offered from the page's content, use this number to contact a web service that describes holdings and, if applicable, availability information for this item, and then integrate this information into the page. Consequently, a user may learn – without any action on the

Figure 3. LibX auto-detects identifiers such as ISBNs and hyperlinks them. A tooltip shows xISBN-derived metadata.

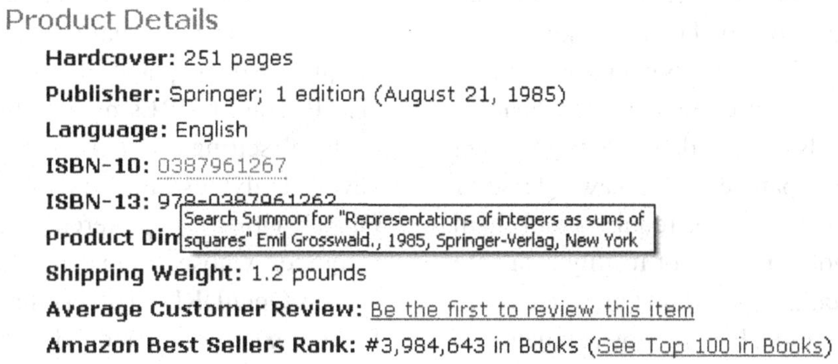

Figure 4. A Libapp in context. LibX checks an item's holdings and availability and integrates the result into the page the user visits.

user's part – whether an out-of-print book offered by an Amazon merchant for an exorbitant price is available on the shelves of their library (or is held and could be recalled).

Other examples of Libapps include embedded tutorials and videos. For instance, an embedded video could introduce the library that is paying for access to a particular resource, and remind users that they can seek reference help and support from the librarians supporting this access.

Motivation for the LibX Edition Builder

Many librarians do not have the skill set, systems knowledge, or time to create intricate and customized software tools such as LibX for their users. The Edition Builder bridges this gap, which requires addressing the two problems of (1) software packaging and (2) software customization.

First, creating plug-ins for today's dominant browser platforms (Firefox, Chrome, and Internet Explorer) is complicated. In addition to writing the software itself, a complex process of packaging and hosting is required. For LibX/Firefox, this process includes the creation of a cryptographically signed manifest to describe the software's origin and compatibility, and specific hosting requirements for the downloadable file and its manifest. For each browser, this process must be repeated whenever the software needs to be updated to provide bug fixes or to be compatible with a new release of the browser. These processes require a tool chain that includes tools that are not readily available on standard installations and that require custom setup and compilation. For our target audience, if we had simply published the LibX software with instructions on how to create downloadable files, we would not have been able to accomplish our goal of wider adoption.

Second, because LibX is highly customizable, the configuration settings for a specific edition are complex as well. For instance, the document type definition (DTD) grammar that describes the structure of a LibX configuration file contains more than 150 elements and attributes, each reflecting a possible setting. These settings include branding information such as icons and logos, configuration preferences that describe which resources and services a specific edition should offer, and metadata descriptions of how to access the configured resources. The most commonly configured resources include Online Public Access Catalogs (OPACs), OpenURL resolvers, and proxy servers that provide off-campus access to licensed resources. Many LibX editions also provide access to free and commercial databases and indices as well as to local digital repositories such as DSpace or Electronic Thesis and Dissertations (ETD) repositories. The librarians who work with these resources daily, and instruct others in their use, are the most qualified to decide which resources to set up for a particular user community that uses a LibX edition.

While librarians know which resources to set up for their edition, they do not typically have the information about how to describe those re-

sources so that they can be configured in LibX. Although the vast majority of these resources are accessible via the HTTP protocol, the specific syntax (i.e., the names and possible values of query string parameters) needed to create HTTP request URLs may be highly dependent on local settings. For instance, many libraries have heavily customized their vendor-provided OPAC with local preferences. Even for those resources whose syntax is relatively uniform, such as OpenURL resolvers that accept requests for context objects encoded via the HTTP transport provided in the NISO Z39.88 standard, the specific syntax to construct a valid OpenURL requires specialized knowledge. Consequently, the LibX Edition Builder pursues the dual goals of shielding librarians from the details of the LibX software packaging and update process and from the details of how to describe and configure their edition's resources.

Our design was guided by two additional, non-technical requirements. First, we did not want to restrict the use of the LibX Edition Builder to only librarians who were officially authorized to represent their institutions. The rationale behind this goal was to avoid red tape or committee overhead before an edition could be deployed. As a result, any librarian with or without the endorsement of their institution could create editions; some of our edition maintainers are staff members or even graduate students. Second, we did not want to require support or buy-in from a library's IT department. LibX configuration and package hosting should not require any hosting or other services, though LibX should be able to use them where provided.

Our discussion of the LibX Edition Builder design focuses on 3 aspects: the design of our user interface, edition and revision management, and the resource auto-detection facilities that help with resource configuration. A complete description of the underlying technology can be found in (Gaat, 2008) and (Tilevich & Back, 2008).

Figure 5. Relationship between end users, edition maintainers, and the LibX Edition Builder

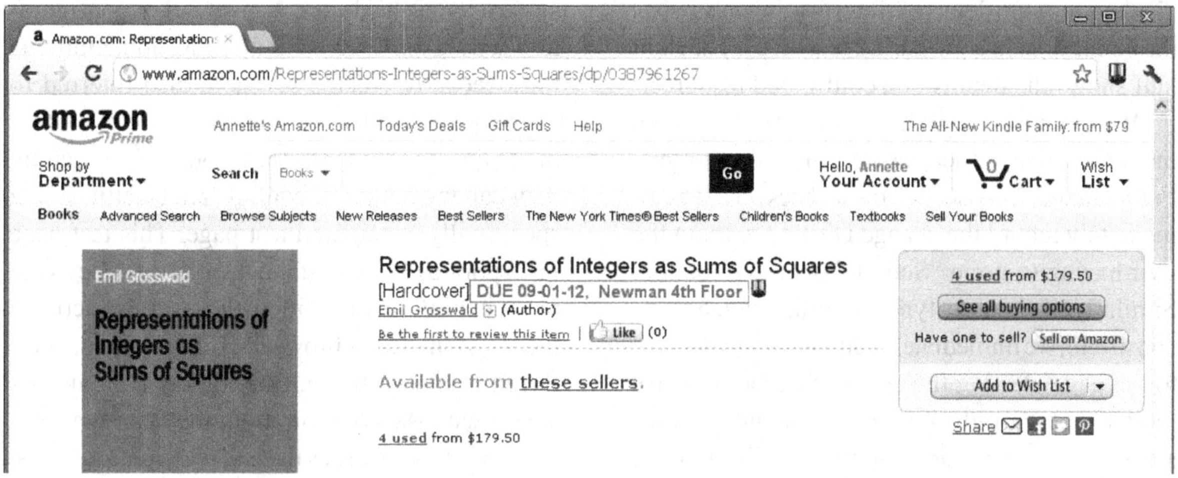

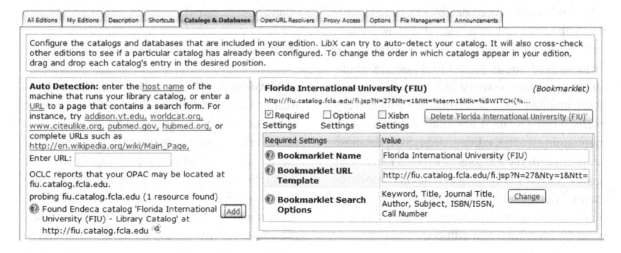

User Interface Considerations

To simplify navigation, we rejected the traditional page-based web application approach and instead chose a single-page internet application style (Mesbah & van Deursen, 2008), which is also referred to as Rich Internet Application (RIA). We chose ZK as our implementation framework (Chen & Cheng, 2007). We used a tabbed display, with each tab providing access to a different group of functionality. These tabs include: a tab to search the list of all editions created by the LibX community;

a tab to see and manage the editions owned by a particular maintainer; and several additional tabs that are used to change groups of settings related to an edition that is currently being worked on. These groups of settings include: general edition metadata, such as a description; the management of shortcuts to be included in the toolbar; the management of the configured catalogs and databases; OpenURL resolver(s); proxies; general options; and the management of branding-specific files such as icons. Figure 6 provides a screenshot of the "Catalogs & Databases" tab that is used to

Figure 6. A Screenshot of the tab-based user interface

configure resources. The left half of the figure displays the auto detection interface, which assists the maintainer in configuring resources. The right half shows the details of a configured resource.

We employed a mode of interaction in which most user input or actions take immediate effect. For instance, if a user selects a checkbox, we immediately record the changed setting without the user having to use a "Submit" or "Save" button. Similarly, if the user types a setting which is inconsistent, we immediately add a warning message to the page alerting the user to that fact. We use modal dialogs only for important interactions, such as when an edition maintainer is about to publish updates to their edition.

We implemented an ample set of online help facilities. In addition to the built-in consistency checking, we placed help icons next to each setting that show tooltips with useful information when the user hovers over them. The tooltips discuss examples of typical settings as well as common mistakes, and provide links to further information. In addition, we integrated a "Help Me With…" option that sends an email to the LibX developers, along with specific information about what the maintainer was trying to do, so they can receive individualized and contextualized assistance in response.

Managing Revisions and Editions

Each LibX edition is owned by one or more edition maintainers, who may take shared ownership. Each edition maintainer can participate in the ownership of any number of editions. A maintainer may share ownership of an edition with other maintainers – a facility that proves very useful when a team of librarians is charged with maintaining a LibX edition. A MySQL database keeps track of the editions to which a maintainer has access. Edition maintainers can choose to make their editions "public." Such public editions can be cloned by anyone, and thus used as a starting point for creating a new edition.

We provide a simple, linear revision control system for each edition. A given set of configuration settings is referred to as a revision. Only the most recently created revision, also referred to as the "test revision," is mutable. As an edition maintainer tweaks the test revision's configuration, they can try out those configuration settings in a specifically configured test page. This test page incorporates the very same JavaScript code as the browser plug-in. In addition, they can also activate the revision in their browser's LibX plug-in. Once testing is successful, and the edition maintainer is happy with the configuration, they can "make it live." This step freezes the test revision's settings, which now becomes the "live" revision for this edition, and triggers the automatic distribution of the new settings to all users who have activated this edition. If there was a previously "live" revision that revision is retained as an archived revision. This model is analogous to the separate staging and production areas provided by some library information systems with which librarians are familiar. However, use of this model requires that edition maintainers become familiar with the concepts of revision and revision history. They must understand the reason for the immutability of the "live" revision's settings and why, consequently, only the settings of the most recent "test" revision in the revision history can be changed. In addition, they must understand the concept of "archived" revisions, which allow an unrolling of changes – if desired.

Resource Auto-Detection and Sharing

We developed sophisticated facilities to aid edition maintainers in configuring the resources they wish to include in their editions. The auto-detection of resources requires two steps: (1) identifying the resources an edition maintainer is likely to integrate and (2) inferring the correct settings for these resources. To identify the resources a maintainer wants to integrate we extensively use registries maintained under the umbrella of the WorldCat Registry (OCLC) by the Online Computer Library Center (OCLC) that list the settings of many libraries' OpenURL resolvers and catalogs. In addition, we designed the Edition Builder to harvest published descriptions using the OpenSearch (Clinton, 2008) standard, scrape search forms, and use server fingerprinting heuristics to detect searchable resources.

When an edition maintainer connects to the Edition Builder, we examine the Internet Protocol (IP) address associated with the connecting client. If this IP address is listed as belonging to an address range associated with an OpenURL resolver in OCLC's OpenURL Resolver Registry, we retrieve the OpenURL resolver settings from this registry. In addition, most registry entries contain an ID for the maintainer's institution in OCLC's institutional registry. Based on this ID, we retrieve the institution's profile, which often contains the URL of the institution's OPAC. Armed with the OPAC URL, the Edition Builder application connects to the OPAC server, examines the response, and applies heuristics to identify the vendor of the OPAC and the settings being used. Though conceptually simple, this process required careful implementation since most resources are not designed for this type of auto-discovery. We have to faithfully mimic redirection techniques and cookie handling implemented by the browsers with which these systems expect to interact.

If the maintainer's IP address is not found in the Resolver Registry, the maintainer may enter an IP address or URL to initiate the auto-detection process. We apply the same set of heuristics as if the domain name had been listed in the OCLC institutional registry. If an institution is not listed in the OpenURL Resolver Registry, it may still be listed in OCLC's Institutional Registry. We provide a search box that allows the user to search this registry and to initiate the resource auto-detection process for any profiles found.

In addition to configuring OPACs and OpenURL resolvers, the Edition Builder can automatically detect settings for many resources that provide search forms. The edition maintainer must copy and paste the URL of the page containing the search form to initiate the auto-detection process. In addition to scraping search forms from these pages, the Edition Builder examines the page for OpenSearch descriptions advertised in the page's metadata. Finally, we consult a shared community database of resources other LibX maintainers have configured in their editions and display any results found therein. This feature is particularly useful if a resource could not be auto-detected, but was already set up manually by another edition maintainer. To avoid cluttering this database with partially configured or nonfunctional resources, we include only those resources that are contained in revisions a maintainer has chosen to make live.

The auto-detection process usually takes a few seconds. To mask latency, we implemented the different heuristics in parallel and results are added to the Edition Builder page as they become known. In some cases, such as when a page uses JavaScript instead of traditional HTML forms, the auto-detection may fail. It also may fail if access to a resource is restricted based on the Internet address or other network topology constraints, which cause the machine running the Edition Builder server to be treated differently by the target resource than machines at the maintainer's institution. Then, the user has an option to report this failure for further investigation. In other cases, a page contains multiple search forms, only some of which may be of interest to an edition maintainer. We display some identifying information we found on the page, such as captions of search buttons. The final decision whether to import an auto-detected resource lies with the edition maintainer. After importing the resource, the maintainer may adjust any settings if needed.

Figure 7. Auto-detection example

```
┌──────────────────────────────────────────────────────────┐
│ ┌────────────────────────────────────────────────────┐   │
│ │ Auto Detection: enter the host name of the machine  │   │
│ │ that runs your library catalog, or enter a URL to   │   │
│ │ a page that contains a search form. For instance,   │   │
│ │ try addison.vt.edu, worldcat.org,                   │   │
│ │ www.citeulike.org, pubmed.gov, hubmed.org, or       │   │
│ │ complete URLs such as                               │   │
│ │ http://en.wikipedia.org/wiki/Main_Page,             │   │
│ │ Enter URL: |                                        │   │
│ └────────────────────────────────────────────────────┘   │
│                                                            │
│  OCLC reports that your OPAC may be located at             │
│  addison.vt.edu.                                           │
│  probing addison.vt.edu (7 resources found)                │
│  ⓦ Found catalog 'Addison' in database (from edition  [Add]│
│     'Virginia Tech Edition'/vt)                            │
│  ⓦ Found III Millennium catalog 'Addison: Search' at [Add] │
│     http://addison.vt.edu 🔒                               │
│  ⓦ Found Bookmarklet catalog                         [Add] │
│     'search.serialssolutions.com with Search' at          │
│     http://addison.vt.edu 🔒                               │
│  ⓦ Found OpenSearchDescription 'Keyword Search'      [Add] │
│     Addison' 🔒 (Click to test)                            │
│  ⓦ Found OpenSearchDescription 'Author Search'       [Add] │
│     Addison' 🔒 (Click to test)                            │
│  ⓦ Found OpenSearchDescription 'Title Search Addison'[Add] │
│     🔒 (Click to test)                                     │
│  ⓦ Found OpenSearchDescription 'Journal Title Search'[Add] │
│     Addison' 🔒 (Click to test)                            │
└──────────────────────────────────────────────────────────┘
```

Figure 7 shows a concrete example of a maintainer's experience when opening the "Catalogs and Databases" tab in the Edition Builder (as introduced in Figure 6). In this example, the IP address was recognized as belonging to an OCLC institution (Virginia Tech) whose OPAC is located at http://addison.vt.edu/. Since the OpenURL Resolver Registry contained a matching entry, the Edition Builder was able to fingerprint the server and identify the OPAC type and URL. In this process, the Edition Builder also detected other possible resources, such as an additional search form embedded in the OPAC interface, and three advertised OpenSearch descriptions. Moreover, the Edition Builder reports that this OPAC is already configured in at least one live edition in the shared community database. The "Add" button on the right allows the maintainer to import any of these resources into the configuration. A maintainer creating a first edition for an in-stitution will often find fewer choices, whereas a maintainer creating additional, specialized editions or consortium editions is more likely to reuse resources already configured.

Edition Builder Evaluation

We performed an extensive evaluation of the Edition Builder, which pursued several goals. First, and most importantly, we were interested in whether the introduction of the Edition Builder led to increased adoption of the LibX plug-in, and at what pace. Second, we wanted to understand its efficiency in terms of how much time edition maintainers will have to invest, and its outcome in terms of success rate and the richness of the configurations built by the maintainers. Third, we wanted to identify which factors influence the acceptance of the Edition Builder application by our community. We used two instruments in this research: a log analysis and a user survey.

- **Log Analysis:** The first prototype of the Edition Builder was released to the public in July 2007. We imported the configurations of all editions we had previously built manually into the system as part of this transition. Initial ownership of these editions was assigned to a dedicated maintainer account we created. Subsequently, when a library for whom we had built an edition contacted us for updates or modifications, we referred them to the Edition Builder. Once the contact person from that library had created a maintainer account with the Edition Builder, we transferred ownership of their edition to this account.

We used the months from July – December 2007 to refine the design and fix bugs, with a particular emphasis on the auto-detection facilities. We also implemented the necessary instrumentation facilities to be able to record the interactions of our maintainers with the Edition Builder. We recorded how much time edition maintainers spent building and testing a revision, we analyzed the complexity of the editions they built by counting the number of included features, and we recorded how often they decided to import resources offered by the auto-detection facilities described earlier. Since we cannot verify if a user attempting to create an edition was successful – because doing so would require individual testing of the created edition – we recorded if an edition maintainer reached certain milestones in their interaction with the Edition Builder. First, we record if an edition is "made live" – that is, if a maintainer decided to freeze its settings, which is a step that is usually taken only after successful testing. Second, we record if an edition is "made public." To make an edition public, an edition maintainer must check a checkbox that is initially off. The edition is then added to the list of editions listed on our website, which also allows other maintainers to locate and copy that edition's settings as a template for cloning.

For the log study, we collected access logs recorded during a 5-month study period from Dec. 7, 2007 to May 7, 2008. During this 5 month period, 562 distinct registered users actively used the Edition Builder. These included both maintainers who had returned to claim ownership of an imported edition as well as new maintainers. Although the Edition Builder allows the creation of editions without requiring registration, a feature we constructed to encourage experimentation, we found few visitors took advantage of this feature. During the study period, 2282 attempts at creating new editions were made. Of those, 693 attempts involved the cloning of an existing public edition. The clone initially contains all settings and resources of the cloned edition, but then those settings can be changed independently. We added the ability to clone to simplify the creation of configurations that are similar to existing ones with respect to the resources they include. 57 of the cloned editions were made live.

The remaining 1589 attempts involved editions created from scratch. In these cases, the edition initially did not contain any resources, and the maintainer had to either add auto-detected resources or manually configure them. From the editions created from scratch, 185 were made live.

Figure 8 shows the cumulative distribution for the amount of time that passed between a maintainer starting an edition from scratch and their making it live. The median is 54 minutes. This time delta does not always reflect continuous engagement with the Edition Builder. For instance, the largest data point reflects an edition that was initiated on January 24, 2008 and made live on March 28, 2008 and on which the maintainer worked only 7 days during that period in sessions ranging from about 1 hour to 5 hours. Unfortunately, our log data does not allow us to determine precisely how long an edition maintainer was using the interface during each session. This lack of precision results because maintainers are not

Figure 8. Cumulative distribution of overall time delta between when an edition was initiated until it was made live, for editions created from scratch

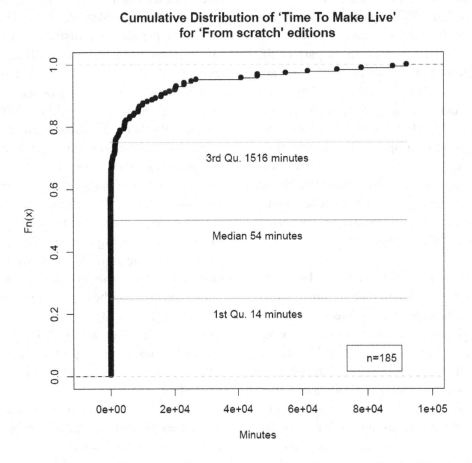

required to "log off," preventing us from reliably determining when they stop using the Edition Builder during a particular session.

For this reason, we performed another analysis to obtain a second estimate for the required effort to create a LibX edition. We estimated the individual session times by recording the time difference between when a maintainer opened a particular revision for manipulation until they made the last change to it during that session, which is an event we log. In this analysis, we consider all editions: those created from scratch during the study period, those created via cloning, and an additional 88 editions that were created before the introduction of the Edition Builder, but

subsequently adopted by maintainers during the study period. The cumulative distribution of these aggregated session time estimates is shown in Figure 9, which exhibits quartiles of 23, 72, and 164 minutes. Combined, these two estimates indicate that the time required to create new LibX editions with the Edition Builder is small.

The small ratio of editions that were made live vs. the overall attempts at creating new editions (242/2282 or 10.6%) does not indicate that the involved edition maintainers were ultimately unsuccessful. At the end of the study period, 46% of all registered maintainers had made at least 1 edition live. This discrepancy likely means that some maintainers simply "started over" in their

Figure 9. Cumulative distribution of estimated total session time for all live editions

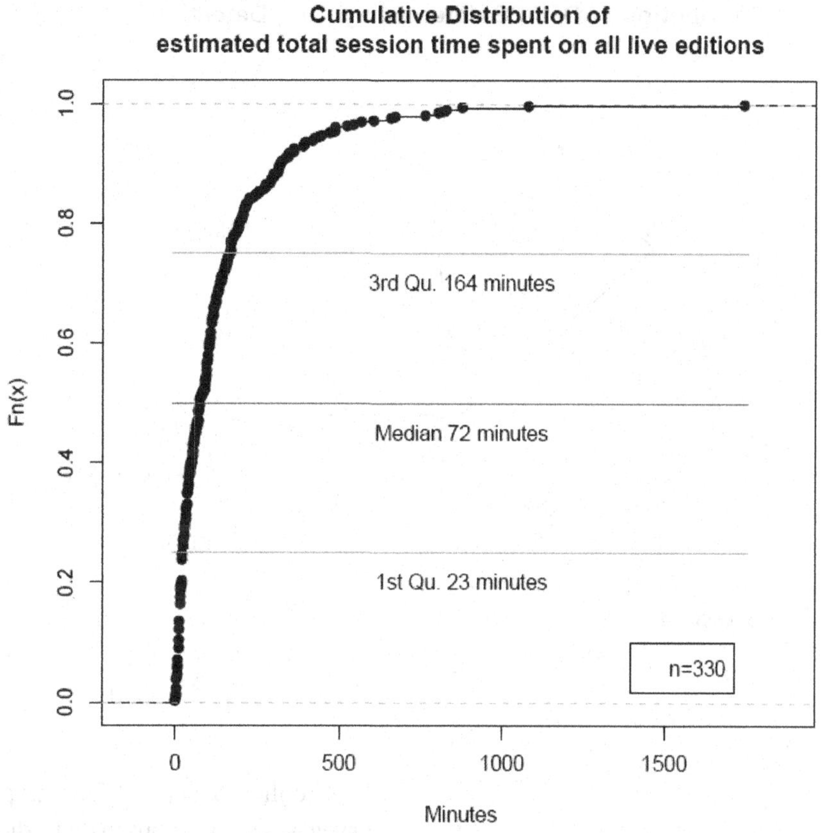

work with the Edition Builder and created a new edition, not realizing that they could have changed the edition they already created to suit their needs.

We further analyzed the complexity of the configurations created for the 330 editions that were made live during the study period. We found that only 3% had failed to add any resource, which indicates a nonfunctional work product. 45% of editions contained 1 resource, typically an OPAC, and 52% of editions had configured more than 1 resource. 50% of maintainers imported auto-detected resources at least once. Figure 10 shows a breakdown of the resources that were imported. Almost half of all resources were of type 'Bookmarklet,' which is the term we use for generic resources, described by a URL template, created from scraped search forms or OpenSearch descrip-

tions. The remainder corresponds to several OPAC vendors for which we provide specific support. The preponderance of III Millennium systems may be explained by the fact that we advertised LibX first in the user group for this vendor's system, as well as by the fact that the auto-detection heuristics for these systems is particularly reliable due to a unique "Server:" header used in its HTTP response.

45% of editions contained a manually-specified proxy configuration, despite the lack of support for auto-detection of proxies. Anecdotal evidence suggests that proxy support is a well-liked feature of LibX. 45% of editions contained at least 1 customized icon or logo; 18% contained 2 or more. Note that such customization is not required to produce a functional edition since we provide a default set of icons.

Figure 10. Breakdown of detected resources by type

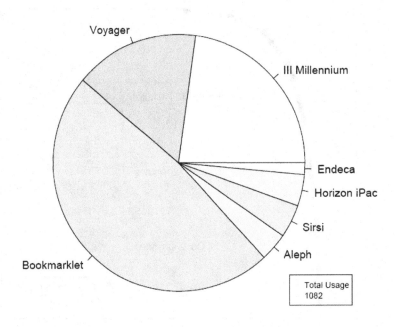

Distribution of OPACS detected by Auto Detection

User Survey

The log analysis described in the previous section measured how maintainers interacted with the Edition Builder, but does not allow us to gauge subjective factors such as perceived ease of use. Hence, we performed a user survey.

We recruited survey participants from willing edition maintainers who had provided a valid email address. Throughout the use of the Edition Builder, maintainers could opt out of contact by clearing a checkbox prominently displayed as part of the user interface. We sent out 810 invitations. 139 respondents participated, yielding a response rate of 17.1%. We conducted the LibX Edition Builder User Survey from June 20, 2008 through July 14, 2008. In contrast with the log analysis study, the audience from which we solicited participants included both maintainers who had been early users of the Edition Builder (during the July - December 2007 time frame) as well as those who had signed up afterwards.

We collected general demographic data, but the survey was anonymous in that it did not collect any personal identifiable information or IP addresses from participants. We also did not ask participants to provide their edition maintainer account, thus we did not link survey responses to a respondent's actual use of the Edition Builder. Survey participants had the opportunity to withdraw from the survey at any time; no data was recorded for those that withdrew from the survey. Participants had the option to skip individual questions, which we recorded as missing values.

The survey contained a total of 33 questions that were divided into seven categories examining different aspects of the Edition Builder. Each section included a brief introduction explaining the type of data the questions were collecting. Part I analyzed the maintainers' overall experience with using the LibX Edition Builder (7 questions); Part II examined maintainers' understanding of essential concepts related to the configuration process (5 questions); Parts III and IV examined maintainers' experiences with the auto-detection features (5 questions) and manual configuration facilities (5 questions); Part V analyzed the experience with the "help" features (3 questions); Part VI addressed the maintainers' impression of using the Edition Builder user interface (5 questions); and Part VII collected demographic information (3 questions). All questions were multiple choice questions; some allowed multiple answers.

- **Statistical Method:** For purposes of the statistical analyses, we created the following variables. 10 questions asked purely ordinal measures, which we mapped to a scale of integers from 0 to N-1. For instance, we asked the maintainers' overall impression with using the Edition Builder and provided as answer choices "Very easy," "Easy," "Slightly easy," "Slightly difficult," "Difficult," and "Very Difficult." We mapped those answers to the integers 0-5, with 0 being "Very easy" and 5 being "Very Difficult." Lower numbers correspond to better ratings. An additional 10 questions contained answer choices from which we could derive ordinal measures. For instance, when asking about maintainers' experience with the proxy configuration, we removed those respondents who had answered that they did not use that feature as their library does not provide a proxy service. We computed the means for each ordinal and derived ordinal variable. The remaining questions asked purely nominal (categorical) measures. For those questions, we derived binomial variables for each answer choice, reflecting whether the answer was chosen (1) or not (0).

We computed Pearson's product moment correlation coefficient for each pair in the above described three groups of observed variables, including ordinal, derived ordinal, and binomial variables reflecting specific answer choices. For each variable, we identified correlations with other variables that exceeded a 0.3 cutoff, ranking the correlations for easier analysis.

Since the absence of linear correlation between the variables derived from a pair of questions does not indicate that the answers to these questions are statistically independent, we also fit an independence model to each pair of questions. In this analysis, we modeled each survey question as a categorical variable. Since the cross tabulations contained many values for which the expected frequency was less than 5, we used Fisher's exact test for this analysis. If Fisher's exact test rejected the null hypothesis that the answers to a pair of questions were independent, and there was also no linear correlation between the answers (either because the question did not map to an ordinal measure for which such a correlation was meaningful, or because the computed correlation did not meet our threshold), then we manually examined the cross tabulation to further characterize this statistically significant lack of independence via additional tests, such as t-tests comparing the means of different subpopulations of answers. Note that we do not draw any direct conclusion from the results of Fisher's exact test for independence, other than using the rejection of the null hypothesis as an indication of the need for further inquiry.

- **Results:** The overall perceived ease of use is a key indicator of the usability of the Edition Builder. Most maintainers surveyed found the Edition Builder "easy" or "very easy" to use (scale 0-5, mean=1.40, sd=0.88, n=138). As shown in Figure 11, only 15 maintainers rated it as "slightly difficult" or "difficult" to use, and no respondent chose a "very difficult to use" rating.

- **Overall Impression vs. Understanding of Concepts:** The overall impression was linearly correlated with several ordinal measures describing the maintainers' understanding of concepts related to the

Figure 11. Overall impression of Edition Builder's ease of use

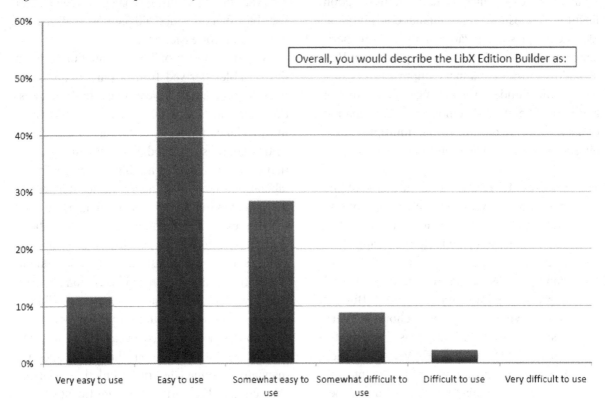

maintenance process. These include the understanding of the distinction between an edition and a revision (r = 0.43[**1]); the understanding of the different types of revisions, i.e., live, archived, and testing (r = 0.44**); and the understanding of the workflow of how to build, test, and revise configurations (r=0.53**). We conclude that the degree of understanding of the underlying workflow abstractions is directly related to the perceived usability.

- **Overall Impression vs. Learning Curve:** We also examined how our maintainers perceived the Edition Builder's learning curve. The perceived learning curve was rated on a 0-5 scale from "Very easy" to "Very difficult." Only 20 out of 138 respondents rated it "somewhat difficult" or "difficult" to learn (mean=1.50, sd=0.93, n=138). Moreover, the perceived learning curve strongly correlated with the overall

impression (r=0.74**). We also asked if maintainers had a favorable initial impression of the ease of use when they started using the system. This initial impression also strongly correlated with the overall impression (r=0.76**), with a majority of respondents having the same initial and overall impression. However, an examination of the cross tabulation revealed that 43 respondents' impression improved, while only 7 respondents' overall impression was lower than their initial impression. Out of the 30 maintainers who initially gave one of the three "difficult to use" ratings, only 15 retained this impression. We conclude that a flat learning curve and a favorable initial impression are strong contributing factors to overall usability.

- **Overall Impression vs. Help Features:** The perceived usefulness of the included help features were also correlated with perceived overall ease of use. These include the usefulness of the built-in help icons (r=0.41**); the rating of the provided FAQ (r=0.36**); the rating of the automatic, live consistency checking facilities (r=0.36**); and finally the rating of the maintainers' experience with the "Help Me With" assistance feature (r=0.37**). Edition maintainers rated their experience with all of these features highly, with a majority agreeing or strongly agreeing that they were useful. Notably, the "Help Me With" assistance, in which a LibX team member assists maintainers with the configuration, was rated on a 0 to 3 scale from "fully answered my question," "mostly answered my question," "partially answered my question," and "did not answer my question" with a mean=0.88, sd=0.91, n=66. We conclude that providing help documentation, automatic live consistency checking, and human assistance when needed all contribute to usability.

- **Overall Impression vs. User Interface Style:** The Edition Builder uses a single page internet application style, rather than the traditional style of web application in which a user must submit forms and navigate between pages. This style was "much" or "somewhat" preferred by the vast majority of respondents (61 and 43, respectively), whereas only 7 respondents preferred the traditional style and 25 respondents did not believe the style matters. Moreover, the overall impression is linearly correlated with this preference when mapped onto a 0-5 scale (r=0.38**).

The second interface choice we made was to forgo "Save" buttons and instead immediately save changes as users made them. When asked about whether this mode of interaction was intuitive, 117 respondents said it was at least somewhat intuitive, whereas only 14 respondents found it counter-intuitive. On a 0-5 scale from "very intuitive" to "very counter-intuitive," we observed a mean=1.32, sd=1.02, n=131. Moreover, viewing this interaction mode as intuitive correlated with overall impression (r=0.37**). We conclude that the use of modern user interface techniques enabled by Rich Internet Application (RIA) technology improves the usability of complex configuration management tools such as the LibX Edition Builder.

- **Overall impression vs. Experience with Auto-Detection of OPAC and OpenURL Resolver:** Since we invested significant effort in the resource auto-detection facilities we implemented, we examined their perceived usefulness and their relation to overall impression. As described earlier, in the ideal scenario, the Edition Builder will have deduced the settings of the primary OPAC of the edition maintainer's community without any input from the maintainer. 45 of 138 respondents had this "zero-effort" experience, an additional 24 respondents reported that the OPAC was detected and the settings needed only slight adjustments to work correctly. Together, these two groups account for 50% of all respondents. For 41 respondents, no OPAC was auto-detected. 7 respondents did not understand how to add the detected OPAC to their edition, and 9 respondents reported that an OPAC was detected, but its settings needed major adjustments.

Fisher's exact test indicated that the overall ease-of-use impression and the experience with the automatic detection of the primary OPAC were not independent (p=0.023). We then compared the means of the overall impression for those respondents whose OPAC was detected and where the settings needed no or minor adjustments after importing (mean=1.20) vs. those whose OPAC was not detected (mean=1.60). A two-sided t-test revealed that the former group had a statistically significant better experience (p=0.014), thus a successful auto-detection experience improved usability.

Since the automatic detection process of the OPAC derives from the automatic detection process of the OpenURL resolver, which is based on OCLC's OpenURL Resolver Registry, we also examined how often maintainers reported that their OpenURL resolver was detected. 44 respondents reported that their resolver was detected and worked when they added it, while 28 reported it was not. When comparing the means of these groups in terms of overall impression, we thought that the group whose resolver was detected had a better impression (mean=1.25) than the group whose resolver was not detected (mean=1.57), but the difference was not statistically significant (p=0.099). This lack of significance may be explained by the relatively large number of respondents who chose an alternative answer that indicated that they were not interested in configuring an OpenURL resolver (52 respondents), perhaps because they were unfamiliar with the purpose of this service. We note that automatic detection of the primary OPAC was effective for about half of our users.

- **Overall Impression vs. Use of Auto-Detection for General Resources:** Whereas the Edition Builder attempts to infer the maintainer's OPAC and OpenURL resolver without any input from the maintainer, the maintainer also has the option to use the auto-detection facilities directly, by inputting URLs to probe for resources. Maintainers were generally successful with this feature as well, with 72 reporting that they were "highly" (28) or "generally" (44) successful in detecting the resources they wanted after providing a URL. 9 respondents reported that they were "rarely" (6) or "never" (3) successful in detecting those resources. Though the number of respon-

dents for which this feature failed is small, their overall impression of the Edition Builder was substantially worse than that of the successful group (means of 2.55 vs. 1.31, p=0.01); in fact, 7 of these 9 respondents found the Edition Builder difficult to use. We conclude that quality of resource detection heuristics significantly improves the usability.

- **Overall Impression vs. Demographics:** We asked three demographic questions: if a participant works in a library, and if so, in which role (librarian vs. paraprofessional staff) and in which department (public services, technical services, or systems). We asked those participants who work in libraries for how long they have worked in the field. Finally, we asked participants about their programming skills.

Table 1 shows the composition of our respondents with respect to their occupation. We did not find any linear correlation above our chosen threshold between membership in any

Table 1. Survey respondents by occupation

Position	Number of respondents
Public Services librarian/faculty	44
Public Services para-professional/staff	0
Systems librarian/faculty	51
Systems para-professional/staff	12
Technical Services librarian/faculty	14
Technical Services para-professional/staff	7
Student - undergraduate	0
Student - graduate	2
Student - other	1
I don't work at a library	7

of the groups and the variables we considered. We compared the means for each group and its complement in the overall impression and perceived learning curve categories. We found that public service librarians found the Edition Builder harder to use (mean=1.70) than the rest of the surveyed population (mean=1.27) at p=0.01. They also found it harder to learn (means 1.77 vs. 1.38, p=0.03). Systems librarians and staff, taken together, found the Edition Builder easier to use than their complement (means of 1.22 vs. 1.56, p=0.02). Without including systems staff, however, the overall impression of systems librarians did not exhibit a statistically significant difference in their mean.

Similarly, we did not observe any linear correlation above the cutoff between the number of years a respondent worked in a library and any other variable. Though respondents with more years of experience found the Edition Builder slightly easier to use, these differences were not statistically significant.

Third, since our motivation had been to enable the creation of LibX editions particularly for librarians with limited programming skills, we asked about their level of programming skills. The majority of survey respondents (89) had at least beginner-level experience in at least one programming language. 30 of these respondents considered themselves to be intermediate or advanced programmers. 48 respondents answered that they do not have any programming skills, but use web applications or their computer extensively. When treated as an ordinal measure, the degree of programming skills was not correlated with any other variable we considered above our cutoff. When comparing the means for the critical overall impression of ease of use variable, we found no statistically significant differences in the means between the groups. This result indicates that prior programming skills are not a precondition for using the Edition Builder efficiently.

- **Maintainer Feedback:** The survey also contained a section for comments on areas in which the LibX Edition Builder could be improved as well as for general comments. The most frequently voiced requests (24 respondents) suggested the creation of additional online help or tutorials. The following comment is representative:

Instructions. Most of the time I was not aware of what I was exactly doing. I appreciate all the work someone did on the edition I copied and used.

As an aside, this comment validated our intuition that providing the cloning feature this maintainer used would ease maintainers' jobs because it allows them to copy and modify an edition, rather than starting from scratch. Despite the large number of respondents that suggested that additional help options would be beneficial, our anecdotal evidence suggests that the existing help information is underutilized, because many questions we received on our "help line" were answered in the online help or FAQ. This experience mirrors the findings of other studies (Novick, Elizalde, & Bean, 2007).

A number of respondents suggested that the Edition Builder should provide an option to retain the authentication of a maintainer across repeated visits, which we have since implemented.

Maintainers also echoed our findings about the importance of understanding the concepts underlying the workflow, as indicated in this comment:

It might be nice to see some links in the FAQ or within the interface to more info about using the Bookmarklet option in Manual Configuration under Catalogs and Databases. Otherwise this is a great interface, I think the only issue is many librarians (including myself) did not start with a clear concept of the software development process, how Firefox plugins work, revisions and archives, etc. The learning curve to these concepts is not at all difficult if one just gives a bit of time to testing plugins and using the Edition Builder interface.

Additionally, some maintainers acknowledged the challenge involved in creating a system such as the Edition Builder, as indicated by the following quote:

It's not super simple configure LibX, but I don't think it's the fault of LibX. It's just a lot of moving parts to deal with. The only thing that really messed me up was a weird caching error with Firefox and older extensions, so that I wasn't seeing updates in the test build. Other than that, you do a great job with a challenging process.

One maintainer commented on the interaction style as follows:

I like the interaction style of the interface, but because there are so few other sites that use it for data input forms like this, it takes a little getting used to (i.e., it's not an established convention yet). Perhaps one improvement would be something prominent that indicates when the data is being saved, as a visual cue that this is happening (like in Google Docs).

We note that the Edition Builder currently does display when data is being saved, but does so unobtrusively by updating a time stamp near the bottom of the interface. This may lead to the uncertainty that this maintainer encountered.

However, there were also a number of comments that demonstrated that not all maintainers were successful in creating a functioning edition, and some felt overwhelmed, as indicated by the following comments:

The help descriptions are still a little too technical for me to understand.

and

I never completed the build because I was unable to find out what was wrong with what I was doing. There is seems to be an assumption that some of the people using the LibX Builder are more computer savvy than they really are. Simplify your instructions so they can be used by a 60 year old librarian and you will have a winning product. I am not 60, but I found the instructions overwhelming and confusing and often difficult to follow.

- **Limitations:** A key limitation of our user study is that participants were self-selected and may not be representative of all potential maintainers. It is possible that the overwhelmingly positive view of our respondents reflected a subset of maintainers that was successful in building editions, and that those maintainers who were unsuccessful did not respond to the solicitation or were underrepresented in our sample.

Modules, Libapps, Packages and Feeds

Similar to how the Edition Builder enables librarians to become creators of software configurations, the Libapp builder enables librarians to become creators of software that runs in a user's browser as they browse the web. To understand its functionality, we will discuss first the programming model used, which consists of Libapps, modules, packages, and feeds.

When designing the end-user programming model for LibX's interactions with webpages, we considered a number of alternatives. First, we considered the use of Greasemonkey user scripts (Pilgrim, 2005). These JavaScript scripts are stored in the browser, and they are invoked when a user visits a URL that matches a given pattern. However, they are difficult to customize and maintain, and they are not modular and make reuse difficult. Most importantly, adaption often requires knowledge of JavaScript, which we did not want to require from all edition maintainers.

Instead, we designed a model based on modules which perform small tasks. These modules are written in JavaScript, but they are created by only a smaller number of JavaScript-savvy adopters, and they are reusable in other contexts. For example, a module might be responsible for extracting an ISBN identifier on a specific book vendor's page. This module can be used for any application in which such extraction is necessary – be it to display availability information, a note to the user, or just a plain link to the catalog. A separate module performs the processing of the extracted information, perhaps to consult a particular backend of a discovery system or OPAC.

Libapps are simply combinations of modules. In programming lingo, Libapps are to modules what programs are to subroutines. However, unlike programs, Libapps's modules are not invoked sequentially, one after the other, to perform their function. Instead, a module is invoked automatically when the preconditions under which it runs are met. We use a tuple space abstraction for this purpose (Gelernter, 1985). As a result, the task of building a Libapp is reduced to combining a suitable set of modules. Modules that produce tuples of a certain pattern must be complemented with modules that consume tuples that match the same pattern.

Packages are collections of Libapps. When related to a computer's file system, a package is to a Libapp what folders are to files. Like folders and subfolders, packages may contain other packages. Users of LibX subscribe to packages. If a user subscribes to a package, their browser will execute all Libapps contained in that package or any of its subpackages. While users can directly subscribe to packages, more commonly the maintainer of the edition they have activated will subscribe all users of an edition to one or more packages that contain Libapps customized for that community. End users have a final say in that they can enable or disable individual Libapps or entire packages via the preferences menu in their LibX plug-in.

If an end user is subscribed to a package, they will receive updates to any module or Libapp contained in that package. This automatic update mechanism is based on the publishing and distribution format in which modules, Libapps, and packages are stored. Rather than invent our own, we chose the widely used AtomPub publishing protocol (Gregorio & de hOra, 2007). Each module, Libapp, and package is stored as an entry in an Atom feed; we use the entry's <id> field as a unique identifier through which to refer to it. The process by which these entries are stored in feeds is hidden entirely from the end users, but it must be understood by Libapp creators and adaptors, because they must organize anything they create in feeds. There are no restrictions on how packages, Libapps and modules must be grouped in feeds. For instance, a user may create a new feed and place a Libapp in it that refers to modules stored in a separate feed (thereby reusing them). The organization of entries affects, however, updates - all entries that are stored in the same feed are updated simultaneously.

The Libapp Builder

The Libapp Builder is a companion web application to the Edition Builder that allows the creation, management, sharing, and distribution of Libapps, which incorporates managing modules and packages. We based its user interface design on the successful Edition Builder design, using the same navigation and interaction style.

At its core, the Libapp Builder allows its users to create and publish feeds. Each feed contains packages, Libapps, and modules, which form a hierarchy (packages contain Libapps or other packages, and Libapps contain modules). Libapp creators can visualize and manipulate this hierarchy using a tree control, similar to the one found in operating systems shells such as Windows Explorer. Each node in the tree can be selected and inspected and its attributes changed.

Like the Edition Builder's ability to share configured resources, the Libapp Builder provides a means to perform a full-text search across all feeds. In this way, users can look for and find modules that may already implement functionality they need. Similarly, the Libapp Builder is integrated with the Edition Builder – edition maintainers can search the Libapp Builder database for packages that are suitable for their users.

Modules and Libapps may accept parameters that control their execution. For instance, a module may be charged with embedding a YouTube video into a page. It is controlled by a parameter 'youtubeid,' which represents the unique identifier for the video. Consequently, if a Libapp creator wishes to embed a YouTube video with a tutorial in a page, they can reuse this module and provide, as a parameter, the desired youtube video id. Requiring parameters raises a robustness issue because the Libapp will fail if a required parameter is not provided. The Libapp builder provides immediate feedback if a Libapp includes a module that requires a parameter, but the parameter is not specified or provided with a wrong type. Similarly, the Libapp Builder checks and ensures that a Libapp is functional by alerting the user when they include a module in an application that requires a tuple that is not produce by any other module in that Libapp. For instance, a module that invokes a service that processes ISBNs will need to be matched with a module that produces ISBN tuples.

FUTURE RESEARCH DIRECTIONS

We may view the LibX Edition Builder and LibX Libapp Builder as meta-design tools that contribute to a shift from "consumer cultures" to "cultures of participation," and "the new understanding of motivation, creativity, control, ownership, and quality" such meta-design enables (Fischer, 2009). But whereas the Edition Builder has been studied and has been successfully deployed for almost 5 years at the time of this writing, the Libapp builder is still in its infancy.

When creating end user development tools (Burnett, Cook, & Rothermel, 2004) it is all too easy to forget the perspective of adopters, who may have limited programming knowledge, and substitute the perspective of programmers instead.

Thus, a key area of future research will be to see if the Libapp Builder can gain momentum. In our opinion, this will require the following – (1) a robust and easy-to-use technological infrastructure, (2) seeding with modules that are reusable in the context of other Libapps, and (3) seeding with Libapps that provide examples that can be adapted by others through suitable parameterization.

CONCLUSION

LibX is an Open Source project that provides software tools for libraries. Since its inception, its use and adoption have steadily grown, reaching now ~200,000 regular users affiliated with over 1,000 libraries. In this chapter, we discussed how LibX can be used to streamline access to resources. The key idea behind LibX is to create a presence for the library and its resources at the point of need, in the user's browser. At a meta-level, LibX is also an example of how to automate and streamline the creation and customization of software tools through interactive, web-based software tools. This chapter provided an in-depth evaluation of the effectiveness and user perceptions of such tools.

We are grateful to Kyrille Goldbeck for her contributions to the Edition Builder survey. We are also indebted to the great team of graduate and undergraduate students who in the past six years have made invaluable contributions to LibX: Michael Doyle, Brian Nicholson, Tilottama Gaat, Nathan Baker, Sony Vijay, Tobias Wieschnowsky, Travis Webb, Arif Khokar, and Rupinder Paul.

REFERENCES

Bailey, A., & Back, G. (2006). LibX - A Firefox extension for enhanced library access. *Library Hi Tech*, *24*(2), 290–304. doi:10.1108/07378830610669646.

Bailey, A., & Back, G. (2007). Retrieving known items with LibX. *The Serials Librarian*, *53*(4), 125–140. doi:10.1300/J123v53n04_09.

Baker, N. E. (2007). *LibX IE. University Libraries*. Blacksburg, Va.: Virginia Polytechnic Institute and State University.

Burnett, M., Cook, C., & Rothermel, G. (2004). End-user software engineering. *Communications of the ACM*, *47*(9), 53–58. doi:10.1145/1015864.1015889.

Chen, H., & Cheng, R. (2007). *ZK: AJAX without the JavaScript framework*. Apress.

Chudnov, D. (2006). Coins for the link trail. *Library Journal*, *131*, 8–10.

Clinton, D. (2008). *OpenSearch 1.1 (Draft 4)*. Retrieved August 13, 2009, from http://www.opensearch.org/Specifications/OpenSearch/1.1

Fast, K., & Campbell, G. (2004). 'I still like Google': University student perceptions of searching OPACs and the web. *Proceedings of the American Society for Information Science and Technology*, *41*(1), 138–146. doi:10.1002/meet.1450410116.

Fischer, G. (2009). End-user development and meta-design: Foundations for cultures of participation. End-user development, 5435, 3-14). Berlin: Springer.

Gaat, T. (2008). *The LibX edition builder. University Libraries*. Blacksburg, Va.: Virginia Polytechnic Institute and State University.

Gelernter, D. (1985). Generative communication in Linda. *ACM Transactions on Programmable Language Systems*, *7*(1), 80–112. doi:10.1145/2363.2433.

Gregorio, J., & de Hora, B. (2007). *The atom publishing protocol*. RFC 5023. 2012. Retrieved from http://tools.ietf.org/html/rfc5023

Mesbah, A., & van Deursen, A. (2008). A component- and push-based architectural style for AJAX applications. *Journal of Systems and Software*, *81*(12), 2194–2209. doi:10.1016/j.jss.2008.04.005.

Nicholson, B. R. (2011). *Libx 2.0. University Libraries*. Blacksburg, Va.: Virginia Polytechnic Institute and State University.

Novick, D., Elizalde, E., & Bean, N. (2007). *Toward a more accurate view of when and how people seek help with computer applications*. Paper presented at the SIGDOC '07: Proceedings of the 25th annual ACM international conference on Design of communication, El Paso, Texas, USA.

OCLC. (2003). *xISBN Web Service*. Retrieved April 22, 2012, from http://www.worldcat.org/affiliate/webservices/xisbn/app.jsp

OCLC. (2005). *Perceptions of libraries and information resources*. Retrieved April 22, 2012, from http://www.oclc.org/reports/2005perceptions.htm

OCLC. (2007). *xISSN (Web Service)*. Retrieved April 22, 2012, from http://xissn.worldcat.org/xissnadmin/index.htm

OCLC. (2012). *WorldCat Registry*. Retrieved from http://www.oclc.org/registry/default.htm

Pilgrim, M. (2005). *Greasemonkey hacks: Tips & tools for remixing the Web with Firefox*. Sebastopol, CA: O'Reilly Media.

Sayers, E. (2010-). Entrez programming utilities help. Retrieved from http://www.ncbi.nlm.nih.gov/books/NBK25500/

Sompel, H. V. D., & Beit-Arie, O. (2001). Open linking in the scholarly information environment using the OpenURL Framework. *D-Lib Magazine*, *7*(3).

Tilevich, E., & Back, G. (2008). *'Program, enhance thyself!': Demand-driven pattern-oriented program enhancement*. Paper presented at the Proceedings of the 7th international Conference on Aspect-Oriented Software Development.

ADDITIONAL READING

Bailey, A., & Back, G. (2009). Rating ILS interoperability: A proposal. *Information Standards Quarterly, 21*.

Blyberg, J. (2005). ILS customer bill-of-rights. Retrieved June 18, 2009, from http://www.blyberg.net/2005/11/20/ils-customer-bill-of-rights/doi:citeulike-article-id:3057267

Burnett, M., Cook, C., & Rothermel, G. (2004). End-user software engineering. *Communications of the ACM, 47*(9), 53–58. doi:10.1145/1015864.1015889.

Cindi, T., & S, P. J. (2010). *Rethinking library linking: Breathing new life into OpenURL* (Vol. 46). Chicago, IL: American Library Association.

Conduit. (2009). Retrieved July 15, 2009, from http://www.conduit.com/

Curbera, F., Duftler, M., Khalaf, R., Nagy, W., Mukhi, N., & Weerawarana, S. (2002). Unraveling the Web services web: An introduction to SOAP, WSDL, and UDDI. *IEEE Internet Computing, 6*(2), 86–93. doi:10.1109/4236.991449.

Datema, J. (2006). Shift happens: LITA faces change at national forum. *Library Journal, 131*(20), 27.

Fischer, G. (2009). *End-user development and meta-design: Foundations for cultures of participation. End-User Development, 5435* (pp. 3–14). Berlin: Springer.

Fischer, G., Giaccardi, E., Ye, Y., Sutcliffe, A. G., & Mehandjiev, N. (2004). Meta-design: A manifesto for end-user development. *Communications of the ACM, 47*(9), 33–37. doi:10.1145/1015864.1015884.

Garrett, J. J. (2005). Ajax: A new approach to web applications. Retrieved July 15, 2009, from http://adaptivepath.com/ideas/essays/archives/000385.php

Giles, J. (2005). Science in the web age start your engines. *Nature, 438*(7068), 554–555. doi:10.1038/438554a PMID:16319857.

Jason, P. (2010). Superpower your browser with LibX and Zotero. *College & Research Libraries News, 71*(2), 70.

Jason, S. P., & Cindi, T. (2010). Sources and resources. *Library Technology Reports, 46*(7), 34.

Mitchell, B. (2007). Library toolbars for use and development. *Library Hi Tech News, 24*(8), 33–35. doi:10.1108/07419050710836027.

Ritterbush, J. (2007). Supporting library research with LibX and Zotero. *Journal of Web Librarianship, 1*(3), 111–122. doi:10.1300/J502v01n03_08.

Schneider, K. (2008). Free for all. *School Library Journal, 54*(8), 44.

Webster, P. (2007). The library in your toolbar. *Library Journal, 132*(12), 30.

KEY TERMS AND DEFINITIONS

AJAX: Asynchronous JavaScript and XML. A communication mechanism that allows web pages to be updated after they are displayed to the user, rather than requiring navigation to a new page.

Application Programming Interface (API): A set of specifications used by software components to communicate with one another, either through direct invocation or via communication through a web service.

Autodetection: Automatic detection of applicable resources or services.

Autolinking: Automatic provisioning of hyperlinks for identifiers such as standard numbers found in an HTML page.

End-User Programming: Approaches that allow users of a system who have no formal training in programming to create software.

Extensions (Add-Ons, Plug-Ins): Software components that extend a base system, such as an off-the-shelf web browser, to add additional functionality.

Open Source Software: Software that is released under a license that grants certain rights, including the right to make changes to the software's source code.

Rich Internet Application (RIA): A style for web-based applications that resemble desktop applications, but run in a browser and do not require any local download or installation.

Web Services: Network-based information services that are designed for machine-to-machine interaction and consumption, typically using the HTTP protocol suite.

ENDNOTE

[1] Throughout the chapter, we use ** to denote statistical significance at the level $\alpha=0.01$, and * to denote significance at $\alpha=0.05$.

Chapter 5
Who Ever Said that ILS Migrations Had to be Stressful?

D Ruth Bavousett
ByWater Solutions, USA

ABSTRACT

Traditionally, migration from one integrated library system (ILS) to another has been an arduous, difficult task; so much so, that libraries may choose to stay with an unsatisfactory ILS longer than they would if a viable and easy alternative were available to them. However, this institutional inertia is not necessary, if the library (possibly in cooperation with a vendor) develops a method for avoiding unnecessary and problems. In this chapter, a process is described that, when implemented, maximizes results, while minimizing pain and stress on the library and its' staff.

INTRODUCTION

As libraries seek ways to save money in this period of declining budgets, one of the factors that may attract attention is the integrated library system. License and support fees for traditional ILSes continue to rise, while functionality frequently does not keep pace. Libraries may look at open-source systems such as Koha and Evergreen as possibilities; these systems are consistently adding new functionality, and both have active developer communities around them. Additionally, there are support vendors available, if the library does not have the technical expertise to "go it alone."

However, if library staff have gone through one or more ILS migrations in the past, they may be reticent to do so again, even with the stated goal of saving significant amounts of money on licenses and support. This reticence may come from a number of factors in the library's institutional history, including frustrations with vendor representatives, lack of openness or robustness in the data-manipulation process, insufficient testing and training protocols, and the slow pace of the work.

Here, we will discuss a process for migrating from a closed-source ILS to Koha, as an example for how to construct a process and team to do so effectively and with minimal stress and cost. It is to the intrepid souls that are facing the new ILS selection and migration process that this chapter is dedicated and to whom it is addressed.

DOI: 10.4018/978-1-4666-3938-6.ch005

ABOUT KOHA AND OPEN-SOURCE SOFTWARE

Koha is a free-for-the-taking, open-source integrated library system, first developed in 1999 for the Horowhenua Library Trust, in New Zealand. Its development is ongoing, centered around a community web site at http://koha-community. org. Koha is in use by libraries around the world, of all types, and new releases are coming out regularly. The community maintains a number of mailing lists to support development efforts and the user community, and knowledgeable people are available on the team IRC most hours of the day (Koha Community). Additionally, there is a useful bibliography of Koha development and history on Zotero, at https://www.zotero.org/groups/koha/items

Koha is licensed under the GNU General Public License, or GPL, Version 2. In short, this means that anyone may take the source code, and use it for any means that they can devise, and modify it in any way that they desire, so long as the modified code is, in turn, given under the same license. No license fee may ever be charged for software under the GPL (Free Software Foundation). This means, of course, that multiple support vendors may arise who are "selling" the same software—in truth, they are not selling the software, but their migration services, support, custom programming, hosting services, and/or training. Libraries may choose one of these vendors, or may utilize in-house expertise to install and implement Koha in their library.

CONSIDERATIONS FOR THE MIGRATION PROCESS

Role and Selection of a Vendor

It behooves you to make a careful, honest evaluation of in-house talent before deciding whether or not to engage a vendor. In the case of Koha, a good understanding of Debian Linux, Perl, and MySQL is needed; the system administrator for the Koha system will need a practical knowledge of installation and configuration of all these components to successfully "spin up" the Koha system. Additionally, if the library is planning to perform their own data migration, then someone with an extensive knowledge of Perl or other scripting languages will be needed, to manipulate the data. This person should have or be prepared to acquire a thorough knowledge of MySQL and the Koha data structures, as well as the structure and composition of the legacy ILS data.

Many libraries either do not have this level of expertise on-staff—or have the available time for those staff members to work on the migration to and maintenance of a new ILS. Enter the vendor community. Obviously, if a library is statutorily required to go through a Request for Proposal (RFP) process, then that is the first step.

As an aside, let's be honest for a moment: No one likes RFPs. Library administrators do not like having to create or process them, and vendors dislike having to answer them. However, they are required in many jurisdictions, so everyone grits their teeth and charges on. You can make this process a bit less painful by making the RFP as brief as possible—only strenuous enough to satisfy your statutory or process requirements. Pages and pages of endless questions about functionality, which may or may not apply to that library, or the proposed solution benefit no one; in the case of an open-source system, the answers will be very repetitive, since the software is the same. Use the RFP to ask the big questions about standards and broad functionality, and about what the vendor can do for the library, rather than trying to make it into a training course on the ILS.

When selecting a vendor, it is common practice to ask the vendor for customer references. When asked, vendors are usually pretty careful to choose customers who really like them, so it follows that this exercise really does not inform the potential customer very well. In the case of an open-source community, it is wise to engage the developer and user community directly. Send an email to a mailing list, or log on to a chat, and ask around.

Obviously, questions about the vendor's customer-support quality are germane, but other questions may also be relevant, including how well that vendor is perceived in the larger developer and user community. Are they regularly and consistently contributing code for inclusion? Are they providing added value to the community, such as sponsorship of events, hosting the community websites, or ensuring that their customers have opportunity to participate in the community directly? Do they support other community players who are not their customers, by being available to answer questions via email or chat? Even if the idealism of an open-source solution is not a driving factor for you, such notions are relevant; the vendor's participation in a robust community of developers gets the library a better ILS, so this is the time to find out about those things.

Another resource that may be helpful in the quest to learn more about vendors is Lib-Web-Cats. Implemented by Marshall Breeding, Lib-Web-Cats is a directory of libraries, with their ILS provider listed, and the data frequently includes which ILS they were on previously; this can be very useful for finding a similar library with which to discuss a vendor or migration process (Breeding).

Finally, your discussions with a prospective vendor may be enhanced by treating it more like a job interview. Theoretically—and, in the eyes of the vendor, hopefully—this relationship will last for many years, it involves a great deal of money and a lot of interaction; how is that different from an employee-hiring relationship? Care should be taken to ask the questions that will answer the question, "How well can we get along with this vendor's corporate culture." Ask to have a phone call with support staff; vendors would rather do that, than have another RFP! Finally, trust your instincts; if you're a hiring manager with a good track record of finding skilled employees, your skills in that will serve you well at this point of the process.

The Hardest Part of an ILS Migration

One of the most-useful realizations you can have in the ILS migration process is that the hardest, most-unpredictable thing you will need to handle is not the data—it is people. The data is really pretty straightforward; while complex, its' behavior is pretty well-defined. The data will not change its' mind on you when it's having a bad day, nor will it take out its' frustrations with training on you, or take a sick day when you really need answers to questions.

The commonest complaint in many projects, from library staff, is "I cannot handle change." Most of the time, however, this is not a complaint about change *per se,* but about unknowns. You see, unless you've stopped growing older, you already deal with change, every day of your life—we get older, our kids get older, our cars age, seasons come and go, the weather changes, politics changes—well, never mind about that last one. The staff member is often, at that point, concerned not so much about change, but about how they will do their job on the day after go-live. It is best to establish a process, either internally or with your vendor, to answer these concerns quickly, lest they fester into animosity and mistrust. The testing and training process described here has also been useful in allying these concerns, as it gives staff members plenty of "hands-on" time to try things out and ask questions so that they will know how things will work after go-live.

Gathering Useful Tools, and Building a Team

If you have elected to go through a vendor, most of the technological tools will be created and used by the company doing the work, and you can spend this part of the project thinking solely about people. If not, you'll find knowledge of Perl useful; see the Additional Reading section for a couple of good Perl books. Commonly-used CPAN modules include MARC::Record, MARC::Field,

Text::CSV_XS, and Getopt::Long. The MAR-CEdit program from Terry Reese is very useful (Reese), as is LibreOffice (Document Foundation), whose Base and Calc modules are Swiss Army can-openers for strange data formats. The Koha community has a collection of migration-related tools and data scripts (Koha Community, Koha Migration Tools repository), and another, less-well-structured repository of code and tools is available, as well (Bavousett, 2012).

For a team, you'll need technology people—or a vendor—and library data specialists within your library. If you are hosting your system yourself, you'll need a system administrator to set up and run the Koha system, as well as someone to script and execute the actual migration. It is useful to have a small team of specialists in the various areas of library services meeting from time to time to talk about data issues. This could include your head of public services, your head of technical services, the director or an assistant, and other department heads, depending on what data you are going to migrate (serials, acquisitions, special collections, etc). Questions from the library staff should be channeled to these people for action, either for direct answers, or referral to the vendor. The people on this team are also the ones who know the collection and library most-intimately in their areas of expertise, and they know where the oddball edge-cases are, like that item that had to be cataloged manually because of some strangeness, or the patron who has a long history with the library and requires special handling. Having them in a group, able to voice their issues coherently, without letting any of them steamroll their plans or hold the entire project hostage to minor matters, can smooth the training and testing process considerably. Include them in meetings with your vendor, and make time in the schedule for them to ask questions, or provide a way for them to log those questions in a way that can be tracked.

It's easy, at this point, to say, "Everyone's important, so we'll include everyone." This is only a useful strategy if your staff is very small, say, five or fewer. Once the group gets larger, things get lots harder, logistically; scheduling meetings becomes difficult, and the meetings can become very disorganized and confusing. A smaller, well-focused team, trusted by other people in their departments, can do a much more efficient job, and can serve as a channel into and out of the team, and to and from the vendor in a much more useful way. If you're using a vendor, this also lowers the stress on your vendor, as a single point of contact is much easier for them to work with.

Process for Analysis and Cleanup of Existing Data

At this point in the process, you may be looking at the mountain of data that is your existing ILS, and wondering where to start. The trick to making the complexity manageable is to "chunk down;" to deal with things that are coherently one type of record, then use it as a building block for something else, and so on. Ideally, exports from the legacy ILS can be in some easy-to-manipulate form—MARC records, or comma-separated values (CSV), for everything non-MARC. In a worst-case scenario, you may have to run reports that do not output CSV, and "scrape" them into a CSV. With these thoughts in mind, start with the collection—bibliographic and item records.

Koha's import tools can easily handle MARC or MARCXML records with embedded item data in 952 tags; look to that format as your goal. Most ILSes can export MARC records, and you can use MarcEdit or Perl's MARC::Record module to transform them. You might need to export item data separately, and "splice" this data in; if that is the case, make sure there is a coherent match point—a bibliographic record number, OCLC number, or some other useful matcher. In the Koha migration tools repository, at migration/Generic/biblio_masher.pl, is a script for doing manipulations of either embedded-item MARCs, or MARCs that have items in an external CSV, which has been successfully used on data from many different proprietary ILSes. Documentation and development for this script is an ongoing work; feel free to ask around in the community for help with it. It will produce MARC records, and if one or more records are too long because of embedded items, it will create a MARCXML file containing those records. It also produces SQL commands to create branch codes, item types, and location and collection codes in Koha. There is a chart in the Koha manual at http://manual.koha-community.org/3.8/en/catguides.html#itemcatguide that you will find useful for building the mapping between legacy data and what Koha expects to find in its' embedded 952 entries (Koha Community).

At a minimum, that can get your bibliographic records and items loaded; however, this is a good time to think about any changes to encoded data that you may have. If you've been thinking about changing item types, by merging them, or deleting item types that are no longer used, or adding or removing collection codes, this is an excellent time to do so, while the data is "out on the table." Rather than make these changes in your legacy system—particularly by hand—plan instead to use the --map directive of the script to shift things around. If you map an item type to the literal string NULL, then the item type will be deleted, as all items must have a type. Likewise, in location and collection codes, you can map values to NULL, and no value will be placed in the record for import. Koha uses ten-character codes for these coded values; it is fairly-standard to use all uppercase for these, and avoid spaces and punctuation, to distinguish them from the descriptions of these values that are displayed in the catalog and staff client.

In a similar fashion, you'll want to export your borrowers, and manipulate them into a CSV that will work for Koha's patron loader tool. As before, coded values can be mapped to new values, as

needed. You may have some manipulation to do with names and addresses; not all systems store thing the same way that Koha does. The script migration/Generic/patron_masher.pl will be of some use in this manipulation.

Other data, particularly checkouts, holds, and fines, should be loaded up in the database using barcodes as the matchpoints; Koha requires that there be only one barcode per item or patron, and those values must be unique in the database, so they make excellent connection points. Here's the minimum you need to make things work:

1. **Checkouts:** Barcode of the item and patron, date of the checkout, date due, and optionally, the number of renewals so far.
2. **Holds:** Barcode of the patron, barcode of an item attached to the bibliographic record, the desired pickup location, and the date the hold was placed.
3. **Fines:** Barcode of the patron (and item, if there is a related item), why this fine has been charged, the date it was charged, and the amount currently due.

Hopefully, by now, you see the general philosophy—take small bites, make sure you have matchpoints, and add each type of record as a building block that sits atop prior work, so you can back out and try again, if you get something amiss.

Coding and Scripting for Data Manipulation

As Koha is written in Perl, most of the community is more-familiar with Perl than with other scripting languages; if you are more comfortable doing manipulations in PHP or some other string-manipulation language, then use that, if you don't mind reinventing some wheels along the way. Your goal is to have the legacy ILS export data in a way that works non-destructively and repeatably, then make manipulations to the data as needed, so that Koha can import that data using built-in tools or simple SQL INSERT operations.

There are a growing number of scripts already in the migration repositories; you might look there first, particularly in the ILS-specific directories, to see if your particular problem has already been conquered. If you need to write new scripts, take a look at existing ones for some coding-style guidance, and try to make your code flexible and readable, and not having things specific to your library hard-coded in, so that you can share your work with the rest of the community. You may think your script is a little bitty, trivial thing, but to a library that comes after you, it can be very useful, so share and be proud of your accomplishment!

Here are some thoughts about coding in Perl that will save you a lot of painful debugging and stress along the way:

1. Include "use strict" in your script. This prevents you from declaring a variable with one name, and using another name for the same thing—all variables must be declared.
2. Likewise, "use warnings" is helpful; it turns on warning messages that might clue you in to an assortment of logical errors in your code.
3. Koha has recently started requiring the CPAN module "Modern::Perl", which encapsulates "warnings", "strict", and several other useful modules; you can use that module instead, as a shortcut, if you're doing your data manipulation on your Koha server.
4. One of the fundamental practices to good Perl coding is to be sure and check your return values from input/output functions, to prevent, for instance, reading from a file that does not exist (which won't read anything, and make you wonder what went wrong). If you will "use autodie" in your script, then any failure in an "open", "close", or "print" statement will immediately cause your script to die with a useful error message.

5. There are a lot of methods in *Perl Best Practices* that can do your code a lot of good, from a readability standpoint; in particular, pay attention to the sections on code layout, naming conventions, and control structures. (Conway, pp. 8-50, 93-131).
6. Install and use "perlcritic" and "perltidy" on the system that you are using to do your scripting; properly configured, these will automate a lot of the coding-style matters, and let you know where you are violating good style guidelines.
7. Code your script with a setting to show what you'll be changing, without changing anything, and/or require that a user enter a setting to allow the script to actually change things in a Koha database. Separate --debug and --update flags will give quite a bit of granularity to the debugging process.
8. You can get a lot of useful help on Perl from the Perl Monks (PerlMonks). They're not a code-writing service, but if you've tried something, and can't quite make it work, a well-defined question there will usually get an answer very quickly. Show the code you've tried, and be sure to describe what's going wrong.

Project-Management Processes and Tips

A quote commonly attributed to Albert Einstein is, "make things as simple as possible, but no simpler." That's actually a paraphrase of a much-longer, and ironically complex statement, but the truth is there—your project should be as simple as possible, without losing anything.

Earlier, there was some discussion of whom to include in a project, and the folly of grabbing everyone indiscriminately, so that need not be reiterated, but on a related note, think carefully about meetings. As an aid to planning effective meetings, use the PACER method, and know these five things before you begin:

1. **Purpose:** Why are we here?
2. **Agenda:** What are the specific items for discussion?
3. **Code of Conduct:** How do we expect participants to contribute?
4. **Expectations:** What do we intend to know by the end of the meeting?
5. **Roles:** Who is leading, and who is taking notes or doing other things the group needs?

If you're working with a vendor, there will usually be a kickoff meeting, to introduce all the people that will be working on your project, and then one or more mapping meetings, prior to the training. After training, there may be meetings to deal with any issues that arose during training, and a final meeting just before go-live. Most other issues can be handled between your local contact, and your vendor's specialist, without a larger meeting being necessary, so think hard about why you're having meetings. "Keeping people informed" is a fine reason to send an email—it's a less-fine reason to have a meeting. If any one of the PACER criteria cannot clearly be defined for a meeting, you need to carefully examine whether or not the meeting should even be held.

You'll need to keep a working document related to your scripting—how you exported data, how you manipulate the data, and how you import it. This should be a "recipe," a list of commands and processes to run. Plan on repeating this process at least twice, possibly more, with the final goal of being able to run the entire "recipe" without interruption or changes on go-live weekend. Box 1 shows part of a recipe, in this case, for a SirsiDynix Symphony migration to Koha. This portion of the recipe covers manipulation and input of bibliographic, item, borrower, and checkout records:

Your final "recipe" will look quite a bit different, in most cases, but note the process: Manipulate data, then load it, then load pieces that use that as a building block. Don't worry about indexing early on (the last line of the script in Box 1); it's easier to just do that from scratch later.

Testing and Training

Testing your migration work is fairly simple; log into Koha, and look at the records you just loaded, spot-checking a few against your legacy ILS to see that you got what you expected. Do this in phases, as it's best to be totally happy with your bibliographic and item records before proceeding to other things. A problem with MARC records found late in the game will require almost everything to be tossed out for reload, so don't be afraid to toss it out early, and start over as needed!

You should examine records both in the staff client and in the public catalog, to make sure that you see how things will look and work for staff and patrons. If you see things that concern you, the earlier you can raise that issue, the better! Have a few key staff participate in this testing; senior technical services staff, for instance, will have the best knowledge of the collection, while senior public-services team members will know borrowers and transactional data.

If your vendor has given you access to a ticketing or tracking system, use it! Be careful not to chain multiple issues into one ticket, instead writing a new ticket for each problem. This makes things easier on your vendor's data migration team, and lets them solve your issues more coherently, without missing something buried in a long ticket. Examples are always, always relevant to this process! The barcodes of items or borrowers that are exhibiting the problem will be greatly appreciated by your vendor. Describe what's wrong, and what the affected record should look like.

Remember that bit earlier, about not involving people you don't absolutely have to have in the process? When it comes to training, throw that out the window! Everyone who will be using the new ILS should receive training, either from your vendor, or from internal sources. The vendor will probably have a test server up and running for your training, and hopefully, you can continue to use it for a while, to give people time to test things out and get comfortable with the new system.

Box 1. Code "recipe" example

```
perl ~/tools/migration/Generic/biblio_masher.pl \
--marc=/home/load/data/biblios_charsetfix.mrc \
--out=/home/load/data/biblios_load.mrc \
--xml=/home/load/data/biblios_load.xml \
--codes=/home/load/data/biblio_codes.sql \
--pricedefault=350a \
--dropfield=852 \
--dropfield=942 \
--dropfield=952 \
--dropfield=999 \
--item=/home/load/src/items.data \
--delimiter=pipe \
--matchpoint="998a:catkey~^a" \
--charset=utf-8 \
--itemcol=homebr:a \
--itemcol=homebr:b \
--itemcol=loc:c \
--itemcol=access:d~date2 \
--itemcol=price:g~div:100 \
--itemcol=barcode:p \
--itemcol=issues:l \
--itemcol=seen:r~date2 \
--itemcol=seen:s~date2 \
--itemcol=price:v~div:100 \
--itemcol=itype:y \
--itemcol=curloc:0 \
--itemcol=curloc:1 \
--itemcol=scheme:2 \
--itemcol=curloc:7 \
--itemcol=call:o \
--itemcol=cat2:8 \
--itemcol=cat1:8+ \
--tally="a,b,c,y,2,8" \
--map=a:/home/load/data/Maps/branch_map.csv \
--map=b:/home/load/data/Maps/branch_map.csv \
--map=c:/home/load/data/Maps/loc_map.csv \
--map=y:/home/load/data/Maps/itype_map.csv \
--map=2:/home/load/data/Maps/callscheme_map.csv \
--map=0:/home/load/data/Maps/discard_map.csv \
--map=1:/home/load/data/Maps/lost_map.csv \
--map=7:/home/load/data/Maps/not_for_loan_map.csv \
--map=8:/home/load/data/Maps/ccode_map.csv
```

continued on following page

Box 1. Continued

```
perl ~/kohaclone/misc/migration_tools/bulkmarcimport.pl -d \
-file /home/load/data/biblios_load.mrc
perl ~/kohaclone/misc/migration_tools/bulkmarcimport.pl -m XML \
-file /home/load/data/biblios_load.xml
perl ~/tools/migration/Generic/modify_items.pl \
--where="location='ONSHELF'" --field=location --val=NULL --update
perl ~/tools/migration/Symphony/By_API/patron_masher.pl \
--in=/home/load/src/users.data \
--out=/home/load/data/patrons_load.csv \
--attrib=/home/load/data/patron_attributes.csv \
--pass=/home/load/data/patron_passwords.csv \
--toss-profiles=DELETE \
--patron-cat=/home/load/data/Maps/patron_cat_map.csv \
--branch-map=/home/load/data/Maps/branch_map.csv \
--cat1=sex \
--cat2=RECIP \
--cat3=TOWNSHIP \
--cat4=DISTRICT \
--cat5=REGLIB \
--xinfo=OCCUPATION:OCC \
--xinfo=LICENSE:DLNUM \
--map=sex:/home/load/data/Maps/gender_map.csv
perl ~/tools/migration/Generic/table_loader.pl \
--in=/home/load/data/patrons_load.csv \
--table=borrowers --update
perl ~/tools/migration/Generic/password_loader.pl \
--in=/home/load/data/patron_passwords.csv --update
perl ~/tools/migration/Generic/borrower_extended_attribues_loader.pl \
--in=/home/load/data/patron_attributes.csv --update
perl ~/tools/migration/Symphony/By_API/charge_loader.pl \
--in=/home/load/src/charge.data \
--dup=/home/load/data/dupe_checkouts.log --update
perl ~/tools/migration/Generic/modify_items_based_on_issues.pl --update
perl ~/tools/migration/Generic/tidy_codes.pl --update
~/kohaclone/misc/migration_tools/rebuild_zebra.pl -r -w -v -v -b -x
```

Timing of training is a matter of some debate; if you do it too soon, it gets "stale" before go-live. Too late, and there isn't time to fix any problems you find. Three to four weeks ahead of go-live seems to work for many libraries, as this gives adequate time for problem-fixing, as well as plenty of time for staff to play around with the system, and learn where things are. A process-based approach works for many library staff. They have a list of things they need to do on any given day, and covering the steps to perform those tasks is a good starting point to get them up and running quickly. Questions likely will arise later, and there should be a process in place from the beginning, well-advertised, for dealing with them either internally, or by passing them along to the vendor.

A training scheduled about three weeks before go-live is also a good time to build some excitement about your new ILS. For some time, possibly as much as four or five months, one or two contacts have been working on this on their own, without a lot of involvement from rank-and-file library staff; having a system ready to show them, with an older copy of their data loaded, can really help boost their enthusiasm for the project. This also carries forward to patrons, as it is time to start promoting that "in a few weeks, we're getting a new library catalog system." Take the opportunity to promote the benefits of your new ILS to your community—new functionality, cost savings for the library, and so forth—and let them know of any changes that will happen, such as password resets, fine amnesties, or changes to how notices will be sent.

Planning for Go-Live Day

Go-live day can be pretty stressful; there's not much way of avoiding the fact that your patrons will need some help with the new public catalog, and the day can be stressful for staff, as well, if they are not thoroughly familiar with Koha, or haven't seen it since training! The go-live process itself, however, should be fairly straightforward, if you have been making a "script" for how to extract, manipulate, and load the data into your new ILS, or if your vendor has been doing so on your behalf.

You'll need to plan your time around when the library is closed. Many libraries are closed on Sundays, so Friday or Saturday night, after the last library branch closes, is a perfect time to start the go-live process. Be sure to give yourself

enough time, as the migration and indexing can take several hours, if your library is large. Working with closed hours only makes it so that you avoid a "cataloging freeze," where items must wait to be cataloged for days or weeks. "Freezes" do not serve your patrons well—if an item is sitting on a cataloger's desk, then perforce it cannot be used by your patrons. More egregious is the notion of a "patron freeze," where changes to borrower information, or registration of new borrowers, are suspended for some period of time before migration. All that "freezes" like these do is create needless stress on staff and borrowers—so don't do it! It's worth staying up late and working a weekend once, in order to prevent weeks or even months of extra stress and nuisances for everyone else.

Extract the data the same way you did for the most-recent test load. It is very, very important that you do this the same way you did before; changes at this point create really amazing headaches for you or your vendor. In Koha, you'll want to TRUNCATE the tables that contain data used in training, but leave behind data tables that contain configuration information; by now, you should have most of your setup and configuration done, so keeping that where it is will be helpful. In the Koha migration tools repository is a MySQL script that will do this cleaning in Koha, at mysql/clean_database.sql.

You or your vendor will then run through the scripted series of data manipulation and loading tools that you have used in your most-recent test, and index the data for searching. A final check to make sure that searches are working, and that things seem to be healthy, and you're ready for the library to open!

FUTURE NEEDS IN OPEN-SOURCE DEVELOPMENT COMMUNITIES

As library budgets continue to tighten, more and more libraries will be looking to open-source software to meet the information-management needs that they have. Open-source software development communities should spend some time and careful effort on:

- Being welcoming to libraries that are considering implementation of their system, and develop means for new libraries and developers to quickly get integrated into the community.
- Developing, documenting, and maintaining tools for libraries to use to assist them in non-vendor-assisted migrations.
- Maintaining and encouraging strong vendor communities, for the libraries that choose to use them.

CONCLUSION

Much of the frustration of the traditional ILS migration process is avoidable, by using these techniques:

- Utilizing open, transparent processes, either vendor- or library-developed, so that all stakeholders may see the work that is being done.
- Ensuring that stakeholders—and, perhaps more importantly, non-stakeholders or marginal stakeholders—are properly identified, and their expertise usefully engaged.
- Proper subdivision of the work.
- A thorough understanding of system data structures of both the legacy and new systems.
- Adequate training and testing protocols.
- A well-planned and promptly-executed zero-day process.

REFERENCES

Bavousett, R. (2012) *Koha migration toolbox*. Retrieved on March 15, 2012, from https://www.gitorious.org/koha-toolbox/koha-migration-toolbox

Breeding, M. (2012). *Lib-Web-Cats*. Retrieved on March 3, 2012, from http://www.librarytechnology.org/libwebcats/

Free Software Foundation. (2012). *GNU General Public License, version 2*. Retrieved on March 3, 2012, from http://www.gnu.org/licenses/gpl-2.0.html

Koha Community. (2012). Retrieved on March 3, 2012, from http://www.koha-community.org

Koha Community. (2012). *Koha migration tools repository*. Retrieved on March 30, 2012, from http://git.koha-community.org/gitweb/?p=contrib/migration-tools.git;a=summary

PerlMonks. (2012). Retrieved on March 28, 2012, from http://www.perlmonks.org

Reese, T. (2012). *MARCEdit*. Retrieved on March 15, 2012, from http://people.oregonstate.edu/~reeset/marcedit/html/index.php

The Document Foundation. (2012). *LibreOffice*. Retrieved on March 12, 2012, from http://www.libreoffice.org

ADDITIONAL READING

Conway, D. (2005). *Perl best practices*. Sebastopol, CA: O'Reilly Media.

Wall, L., Christiansen, T., & Orwant, J. (2000). *Programming perl*. Sebastopol, CA: O'Reilly Media.

Chapter 6
Chatbots:
Automating Reference in Public Libraries

Michele McNeal
Akron-Summit County Public Library, USA

David Newyear
Lakeland Community College, USA

ABSTRACT

The authors discuss their experience with using artificial intelligence and chatbots to enhance their existing web sites and information services in public library settings. The chapter describes their budget driven motivations for embarking on this project and outlines the development and implementation of the bots in their library settings. They show how the bots are positioned to enhance existing services and describe the various reactions to the bots from their patron base, and staff. Different implementations of the bots are highlighted (text only, animated talking avatar, mobile site, desktop help icon) as well as the differing levels of complexity of these different implementations. They address the oft posed question "Does AI spell the end of Reference?" and describe the InfoTabby code sharing project.

CHATBOTS TO EXTEND LIBRARY INFORMATION SERVICES

History

Our journey toward the implementation of AI chatbots began at the Mentor Public Library (MPL), following then Ohio Governor Ted Strickland's June 19, 2009 proposal to cut state funding for

DOI: 10.4018/978-1-4666-3938-6.ch006

public libraries (Office of the Governor, State of Ohio, 2009). At that time the proposal included cuts which would have removed about 50 percent of the Ohio's funding for public libraries between 2010 and 2012. This was on top of a 20 percent reduction already faced by Ohio's public libraries due to decline in the state's General Revenue Fund (Sun News, 2009). As the Sun News (2009) reported, "Nearly 70 percent of the state's more than 250 public libraries rely solely on state funding. A reduction of this magnitude could mean many will close branches or reduce hours and services."

Mentor Library's Board of Trustees responded to Governor Strickland's proposal with a resolution directing the library administration to seek and implement technology to make up for the budget shortfall (Mentor Public Library Board of Trustees, 2009, p. 2). To this end, self check-out stations were installed in the Main Library during the summer of 2009. As he watched the installation of these stations, and faced with the potential of unstaffed reference desk time, it occurred to David Newyear (at that time Manager of Adult Information Services at MPL) that there might be a way to create self-serve information stations as well.

Having seen successful implementations of AI Chatbots in corporate and public sectors, David did some research and began creating a chatbot using SitePal (a company which provides animated avatars) in June of 2009. Emma, the Mentor Public Library's virtual agent, made her debut on the library's website on November 19, 2009. This first iteration of Emma provided answers to a list of twelve "frequently asked questions" or about the library's services and policies. In January 2010, Sitepal's artificial intelligence component was enabled allowing Emma's programming to begin to evolve to address more and more complex questions. During the month of February, Emma answered nearly 10,000 questions posed by patrons.

From March to May of 2010, Emma's "brain" was migrated from SitePal to the Pandorabots website, and Emma was rebuilt to take advantage of Pandorabot's Superbot 2.0 base AIML files. At this point Emma was answering approximately 300 questions each week with a correct response average of 60%. In July 2010, David invited Michele McNeal, the Web Specialist for the Akron-Summit County Public Library, to collaborate on the project and they began to work together to expand the scope of the bot's capabilities and improve the correct answer ratio.

Measuring Correct or Meaningful Responses

User conversations may be divided into three general categories:

1. Questions about the library.
2. Catalog or information searches.
3. Chat.

Determining the accuracy of responses in each of these categories is based on an examination of the conversation logs. This is fairly straightforward in the first category, which accounts for around 40% of user questions. For example, if a user asks how to renew their eBooks, the bot responds that eBooks cannot be renewed. This is counted as a correct or meaningful response. Ideally, every question should be answered in a single response. This is not always the case, so Emma will guide or prompt users to clarify their questions. If Emma can lead the user to the answer through this guidance, the response is counted as correct. If users abandon the conversation before arriving at the answer, the response is counted as incorrect.

Passing queries to the catalog or to other online resources is the second category and accounts for approximately 40% of user queries. In this case, sending the query to the appropriate resource and returning the result is counted as a correct response. In these searches, the lack of follow through on the user's part becomes more of an issue. About half of catalog or database searches are abandoned before completion.

General chat makes up the remainder of conversations. Since the primary purpose of the bot is to provide information about the library and to assist with its use, correct or meaningful responses to general chat have not been tracked. The majority of the bot's chat capacity comes from the SuperBot 2.0 AIML files, which have been edited to remove categories in conflict with library information. Some additional content has been written to smooth out conversation or to add to the bot's personality.

Passing Searches

Emma's knowledgebase contains over 33,000 categories. When input fails to match one of these, Emma will initiate a process that will pass the input to the library catalog or to another resource. The number of abandoned searches has been noted above; in order to reduce this number, changes have been made to the way in which the bot processes unmatched input. In her earliest version, Emma would inform the user that she needed to ask some questions to help them. The number of questions was unspecified, and the bot gave no indication of the progress towards an answer. This was changed to the "Three Question Game." Emma told the user that she needed to ask them three questions to help them and assured them that it would be fun, "Like 20 questions, but a lot shorter."

The three questions were:

1. "Yes or no, is this something you want to check out from the library?" If "yes," the query was passed to the library catalog, if "no," the bot responded,
2. "O.K. Second question. Yes or no, is this something you want to know *about* the library?" If "yes," the query was passed to a site search, if "no," the bot responded,
3. "This is the last question and it's a *little* bit more difficult. Is your question best answered by words or numbers?" If "words," the query was passed to the Ohio Web Library, if "no," the query was passed to Wolfram|Alpha. Any other input caused the query to be passed to Google.

Informing the user of the number of steps and the search progress was an improvement, and the number of abandoned searches dropped below 50%. Although this was an improvement, users

sometimes refused to respond to the bot's prompts with a "yes" or "no." Additional code was written redirecting such responses to the search process. Many users were also unsure how to answer the third question. The entire process has recently been rewritten using a single question, "Is this something you want to check out, like a book or CD or DVD?" If the user answers in the affirmative, the query is passed to the library catalog, with an option to "keep looking," which will pass the query to OhioLINK catalog (a consortia of Ohio public and academic libraries). If the user answers in the negative, the query is broken down into keywords and sent to the most likely resource based on those keywords. For example, words like "population," "how far," "how many," etc., will send the query to Wolfram Alpha. Words like "history," "who was," "who were," will send the query to Dbpedia. This portion of the bot's knowledgebase is under continual refinement.

Patron Reaction

Patron reaction to Emma has been overwhelmingly positive, especially among teens and young adults. She regularly receives marriage proposals and many patrons return to Emma repeatedly simply to chat. In addition to the local library community, Emma now receives inquiries from people across the United States and from as far away as South America and Russia. A number of users in Estonia have become especially fond of her and engage her in regular conversations. In 2011, users had a total of 7116 conversations with Emma, and asked 2223 questions about the library and its services and policies. 2551 user queries were passed to our catalog, databases, or other online resource. Concurrently Emma's correct response rate for questions about the library and library services improved from about 80% in January 2011 to an average of 90% during November 2011.

Emma has, for the most part, avoided some of the negative responses described in the DeAngeli study (2001). This study used transcripts of conversations as well as user comments to evaluate the social interactions of human users with the chatbot ALICE (Artificial Linguistic Internet Computer Entity). The study indicated that introducing human-like traits to chatbots, could generate negative reactions among users. DeAngeli (2001) indicated that users sometimes react to chatbots in a dominant way, feeling they can "punish" the bot by turning if off and/or demonstrating a competitive attitude toward the bot by trying to test its knowledge or limits. Emma's avoidance of these problems may be related to the fact that users who address questions to Emma are specifically selecting to use the bot on the help page rather than being automatically directed to the interface either in a pop-up box, or on the homepage itself (as is sometimes the case with corporate installations). Also, most users approach Emma with a specific task or need in mind, rather than simply to "chat". Interaction with Emma is therefore both optional and task directed. These two aspects our user's experience with Emma may help focus the user's attitude toward the bot onto the success of the task or fulfillment of the need with which they arrive.

In addition, Emma has mostly avoided the problem of users thinking there is really a human being providing the chat interaction, another issue mentioned in the DeAngeli study (2001). This may be due to the avatar being nonhuman (a cat) and to the introductory data on Emma's interface screens which clearly indicate she's an artificial intelligence.

Staff Reaction

Staff reaction to the implementation of the bot has been much more mixed. Some staff welcome the bot's presence. They see Emma as a way to streamline and focus their reference work load and take up some of the burden of repetitious hours, directional, library policy, and catalog function questions.

These staff members have seen some of the benefits documented in a case study conducted by SitePal (2008). This study of one of SitePal's clients documented a 50% decrease in the average time spent on each customer service telephone interaction, and a 30% decrease in the length of their "in person" consultations. The documented changes were due to three chatbot outcomes:

- Customers were better informed about the variety of available resources and services.
- Customers had a better idea of what they wanted when they spoke with a human agent.
- The client was better able to monitor the information needs of the customers and tailor the bot's knowledge base and other supporting materials to those needs.

Such a decrease in time spent dealing with routine questions meant that the professional staff could focus more on doing other professional tasks such as responding to reference/research questions, developing the library's collections, creating primary content (using blogs, wikis, etc.), providing programming, developing community relations, and doing outreach.

Some staff members, however, are concerned that the library administration will use the presence of the chatbot to cut staff or reduce the number of hours the reference desk is staffed. In addition, there is, as always with all new technology and change, a certain level of resistance.

In our case, this resistance can be seen to come from a number of related concerns. First, there is resistance to releasing control of a component of customer service interactions. There is also a fear that users may be unhappy or uncomfortable with the information they receive from the bot, or that the bot will provide incorrect information. As librarians we are constantly trying to be sure that our customers receive the best, most relevant, and most up to date information we can provide. For many staff members it is difficult to entrust a computer program with providing such information.

Emma and the Reference Department

Emma's role in the reference department is really to assist users in locating basic information about the library's physical plant, its policies and its services. These directional and basic information questions actually account for a large percentage of questions posed at the reference desk. In addition, Emma has been programmed to assist off site or after hour's users to locate appropriate information resources.

For example, in a three month survey of all business at the Mentor Library's reference desk, statistics indicate that reference questions accounted for less than twenty percent of the total number of questions posed. (19.12% of all transactions and 17.23% of in-person transactions). This can be compared to requests for computer help which accounted for 35.52% of transactions, requests for

assistance placing holds which provided 25.74% of transactions, and "other" requests (21.51%).

Mentor Public Library is not alone in this trend. Zabel (2005) noted that reference transactions began to decline in 1998 and have continued to decrease each subsequent year.

Similarly statistics from "virtual" reference interactions (such as SMS text messaging and IM/chat sessions) indicate that requests for immediate information about library hours and for specific books or other materials/resources predominate (Radford et al, 2010). These are exactly the kinds of questions that a chatbot is particularly well suited to answer.

Finally, Banks, and Pracht (2008) found an increased emphasis on developing Web-based reference services, a focus on the Internet, and a decreased number of questions asked at physical reference desks. This is definitely in keeping with the creation of Chatbots which provide a natural language interface to assist users in selecting and accessing information services.

Interestingly, in the course of strategic planning, Mentor Public Library's community focus

Figure 1. Reference questions compared to other service requests

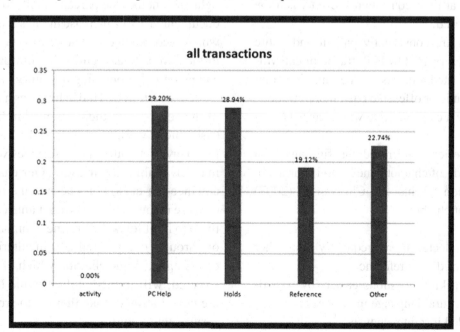

group ranked reference services next to last in their desired service responses. Their top five desired services were:

1. **Create Young Readers:** Early Literacy.
2. **Understand How to Find, Evaluate, and Use Information:** Information Fluency.
3. **Satisfy Curiosity:** Lifelong Learning.
4. **Stimulate Imagination:** Reading, Viewing, and Listening for Pleasure.
5. **Connect to the Online World:** Public Internet Access.

Benefits of a Chatbot

As previously mentioned, chatbots can ease the burden of routine questions so library staff can make better use of their time. In addition to improving the productivity of existing staff as discussed above, they can also help fill voids in staffing left by budget cuts or redeployment of staff. They can be deployed simultaneously at multiple locations and during times when then the reference desk is unstaffed. This enhanced availability: twenty four hours a day, seven days a week, makes it possible for users to immediately access needed library information at their convenience even when the library is closed.

Chatbots are consistently patient and polite and cannot be ruffled by high traffic or call volume, by repeated requests for the same information, or by rude or offensive customers. In short, they provide a cost-effective way to answer routine questions.

Our experience mirrored the findings of a German study of chatbot benefits to libraries. In her discussion of chatbots, Christansen (2007) suggests that chatbots:

- Were selected more frequently than other forms of digital reference.
- Made asking questions easier (by providing a natural language interface).
- Provided instant responses.

- Were anonymous (which encouraged shy users or those who thought their questions might be "stupid") .
- Provided a marketing tool for reference services.

It is important to note Chistensen's final suggestion that chatbots can serve as a marketing tool for our resources and services (2007). We enhance access to and knowledge of these resources by embedding links to our various resources and phone numbers for our service desks, as well as by creating dialogs which pass searches to our catalog, online databases, or other related products. Customers who query the bot simply looking for a definition or a unit conversion, receive this information; but these customers are also introduced to a gamut of useful resources through our linkages with Merriam-Webster Online, Wolfram Alpha, and the Ohio Web Library. The bot also links to the library's Facebook page, and can pass queries to Weather.com, Food Network, and IHeartRadio.

In addition, chatbots allow us to personalize of the user experience, an aspect of their value which cannot be overemphasized. The importance of welcoming and putting the user at ease is highlighted in another Sitepal case study (2010), which documented a 40% increase in sales at a company which incorporated a chatbot to welcome users to their website and guide them through a vast and potentially confusing inventory of products.

The situation we find in libraries mirrors that of the company in the previous SitePal (2010) case study: we have a vast and often confusing array of resources and services we want to make available to our users. Our chatbots can enhance our customer's access and navigation of these resources by making it unnecessary for them to hunt for what they need on our website or through our printed documentation. They can simply ask the chatbot, which, if well constructed, will immediately provide them with the information or take them to the resource or service they require.

Chatbots also provide an opportunity to connect the library with new groups of users. Chatbots appeal to both the most and least computer saavy users. To those on the cutting edge of technology, they provide an opportunity to quickly access information without waiting for assistance. We've seen the "cool" appeal of the chatbots among our teenage users and, in general, conversing with Emma is viewed as being fun and useful.

But chatbots are particularly helpful to our least computer literate users as well. For these patrons, as well as those who are shy or uncomfortable making face to face information requests, chatbots provide a private, nonthreatening, nonjudgmental interface where they can seek the information they need. This is due to their employment of Natural Language Processing (NLP).

Natural Language Processing

Warschauer and Healy (1998) define Natural Language Processing, as the process of a computer extracting meaningful information from natural language input and/or producing natural language output. Normal searching, even in its most simple form, requires the user to compose a search for the information they need. Chatbots on the other hand, allow the user to present an inquiry as they would to another human being. The burden of finding the needed information shifts away from the user and onto the bot's designer or programmer. A well designed chatbot will guide the user to the information they need by engaging them in a dialog.

The chatbot designer creates a paradigm which leads the user through a question-and-answer conversation to discover their information needs. The designer then responds to these needs by programming the chatbot to provide the needed information or links to additional resources. This process addresses the underlying problem of customers not using the library terminology or jargon with which we are familiar.

At the same time, conversation log review allows the chatbot designer to continually monitor the types of questions and terminology being used, and to update the information provided by the chatbot and the language it will recognize. This is why the chatbot is particularly convenient to those patrons who are least familiar with the library, its services, and resources.

The Many Lives of Emma

One of the most exciting things about chatbots are the multiplicity of places and ways the can be implemented. We've made use of this to introduce the chatbot's functionality to a variety of places in our library's virtual and physical environment.

At its simplest, Emma can be accessed as a text only interface within a library's III WebPac. Here Emma can be called on to explain simple tasks (renewing items or placing a hold) or to give information about the library's policies or resources (how to get a library card, where the microfilm is located, how to check out magazines, etc.)

In addition to this very simple interface, Emma is fully deployed on the help page for the Mentor Library system. Here she is an animated, talking avatar. Users can ask questions about the library, ask for information, or other resources from the library's collections or from the Ohio Web Library or Wolfram Alfa databases. Here users are reminded to turn off their pop-up blocker for the mentorpl.org site in order to capitalize on Emma's ability to search additional sites.

Also on the library's homepage, the Emma's avatar can be used to provide a pre-recorded message about upcoming programs or highlighted resources. In this implementation we capitalize on the branding associated with the bot to call attention to the information we wish to advertize. The bots text to speech capability allows us to customize the message or create pass-through links to pertinent resources.

Figure 2. Text only interface embedded in the library catalog

Figure 3. Full speaking avatar implementation on library help page

Figure 4. Mini avatar implementation on homepage to advertise special programs/services

We have also created a mobile access page for our chatbot. Here we've used the text-only interface with a simple thumbnail image to maintain the branding of the site. Users with smart phones which provide speech-to-text capability can simply speak their question into the phone. Users with QR code scanners can simply scan a QR code to reach the interface.

The four implementations described above have focused on the needs of our users who are outside the library itself. We have also created two special resources for in-house users.

The first is an information kiosk similar to those you may have seen in large office buildings or malls. Here we've identified the questions most frequently asked by in people first entering our building. Using these questions we've created a touch screen interface to help users quickly find the information they need. However, there are always questions for which we don't have "quick buttons." To answer these we've implemented a special version of the chatbot with knowledge tailored to the needs and resources of our Main Library. This includes answers to such questions as "Where are the restrooms located?" "Where can I get my parking ticket validated?" and "Where are your audiobooks?"

In addition, we sought to make the chatbots easily accessible to patrons at the library computers. To this end, we are developing a desktop icon connected to a window, open script which will allow users to click the icon and immediately access the full animated interface from their screen. We see this as being particularly useful for computer use related questions like, "How do I print?"

The End of Reference?

Chatbots cannot replicate the complexity of the reference interview, nor replace the personalized instruction and assistance user's associate with professional librarians. However, we need to recognize the fact that virtual agents are here today. NextIt, VirtuOZ and Zabaware are only a

Figure 5. Mobile text based implementation with static image. QR code for easy access to mobile site for phones or other mobile devices with QR scanning software.

Figure 6. Information kiosk implementation

Figure 7. Desktop icon with window open script allows quick access from patron computers

few of the companies vying to create these agents for public sector companies. In addition to these company specific resources, there is a new wave of "personal assistant" agents like Ultra Hal (2012) for windows computers, Siri (2012) for the iphone, and Evi (2012) for Android devices which will perform meta-search functions. With the success of these programs, virtual agents are likely to become more prevalent in the future.

Indeed, a VirtuOZ/CCM Benchmark Group study projects a "400% increase in virtual agents by 2014" (VirtuOz Inc., 2011). Virtual Agents are now gaining popularity in libraries. From the German Bots created in the 1990s to Emma, Pixel, and Ella, bots are enhancing services in physical libraries and providing assistance to users and are providing this assistance in an extremely cost-effective manner. As noted above, Emma answered a total of 4774 library related questions in 2011; the cost of providing this service was $0.14 per

use. As library funding continues to erode and chatbots become more intelligent, automated reference services will become an increasingly attractive option, if not a necessity.

In addition, virtual information agents are central to providing information services to users of virtual libraries, such as the Neil A. Armstrong Library and Archives on secondlife.com (Bohle, 2010). This virtual collection, like Curiosity AI, (the second place winner of the 2011 White House/ Department of Defense Federal Virtual Worlds Challenge,) is an entirely virtual world where the user himself is represented by an avatar which interacts with the virtual agents and resources.

We believe the correct question we need to ask as professional librarianship is, "What happens if we ignore this technology?" If we don't embrace and attempt to direct this technology and its implementation, we are likely to have it thrust upon us in ways we cannot direct, either by library boards or administrations whose bottom line is the dollar cost per question answered. If we use Virtual Agents to enhance and streamline our information services, and as a marketing tool for both our traditional reference services and our online resources, we can reap the benefits of this technology and, at the same time, position our professional librarians to provide those value-added services to our users at which they alone can excel.

The InfoTabby Project

Finally, we'd like to introduce you to the InfoTabby, the Ohio Virtual Reference Project. This is a growing partnership between a number of Ohio public libraries and librarians which invites libraries to get a head start on developing their own virtual agents by editing a base repository of code to conform to the needs of their own locations. This open source project is hosted on Google Code and distributed under an Apache 2.0 license. In addition to the Google code base aiml files, the project employs Tortoise Hg, a Mercurial client, to allow multiple users to work on the same project, and then merge the results of their independent work into a unified result.

There are two ways to participate in the project.

On the "Non-Sharing Path," the developer downloads a zip file containing the InfoTabby .aiml files (and some customized .html files) to their local computer. Using this copy of the code, they simply edit the files to reflect their own library's hours, location, staff, and resources. Once they have created their customized content, they need only create a new bot at Pandorabots.com. This site offers both a free Community Server "sandbox" workspace and a secure shared site for bot creation. They also offer an enhanced suite of precoded aiml patterns called the AIML Superbot, which can speed the creation of your bot.

Once the bot is created, the developer uses a template of bot properties to define basic personality characteristics of the bot. This information creates a description of the bot similar to the "profile" information users can enter into their Facebook or Twitter accounts. The greater attention given to the customization of these properties files, the more lifelike characteristics the bot will have and the more realistic information it will have to draw upon to converse with users. At this point the bot can be deployed for testing. The developer needs to ask the bot the types of questions it is to answer, and evaluate its replies both in real time, and by using the conversation logs.

In addition to the above described process, we strongly encourage bot developers to become sharing members of the InfoTabby community. This "Sharing Path" has two special advantages:

- First the developer receives alerts of and has access to the latest changes to the base AIML code made by other members of the library community.
- Second, the improvements made by the developer are easily shared with all community members.

By choosing to participate as full sharing members of the project, bot developers help to bring the power of open source computing to the task of improving the bot's language recognition, knowledge base and searching/linking capabilities.

To implement this path, the developer creates a code Mercurial code "repository" on their computer. Changes to this repository are tracked and can be shared with the rest of the community. This process has the added benefit of allowing the developer to easily monitor changes to the bot through various "versions" and more quickly identify and resolve problems created by specific alterations to the base code.

REFERENCES

Banks, J., & Pracht, C. (2008). Reference desk staffing trends: A survey. [from *MasterFILE Premier*. EBSCO. http://web.ebscohost.com.proxy.oplin.org]. *Reference and User Services Quarterly*, 48(1), 54. Retrieved March 8, 2011

Bohle, S. (2010). The Neil A. Armstrong Library and Archives: That's one small step for a virtual world library, one giant leap for education! *Journal of Virtual Worlds Research, 3*(1). Retrieved January 23, 2012, from http://journals.tdl.org/jvwr/article/view/1600

Christensen, A. (2007, August 28). *A trend from Germany: Library chatbots in digital reference.* Paper presented at the International Ticer School, Digital Libraries à la Carte, Module 2: Technological Developments: Threats and Opportunities for libraries. Tilburg, Netherlands. Retrieved March 2, 2012 from http://www.slideshare.net/xenzen/a-trend-from-germany-library-chatbots-in-electronic-reference-presentation

De Angeli, A., Johnson, G. I., & Coventry, L. (2001). The unfriendly user: Exploring social reactions to chatterbots. In M.G. Helander, H.M. Kalid, & T. Ming Po (Eds.), *Proceedings of the International Conference on Affective Human Factor Design, London: ASEA*. Retrieved from http://disi.unitn.it/~deangeli/homepage/lib/exe/fetch.php?media=references:deangeli:theunfriendlyuser.pdf

Mentor Public Library Board of Trustees. (2009, September 16). Board of trustees meeting minutes. Resolution #09-104.

Office of the Governor. State of Ohio. (2009). Governor Strickland's proposed framework to balance FYs 10/11. Retrieved from http://obm.ohio.gov/document.aspx?ID=d9f88041-a555-4840-a211-23f69bbf1829

Radford, M., et al. (2010, July). *Taking the library with you: VR going mobile.* Paper presented at the ALA Virtual Conference. Retrieved from http://www.amandaclaypowers.com/2010/06/28/taking-the-library-with-you-vr-going-mobile/

Sitepal. (2008, May). *Virtual meeting with Kathleen improved sales productivity by 50%.* Retrieved January 23, 2012 from http://www.sitepal.com/pdf/casestudy/loftusphotography.pdf

Sitepal. (2010, February). *Daughter nature increased online sales by 40%.* Retrieved January 23, 2012, from http://www.sitepal.com/ccs2/oddcast/casestudy/14.pdf

Sun News Staff. (2009, June 24). Governor Ted Strickland's proposed budget cuts endanger future of public libraries. *Sun News.* Retrieved March 13, 2011, from http://blog.cleveland.com/garfieldmaplesun/2009/06/governor_ted_stricklands_propo.html

VirtuOz Inc. (2011, May 10). VirtuOz/CCM benchmark group study predicts a 400 percent increase in ecommerce adoption of intelligent virtual agents by 2014. Retrieved January 23, 2012, from http://www.virtuoz.com/company/news-events/press-releases/virtuoz-ccm-benchmark-group-study-predicts-a-400-percent-increase-in-ecomme/

Warschauer, M., & Healey, D. (1998). Computers and language learning: An overview. *Language Teaching, 31*, 57–71. doi:10.1017/S0261444800012970.

Zabel, D. (2005). Trends in reference and public services librarianship and the role of rusa part one. [from *MasterFILE Premier*. EBSCO. http://web.ebscohost.com.proxy.oplin.org]. *Reference and User Services Quarterly*, 45(1), 7. Retrieved March 8, 2011

ADDITIONAL READING

Evi. (n.d.). Website. Retrieved from http://www.evi.com/

InfoTabby: The Ohio Virtual Reference Project. (n.d.). Website. Retrieved from http://www.infotabby.org/

Pandorabots. (n.d.). Website. Retrieved from http://www.pandorabots.com

SightPal. (n.d.). Website. Retrieved from http://www.sitepal.com/

Siri. (n.d.). Website. Retrieved from http://www.apple.com/iphone/features/siri.html

Ultra Hal. (n.d.). Website. Retrieved from http://www.zabaware.com/assistant/

Chapter 7
Managing Automated Storage in the 21st Century Library

Carolyn Adams
Boise State University, USA

ABSTRACT

The Mathewson Automated Retrieval System (MARS) is the second largest automated library storage system in the world. Housed in the University of Nevada, Reno's spectacular Mathewson-IGT Knowledge Center, MARS provides storage for half of the print collection, and nearly all government documents, special collections materials, and multimedia equipment. This chapter will explore automated library storage management, including maintenance and care of the equipment, safety, stewardship of the collection, and how automated storage challenges our beliefs about the purpose and function of libraries.

INTRODUCTION

The Mathewson-IGT Knowledge Center opened August 11, 2008, on the University of Nevada, Reno campus. UNR is a large academic institution with 18,000 undergraduate and graduate students and nearly 1,000 full and part-time faculty. The University Libraries include three campus locations—the Mathewson-IGT Knowledge Center, which is the main library location and serves the majority of academic disciplines, the DeLaMare Science and Engineering Library, and the Savitt Medical Library. The Knowledge Center serves University faculty, staff, and students through

electronic and physical library collections and extensive technology resources. Named in honor of a $10 million joint donation from former Chief Executive Officer Charles N. Mathewson and International Game Technology, the Knowledge Center is considered to be one of the most technologically advanced libraries in the country. The design and functionality of the building hinges on the flexible spaces available to students for a variety of learning activities. The ability to provide learning spaces such as numerous group study rooms, café style booths, and over 300 computer workstations throughout the 295,000 square foot structure, is dependent on efficient storage of library collections. By including an automated storage and retrieval system (ASRS) the architects

DOI: 10.4018/978-1-4666-3938-6.ch007

were able to save approximately 100,000 square feet of space that would have been filled with traditional shelving.

The Mathewson-IGT Knowledge Center is the physical manifestation of the vision of former Dean of University Libraries and Vice President of Information Technology, Dr. Steven D. Zink. What is a Knowledge Center, and how does it differ from a library? Dr. Zink describes this distinction best in the case statement written to secure funding for the $75 million project.

The single greatest intellectual force and competitive advantage in the 21st century is the rapid assimilation of new knowledge to fuel innovation. New knowledge, applied to existing tasks, results in increased productivity; new knowledge applied to new challenges and tasks is fundamental to innovation. Recognizing this critical interplay between knowledge and innovation, the University of Nevada, Reno has established one of the first centers in the nation built specifically to embrace these dynamics of the 21st century. The Knowledge Center will encompass all facets of the digital age in a single, synergistic complex. Computing and information technologies are combined with the latest in graphical design technologies and the resources of the university library in a physical environment designed to maximize learning, nurture the production and distribution of new knowledge, and stimulate and sustain innovation. (Zink, 2010, p. 16)

In order to create Dr. Zink's vision, it was essential that the design of the building supported a wide variety of user spaces, and that those spaces were flexible and adaptable to keep pace with a rapidly changing learning environment. The building is, however, also a library, and the one million volume collection needed a home. Automated storage was familiar in the state of Nevada, owing to the system in the Lied Library at the University of Nevada, Las Vegas (UNLV), and several universities in California. After careful consideration

of the size and anticipated growth of the library's print collection, the automated storage system was designed with enough space to realistically accommodate the current collection and twenty years of growth. The Knowledge Center offers a broad range of support for media design, poster printing, computing, instructional design, and high-end mathematical, statistical, and geographic information systems software. The look and feel of the building is comfortable, yet elegant, with group study rooms equipped with large monitors and laptop connections, and a mix of study carrels, tables, and comfortable seating. In order to accommodate ever increasing mobile computing such as laptops and smart phones, study tables and carrels contain multiple electrical outlets. This first floor of the building, @One, is distinct in design and function, with a focus on high-end computing, multimedia design, large format poster printing, and multimedia checkout. The name @ One originates from the concept of having all of the production and software resources, including a sound recording booth and green-screen special effects studio, plus traditional library research help, a 163-seat auditorium, computer training lab and video conference room, @One place, on floor one. There are traditional book stacks on floors three through five, housing approximately 50% of the library collection, with the remaining half stored in MARS. Daily operations of the system are managed by the MARS and Stacks Maintenance supervisor and the general collection is stored and retrieved in the operator area located in the Library Services department. The department is a combination of circulation, interlibrary loan, print and electronic reserves, and MARS and stacks maintenance. Everyone in the department uses the system regularly.

Automated storage and retrieval systems provide libraries with an opportunity to optimize space and store their collections on-site. By repurposing a technology well known in private industry, libraries win space and efficiency, but must learn to adapt to an industrial workplace typically unfamiliar to librarians and library staff. This adjustment can be daunting. Patron response is varied, and often met with resistance in the beginning. The planning phase is long and complex. Depending on the size of the initial load, it can take several months to complete the storage process. When the system is loaded and fully functioning, the work continues with regular maintenance, troubleshooting, and safety requirements. This chapter will provide an in-depth exploration of the automated storage system at the Mathewson-IGT Knowledge Center. We'll examine system design, the logistics of the initial load, safety protocols, staffing needs, and use. User reactions to the system will also be discussed, and the ways users and staff interact with the equipment on a day to day basis. We'll consider the impact of an ASRS of this size. As more and more information becomes available in digital format, is the cost of building an automated storage system justified? What opportunities can automated storage provide? Finally, what assumptions and values held by librarians and users are challenged when we hide our collections away from the public view? With the digital age upon us, and budgets shrinking, it is vital that we transform our buildings and services to connect users with information efficiently and provide learning spaces that encourage innovation and knowledge creation. Automated storage provides a unique and exciting opportunity to redefine our relationship with library spaces and collections. This chapter provides a roadmap for librarians considering, planning, or maintaining an ASRS, and explores opportunities created by redefining our relationship with our physical library collection.

LITERATURE REVIEW

Automated storage and retrieval systems offer libraries a space saving solution to the book storage challenge. As libraries evolve from book

repositories to spaces which support collaborative work, technology, community events, and innovation, automated storage allows book collections to remain intact and immediately available to users. Student study space, group study space, and hosting events is challenging when thousands of square feet must be devoted to shelving. Off-site storage poses additional challenges, including the slow delivery to the customer, which is inconvenient and may lower the use of important materials. By relocating collections into the high-tech and highly efficient ASRS, printed materials are preserved without sacrificing ease of use or valuable space.

Serious consideration of automated storage makes its debut in the library literature in the mid- 1980s, when the vast and expanding California State University (CSU) system considers an ASRS as a solution to library space concerns. Creaghe and Davis (1986) describe the benefits associated with ASRS at the American College and Research Libraries Fourth National Conference, sharing details of the proposed installation of an "Automated Access Facility (AAF)" (p.496) at the California State University, Northridge (CSUN) campus. The first ever automated storage system installed in the United States was the Randtriever at the Ohio State University Health Sciences Library in the early 1970s, along with a few other installations in that time period. The equipment, however, was unreliable, and did not re-enter the library world until the CSUN project, completed in 1991 (Bullard & Wrosch, 2009). In the twenty-year period following CSUN, libraries across the world have installed ASRSs to store their collections. It is difficult to obtain a comprehensive list of libraries utilizing the technology, although a rough estimate hovers in the twenties. The 2008 installation at the Mathewson-IGT Knowledge Center was, at the time, the largest automated library system in the world. This distinction was recently surpassed by the University of Chicago, when the Joe and Rika Masueto Library finalized the move of their 3.5 million volume collection to a spectacular underground automated storage system in fall 2011.

John Kountz (1987) provides a detailed overview of the benefits of automated storage in libraries. At the time, he was the Associate Director for Library Automation in the CSU Chancellors Office, and oversaw ASRS installations at CSUN and California State University, Long Beach. At the onset of every ASRS installation, there is an outcry from some library users that librarians are hiding the books away, and that research will be significantly crippled by the loss of serendipitous discovery in the open stacks. These conversations occurred at UNR as they have at other institutions planning for automated storage, including the Bruce T. Halle Library at Eastern Michigan University (Shiarto, Cogan, & Yee, 2001). Kountz offers two distinct advantages of automated storage that eloquently contradict this common concern. First, by selecting low-use materials for storage in the ASRS, the items left in the open stacks are the more popular, higher use items, thus "reducing the number of dead and dying items that must be winnowed to satisfy a search" (p. 67). Second, open stacks are subject to the ravages of other users. Items may be mis-shelved or misplaced, making it difficult or impossible for users or staff to locate the material at the time of need. Kountz asserts, "The ability of the user to request specific titles with confidence in their timely availability is an inestimable service enhancement to any library" (p. 68).

From an architectural standpoint, there is an aesthetic decision that must be made when designing a library with an ASRS; that is, should the system be visible to the public? At Eastern Michigan University, the system is hidden from public view. Shiarto, Cogan, and Yee (2001) consider what they might do differently if they built their library again, and recommend making the ASRS "clearly visible from the circulation desk"

(p. 258). At Grand Valley State University, another Michigan university on the Western side of the state, "patrons can order a book online and then see it retrieved by the robot and brought directly to the circulation clerk – something that is quite a showpiece" (p. 258). The Lied Library at UNLV built a storage system surrounded by glass walls on three sides so the system is visible from each side, and even from the floors above. Haslam, Kwon, Marilyn, and White (2002) argue the visibility of the system, named LASR, helped overcome resistance encountered by other libraries. "LASR is a focal point of the Lied Library….designed to be visible to the public, creating enthusiasm among users and winning user acceptance" (p.89). Visibility is a two-edged sword, however, in the long run. As systems age, their sparkle may fade, thus with the decision to let users see the nuts and bolts of an ASRS must come the commitment to include cleaning and polishing to the regular maintenance schedule of the area.

In 2011, Library Journal published a new ASRS article, David Rapp's "Robot Visions." Amazingly, a quarter of a century later, the content is virtually identical to Kountz's 1987 account of automated library storage. Greater emphasis is placed on marketing and creative learning spaces, but the basic equipment design and function, cost savings, and the overwhelmingly positive response from librarians, have remained unchanged. What is lacking from previous research, however, is a realistic account of day-to-day life managing an ASRS. The literature provides detailed design specifications and accounts of the initial load of materials (Haslam, Kwon, Marilyn, & White, 2002), cost estimates (Rapp, 2011), and early user reactions (Shiarto, Cogan, & Yee, 2001), but when the dust settles, things can change dramatically. This chapter will delve deep into daily operations at the Mathewson-IGT Knowledge Center, providing insights into managing automated library storage that is lacking in previous literature.

TAKE A TRIP TO MARS

System Design

The Mathewson Automated Retrieval System, MARS, was built by HK Systems, Inc., now Dematic Group, the preeminent manufacturer of automated library storage and retrieval systems. While other automated storage manufacturers dabbled in libraries in the last four decades, HK Systems emerged as the clear leader specializing in automated storage customized to the library environment. The acquisition by Dematic in fall 2010 (Gallagher, 2010) left library clientele uncertain about the future of their systems, but the quality of products and service has remained at the remarkably high and individually personalized level HK Systems customers came to know and appreciate. Dematic promised the merger would retain all of the best characteristics of both companies, while expanding their reach in the global material handling market, and they have kept that promise. Working with Dematic sales, customer service, project management, and software development staff is always enjoyable. Their product is well-designed and reliable, and their staff is skilled and supportive throughout the design, implementation, and ongoing operation of the system.

MARS is comprised of six mini-load storage and retrieval machines (SRMs) which pick and store 25,338 containers located in 3-story racks on each side of the aisle. The containers are 24" wide and 48" deep, with a variety of heights: 8", 10", 12", 14", and 18". The majority of bins are 10" and 12" high containing primarily monographs and serial publications respectively. There are 17 computer workstations distributed on 3 floors, with 11 workstations, and the majority of activity, occurring in the second floor Library Services department. Book and periodical checkout is managed at the Library Services desk, however the first floor has access to one aisle of the ASRS behind the Multimedia desk, and the Special Collections and Archives department has access to two aisles on the third floor. Access to the system from distinct, secure areas on three floors is one of the unique features of MARS that distinguishes it from all other automated library systems. The Special Collections and Archives department has custom-designed workstations that are slightly lower than the standard configuration, allowing for easier storage and retrieval of heavy manuscript boxes.

Each item barcode is associated with a bin location indicating the aisle, left or right side of the rack, vertical location, horizontal location, and the section of the bin where the item is located. Cardboard dividers create a variety of bin configurations and all materials are stored randomly by height and size, without consideration for call number or subject area. When an item is requested, a graphical representation of the number of sectors in the bin is displayed on a computer monitor. Trained library staff and student assistants login to the ASRS software program, and use the graphical display and item barcode information to locate the requested item within the bin. All barcodes are placed on the front of the book to make visual identification easy. The item is removed from the bin, the barcode is scanned, and a receipt is printed with the name of the individual who requested the item.

The initial plan for books retrieved from MARS was to place them on a bookshelf in a public area near the Library Services desk. Storing requests where users could retrieve them at will mimicked the experience of browsing the shelves. This decision came partly from initial concerns over hiding the books away. There was reluctance to require users to check out books they requested from MARS. Staff involved in the planning phase had seen similar setups at libraries they visited, and liked the idea of putting MARS requests out in the open instead of using the traditional hold shelf behind the service desk. This idea never materialized, however, because the receipts printed included identifying information about the requestor, including first and last name and library barcode number. Editing the format and content of the receipts was expensive, and there was little interest in spending money for this purpose. The requested items, averaging 60 to 80 per day, continue to be stored on the Library Services hold shelf. Shelving from the original MARS bookshelf was relocated to increase storage capacity behind the service desk.

Most user requests are initiated from the online library catalog. UNR affiliates click a "Request from MARS" button and log in with their university ID and password. When the request has been processed, the system displays a message to the user stating that their request was successful and the item(s) may be retrieved at the Library Services desk in 10 minutes. At this point, the item information is transferred from the library catalog system, powered by Innovative Interfaces, Inc., and the appropriate crane receives the command to pick the bin containing the item and deposit it at the operator workstation. Part of the beauty of the online request process is that the user can initiate the request at any time, and wait up to ten days to visit the Knowledge Center to pick up the item. Users appreciate this convenience, and frequently request a similar paging process for materials housed in the open stacks. Although we do not page items at this time, it is likely we will in the near future.

Unlike other items saved on the hold shelf, such as interlibrary loan requests, books from our consortial borrowing program, and other traditional holds, users are not required to check out MARS items to view them. Users are given the option to browse the material, photocopy, or read it at their leisure without checkout. In order to make identification of these materials easy for checkout, sorting, and re-shelving, a transparent green sticker is placed over the barcode on the front of the item when it is retrieved from MARS. In addition, the receipt with the user's name is left in the book so it can be placed in alphabetical order on the hold shelf. Most users

check out the books they request from MARS, however some browse the material quickly at the desk, and then decide whether to keep or return it. Allowing users to view materials without requiring checkout is a special courtesy to our community users. We did not want to unfairly limit their access to MARS items by requiring that they purchase a Community Borrower library card, which comes with a $25 annual fee, to view items stored in the system. The process works well, users appreciate the choice, and we have no indication that the policy is abused. If a researcher needs access to a large number of items, such as multiple journal volumes, we place the materials on a book cart and store it in the Library Services staff area for as long as it is needed. This service is requested rarely and met with hearty thanks from researchers.

One disadvantage of allowing users access to MARS books without checkout is that use statistics cannot be tracked by circulation numbers alone. During the first eight months of operation, loading statistics were tracked, but we were unaware that other data such as items retrieved, stored, and audited, was purged from the SQL database after six days. This loss of data was a profound disappointment, and one of the difficult lessons learned in our first year of operation. The period of time that data is stored is easily adjusted by the manufacturer upon request. In response to this loss, the Applications Development Librarian created a partial backup for pick, store, and audit data. By creating queries in Microsoft Access, we can track total items stored, top 100 requested items by title, top 100 stores by user, and average, high and low picks per day. This backup is useful for the time period from January 2009 to June 2011; however the software and server upgrade in July 2011 changed the data location so statistics are available in another format. With the software upgrade, we requested one year of data storage from the manufacturer, and will work in summer 2012 to consolidate both datasets into one searchable archive.

The Initial Load

In preparation for the move from the Getchell Library, built in 1962, to the new Knowledge Center, librarians and technical services staff created a list of items destined for MARS. Staff and student assistants spent countless hours placing blue dots – small dark blue circular stickers – on all of the titles headed for the ASRS. This visual distinction was necessary for hired movers to properly identify items that would be moved to traditional shelving in the new building. Simultaneously, the cataloging department was furiously creating 80,000 brief serial item records and changing location codes. The Business and Government Information Center measured 10,000 linear feet of uncatalogued government documents and placed yellow cards with a barcode assigned to each foot of material planned for one sector in a MARS bin (Adams, 2010, p. 20-24). The Life and Health Sciences library organized their materials as well, and Special Collections and University Archives, and the Basque Library each prepared for the new building located a brisk ten-minute walk up the hill. The move to the Knowledge Center, adjacent to the newly built Joe Crowley Student Union, shifted the focal point of campus from the university's historic quadrangle to the mid-campus area; halfway between the oldest buildings to the south, including Morrill Hall built in 1887, and the medical campus to the north.

Due to construction delays, the initial move of materials to the ASRS began later than planned, and the movement of books continued for seven months following the building opening in August 2008. Loading began in June of that year, when the first of six aisles was complete. The libraries offered limited service during the slow summer, grateful for the relatively quiet campus. Books were paged for users, while staff scurried between two libraries, managing a split collection in constant flux, as moving vans loaded with large wooden book trucks made the trek back and forth from Getchell to the "KC." This back and forth

movement continued throughout the fall semester into spring, led by the MARS and Stacks Maintenance supervisor and his crew, comprised of one full-time temporary employee and 6-10 part-time student assistants. The Supervisor coordinated with the technical services department. Batches of item records were loaded into the ASRS database and the corresponding ranges of books were delivered to the Knowledge Center, sorted by size, and loaded one-by-one into MARS. Staff volunteered for the loading schedule, but the moving process was inconsistent at times, and occasionally there were no books to load. When the books were ready, email announcements declared "Calling all Martians!" and volunteers from library and IT departments flooded the Library Services department to load MARS. From June 2008 to February 2009, 550,000 items were stored in MARS at a peak rate of 20,000 to 25,000 items per week.

Items identified for the initial load include the following: books published before 1995 that had been checked out 7 or fewer times in the last 20 years, journals older than one year, most Special Collections and University Archives materials, and most of the government documents collection. The items generally excluded from MARS are newer, high use books, pictorial works, music scores, books on Nevada history, and fictional literature published after 1980. The original intention was to move any items requested frequently from MARS back to the open stacks, and to move older journals each year into MARS. The decision was made fairly quickly to change the serials retention policy and leave all journals in the open stacks published in 2008 or later, rather than processing a yearly move to MARS. Because of ample space in the stacks, and the reality that many print journals had been cancelled or replaced by electronic versions, it was easier to leave newer journals on the shelves and revisit the policy if shelf space ever became an issue. Moving frequently requested items out of the storage system has never become a priority. Without users complaining about requesting items from MARS, there is little motivation to remove items. The more frequent request from our undergraduate users is for us to move all of the books into MARS because it is so simple and convenient to obtain the materials.

Immediately following the 550,000 item push, the remote storage facility in Stead had to be emptied. Nearly 100,000 books and journals and several hundred Special Collections boxes were moved to the Knowledge Center by June 2009. Many of these older materials required technical services processing before they could be stored in MARS. The professional movers who filled the open stacks had left a large section of empty shelving on the fifth floor. We utilized this space as in-house storage for nearly two years while technical services staff tirelessly updated more than 35,000 catalog records and slowly moved the items to MARS. In August 2009, the Teaching and Learning Technologies department relocated to @One. TLT utilized the first floor access to MARS for the first time, storing a wide variety of multimedia equipment available to university faculty, staff, and students. In spring 2011, the final substantive relocation of materials involved 60,000 journals moved from the DeLaMare Science and Engineering Library to MARS. As of April 2012, MARS contains more than 733,000 items and is approximately 49% full.

Despite careful planning, there are always surprises with a project of this magnitude. Construction delays seriously impacted the timing of the move and initial load of materials. In addition, we began the move without a specified deadline. That changed with little notice, however, in October 2008 when we were informed that the heating system would be shut down at the old library and all archival materials had to be moved immediately. At that time we were given four months to remove the remaining print collection, approximately 150,000 items, hundreds of Special Collections and University Archives manuscript boxes, and the entire collection, 10,000 linear feet, of uncatalogued government documents. Another unforeseen challenge was finding space in the

Knowledge Center for sorting the books by size. The sorting process takes time, and there wasn't an adequate staging area in the new building. Fortunately, the Building Operations Manager stepped in an offered an unfinished room in the basement of the building where he erected temporary shelving to hold materials as they were sorted and loaded into MARS. There was sufficient funding to pay a moving company to bring the remaining books to the Knowledge Center, fill the temporary shelving, and allow time between loads for sorting. Meanwhile, the library moving crew focused attention on Special Collections and government documents. The plan worked, and the deadline was met with one week to spare.

Staffing, Maintenance, and Safety

Because MARS is designed with access in three distinct departments, nearly 100 library staff and student assistants, referred to as operators, use the system regularly. There are also a handful of technicians who are qualified to work on the cranes directly. The MARS department was initially staffed with two dedicated full-time employees. Each had additional duties beyond managing the system, but it was their primary responsibility. The supervisor worked a standard day schedule, and the assistant worked an evening shift in order to resolve any problems that occurred after business hours when the building was still open. When the MARS and Stacks Maintenance supervisor retired in summer 2011, the position was eliminated because of budget constraints. The assistant was promoted and moved to a day schedule, and a small portion of the evening circulation supervisor position was shared with MARS to help with evening issues. Managing the system with just one full-time employee is difficult, but the talented and dedicated Building Operations Manager has donated a significant portion of his time to MARS maintenance and troubleshooting. The Head of Library Services manages vendor relations, including communication with Dematic, contract negotiations, and project management. Student shelving assistants store returned materials and continually audit the contents of full bins to ensure accurate inventory. Unfortunately, a complete audit of the system will take many years, since student hours have been reduced by nearly 50% due to library budget cuts.

One of the best decisions made during the first year of operation was to invest in a week-long maintenance training session provided by an HK Systems technician. The MARS Supervisor and assistant, the Building Operations Manager and assistant, two electricians and a mechanic from the campus facilities department, and the Head of Library Services participated in the training. The initial training, coupled with several years of experience, allowed money planned for the yearly maintenance contract to be spent on a software and server upgrade in 2011. Monthly, quarterly, semi-annual, and annual preventative maintenance, and troubleshooting, are all performed by Knowledge Center faculty and staff. Maintenance records are stored in a keyword searchable online department workspace, using FootPrints ticketing software by Numara, Inc. Mechanical errors encountered throughout the regular course of operations are also tracked, and details regarding any part replacements or extended troubleshooting. Extensive recordkeeping allows technicians to track patterns in equipment behavior and regular wear and tear that might otherwise be missed. Cleaning the area is also a priority. To facilitate the removal of dust from the concrete floor, the area was coated with a wax sealant. This not only minimizes dust that can increase wear and tear and damage the sensitive electrical and computer components in the crane, but makes the floors shine, adding to the visual appeal of the area. Desks with standalone computers were installed at the end of each aisle, providing a workspace and easy access to the end-of-aisle software often used for troubleshooting and diagnostic purposes.

Adjusting to the industrial requirements of an ASRS can be a daunting task for librarians unfamiliar with complex maintenance and safety protocols. Although the system manual provides generic safety procedures, it is the responsibility of every site to develop a customized safety program. This process can be overwhelming, and if safety rules are not created before the system is use, it is especially difficult to win staff support and buy-in after they have become comfortable with the machinery. We spent several months developing a MARS technician safety manual, and negotiating appropriate rules with the affected staff. The final document was approved by the university Environmental Health and Safety department. The negotiation process with staff is vital. The key to a safe work environment is creating a culture of safety, where all staff believe in the importance and value of the policy. As a supervisor, it is impossible to be present at all times to enforce safe practices. The supervisor must be able to trust that staff will follow every safety procedure, every time, without bypassing steps and creating shortcuts. The single most important part of managing an ASRS is creating consensus regarding safety policies, promoting and rewarding safe work habits, and reinforcing an environment of trust, to ensure employees working on the machinery always protect themselves from harm. Operators loading and retrieving books are not subjected to the level of risk technicians face, however, any contact with the system warrants explicit, detailed safety training. Operators are provided with personal safety training before they are able to work with the equipment, and all student assistants must recertify in safety policies once a year. The department uses Blackboard online course management software to post documents and procedures, and an online quiz was created to test students' knowledge of safety practices annually. The most important safety rules are posted in the work areas to remind employees to think about safety throughout their daily activities.

Marketing MARS

The Mathewson Automated Retrieval System was named by a talented student who submitted the winning entry in a campus-wide naming competition. Early communication about the upcoming Knowledge Center and its futuristic robotic storage system was an important step in preparing users for the change. Subject liaison librarians visited departments and communicated with faculty across campus. Although there was some concern voiced by faculty, the transition was relatively smooth. There is an ongoing policy to remove items from the system at the request of faculty, and in nearly four years we have received only a handful of requests.

The system is not visible to the public. The Knowledge Center is visually stunning and elegant, and the industrial, mechanical look of the ASRS did not fit with the architects' vision for the building. However, the sheer size of the system, and its unique presence in libraries, consistently delights and amazes visitors. The first year of operation, tours of MARS were provided upon request. Library Services staff escorted visitors to the operations area behind the service desk to show off the system. Although the number of tours provided during this year was not recorded, there were multiple requests per day, as many as 15-20 at peak times. Eventually, we created a dummy item in a bin at the very back of the rack specifically for tours. This allows us to track tours by counting the number of requests for the "MARS Tour" item, and shows the system at its best by commanding the crane to move the entire distance of the aisle at top speed. Because the system invokes strong emotional reactions in all its visitors, it is a powerful marketing tool. Much to the relief of staff, tours on demand were discontinued after the first year. There are exceptions to this rule, however, for potential donors brought through by the Director of Development, the Dean of Libraries, and prospective students or other key visitors as approved by the Head of Library Services. Our

beloved late University President, Milton Glick, was excited and proud of MARS, and was known to stop by the Knowledge Center frequently to show off his favorite building and its magnificent robot to his friends, family, and honored guests.

Early in the initial loading process, the campus videographers created a clever video that is looped on a large monitor near the service desk. The video was created by placing a camera in a bin, and the viewer can see the system in operation from the perspective of a stored book. Campus tour guides from the Office of Prospective Students always include the MARS video in their trek across campus. Visitors who request a tour of the system are referred to the video and anyone new to the building stops to watch the second largest automated library storage system in action. Search "UNR Knowledge Center MARS cam" online to watch the video.

In order to generate continued excitement about the system, this year we hosted a MARS Open House once per semester. Advertising included posters in the Knowledge Center breezeway and promotional displays on large information monitors, flyers at all public service desks and in the Student Union, and postings on the library webpage and blog. Visitors are given "Trip to MARS" buttons as a reminder of the experience. The open houses are a success, with nearly 250 visitors this year. The second open house ran for two hours a day on two consecutive days, and had nearly twice as many visitors as compared to the initial one day, four hour event.

Automated Storage and the Future of Libraries

The Knowledge Center was designed to allow for twenty years of print collection growth. The original estimated capacity of MARS was two million volumes, however after the first 550,000 items were stored, the capacity estimate was adjusted to 1.5 million volumes. There are several possible reasons for this adjustment. There are a many large items stored resulting in a bin containing only 1-3 items instead of 80 used in the original estimate. Because of the way uncatalogued government documents were stored, we have 10,000 linear feet of material represented by only eight items per bin, despite the fact that each sector is filled with numerous individual items. Even with smaller estimated ASRS capacity, there is no concern we will ever run out of space for books in the Knowledge Center. At this time, all of the large loading projects are complete, and adequate space remains in the system to store all of the materials in the open stacks. The print collection grows very slowly each year, and the rate of growth will likely decrease in the future as digital format overcomes printed volumes in affordability and user preference. Although users generally accept and prefer digital journal articles, that is not yet the case for electronic books. As Shelburne (2009) discovered in her survey of University of Illinois faculty, graduate students, and undergraduate students, most users still rely on both print *and* ebook content because of differing strengths and weaknesses of each format. Until issues of digital rights management, ease of use, accessibility, the ability to mark-up and highlight electronic books, and eye strain with standard computer screens are resolved, not to mention historical content that may not be digitized for decades, printed books will continue to have value and purpose in the academic library.

With thousands of feet of traditional shelving and room for at least 800,000 additional items MARS, the Knowledge Center is a library that does not have to weed its collection. Typically, space concerns in libraries are paramount, and regular weeding projects are unavoidable. Off-site storage can alleviate some space issues, but it is costly and typically requires users to wait days for requested materials to be delivered. Compact shelving is costly, and is often installed after a library is nearing capacity, and regular weeding is still required. Automated storage provides the best of both worlds; users have virtually instant access to the materials, and librarians no longer have to make tough weeding decisions. This luxury seems almost unimaginable to most librarians. Looking forward, the problems with ebook format and delivery mechanisms will eventually be resolved, and few would argue ebooks will likely overtake printed books as the preferred format. The ability to perform keyword searches, ease of transport, and 24/7 access make ebooks a powerful research and reference tool for academics. I predict printed books will become increasingly rare in the future, and the archives of this medium will likely increase in value. For this reason, automated storage provides a secure, long-term storage solution that allows libraries to maintain their historical print collections.

Libraries' primary function has been shifting for several years from book repositories to learning and community spaces. Modern libraries provide room for collaborative learning, community events, and social gatherings. Libraries provide technological support for learning, not only by containing computer workstations for student use and laptops for checkout, but by housing experts who can assist with media production, high-end statistical processing, and personal computer support. The ability of libraries to provide a place for students to learn and play is seriously reduced when miles of book stacks fill the building. The Knowledge Center saved 100,000 square feet of student learning space by building a 12,789 square foot automated storage and retrieval system. By resolving the issue of book storage before the building opened, there is automatic flexibility in the configuration and purpose of learning spaces. As student learning, research, and campus community evolve, spaces can evolve to meet changing needs. This is already happening at the Knowledge Center. The majority of the building is designed for collaborative work, and only one of five floors is designated quiet study space. One area of shelves on the quiet floor was exposed to direct sunlight during most daytime hours. This preservation concern led to space configuration discussions. The rows of shelves blocked most of the natural light on the floor, making it appear gloomy and isolated compared to other areas in the building. By removing several rows of shelving in the center and edges of the section, the books exposed to sunlight were relocated, and additional tables were added between the stacks providing a private, quiet ambiance which suited the purpose of the area. The decision was undisputed, because removing shelving had no impact on the collection. Several rows of shelving were removed on the fourth floor as well, to create a new study space and provide better visibility for the Instructional Design department. The area was filled with whiteboard tables; ordinary tables painted with whiteboard paint to allow students to draw on the surface. The whiteboard tables are a huge hit with the students, as evidenced by a note left for library staff, "We love this table! Thx." With room to adjust and retool, libraries can respond quickly to user needs without sacrificing their commitment to supporting research and teaching at the university. Books and people are no longer in competition for the most sought after and valuable commodity in any modern academic library: space.

CONCLUSION

There is a basic human desire to see into the future and attempt to predict how life may change as time passes. Scholars and science fiction writers alike have discussed how the world will change, and how societies and technology will function, in years to come. In a fascinating review of scholarly and science fiction predictions describing the library of the future, Steele (1987) ponders how technological advances may change what libraries look like, and represent, in the 21st century. Amazingly, many of the concepts he explores ring true. For example, this question posed by an archivist speaking to a user in Phillip Dick's *Counter-Clock World* (1957) is eerily relevant today:

Who controls the storage and the retrieval? To what extent is the material there for anyone who wants and needs it, and to what extent is it 'there' only for those who have the information that it is there, the education to obtain that information, and the power to get that education? (Steele, 1987, p. 31)

From a literal viewpoint, hiding books from the users' sight in an automated storage system relates to the Archivist's musing. However, the complexity of information available on the Internet and controlled by publishers and database vendors who limit access to libraries, and individuals with money, certainly echo his viewpoint. The ever-increasing complexity of evaluating and accessing information is changing the way librarians teach and discover research materials. The nature of information is changing, as social media and multimedia formats dominate, and user-created knowledge rises in authoritative acceptance and social relevance. The challenges librarians face teaching students to find reliable information online, and managing the exploding information landscape, are clearly reflected in this forecast over a half century ago.

As Brian Mathews (2012) asserts, libraries must look to the future with a broad view, searching for new ways to support 21st century learners instead of making incremental changes to the services we already provide. He calls for innovative methodologies, and "breakthrough, paradigm-shifting, transformative and disruptive ideas" (p. 1) to ensure libraries will stay relevant and necessary.

We have to face the future boldly. We have to peer upwards and outwards through telescopes, not downwards into microscopes....[W]e need to implement big new ideas, otherwise the role of the library will become marginalized in higher education....We'll become just another campus utility...rather than the intellectual soul of the community. (Mathews, 2012, p. 1)

Robots in libraries are not the beginning of this transformation. In fact, they are an overdue step forward using a technology accepted in the business world four decades ago and improved over time. By incorporating ASRS technology into library buildings, we are coping with the library's traditional role as a warehouse and storage facility for printed information by maximizing space utilization and efficiency. Collections are available and usable, but no longer impede the evolving role of libraries as a center for learning and knowledge creation. This evolution is occurring at an increasingly rapid pace. Librarians are grappling with ways to keep up and stay relevant in a world of instantaneous access to the digital information explosion. In order for the Mathewson-IGT Knowledge Center to fulfill Dr. Zink's vision and for libraries to continue to be necessary to communities that we know need us - but may not realize they do – we must become change agents and purposefully build the libraries of the future.

REFERENCES

Adams, C. (2010, September). *New managers of automated library storage: What to know*. Presented at the 25th annual Material Handling and Logistics Conference, Document Management Track, Park City, UT. Retrieved April 20, 2012, from http://www.hkplanet.net/resources-mhlc-presentations.cfm?year=2010

Bullard, R., & Wrosch, J. (2009). Eastern Michigan University's automated storage and retrieval system: 10 years later. *Journal of Access Services*, *6*(3), 388–395. doi:10.1080/15367960902894187.

Creaghe, N., & Davis, D. A. (1986). Hard copy in transition: An automated storage and retrieval facility for low-use library materials. *College & Research Libraries*, *47*(5), 495–499.

Gallagher, K. (2010, August 4). HK systems to be acquired by Dematic. *Milwaukee, Wisconsin Journal Sentinel*. Retrieved April 23, 2012, from http://www.jsonline.com/business/99969954.html

Haslam, M., Kwon, M., & Pearson, M. (2002). The automated storage and retrieval system (ASRS) in Lied Library. *Library Hi Tech*, *20*(1), 71–89. doi:10.1108/07378830210420708.

Kountz, J. (1987). Robots in the library: Automated storage and retrieval systems. *Library Journal*, *112*(20), 67.

Mathews, B. (2012). Think like a startup: A white paper to inspire library entrepreneurialism. *VTechWorks Institutional Repository, Virginia Tech University*. Retrieved April 21, 2012, from http://hdl.handle.net/10919/18649

Rapp, D. (2011). Robot visions. *Library Journal*, *136*(15), 20–24.

Shelburne, W. A. (2009). E-book usage in an academic library: User attitudes and behaviors. *Library Collections, Acquisitions & Technical Services*, *33*(2-3), 58–72. doi:10.1016/j.lcats.2009.04.002.

Shirato, L., Cogan, S., & Yee, S. G. (2001). The impact of an automated storage and retrieval system on public services. *RSR. Reference Services Review*, *29*(3), 253–261. doi:10.1108/EUM0000000006545.

Steele, C. R. (1987). From punched cards to robots: Our ascent into technology. *Wilson Library Bulletin*, *62*(2), 29–32.

Zink, S. D. (2010, September). *A building for the post-Gutenberg era: The Mathewson-IGT Knowledge Center at the University of Nevada, Reno*. Presented at the 25th annual Material Handling and Logistics Conference, Document Management Track, Park City, UT. Retrieved April 20, 2012, from http://www.hkplanet.net/resources-mhlc-presentations.cfm?year=2010

Chapter 8
Development of Academic Library Automation in Brazil

Michelângelo Mazzardo Marques Viana
Pontifical Catholic University of Rio Grande do Sul, Brazil

ABSTRACT

The automation of university libraries in Brazil underwent a restraint of trade on computers and software, which took place in the country between 1980 and 1990, restricting the initial use of automation systems. However, they were often developed in creative ways: systems and applications were created and used in various universities, some as free software, others based on the ISIS platform from Unesco, in addition to using modern foreign systems, which only occurred in the 1990's. This chapter provides a historical overview of the development of automation in the country's university libraries, from the moment in which Brazilian researchers began to disseminate information technology, creating an automation culture in higher education institutions. Many people and institutions have also contributed to promoting and implementing automation in university libraries. This paper is on future perspectives of academic library automation in Brazil with discovery tools, next generation cloud-based systems and library automation equipment. Some possible future developments are also presented.

INTRODUCTION

What does library automation mean? I would like to introduce this chapter by assuming the position by Rowley (1994) and Barsotti (1990). Professor Roberto Barsotti (1990), an Italian naturalized Brazilian, former professor of Librarianship at the University of São Paulo (USP) and the Teresa D'Ávila Integrated Colleges (FATEA) in his book *Computers in librarianship and documentation* states:

DOI: 10.4018/978-1-4666-3938-6.ch008

When we speak of library automation, we mean the automation of the library's technical processes. Basically, acquisition, cataloging and/or indexing and circulation. This automation is often confused with the creation and exploration of databases containing the library archives. They are different things, with different focuses and results, involving different software. (Barsotti, 1990, p. 65)

I believe that when computers were first being used in Brazilian libraries, there was a bit of confusion on the part of students, professors, librarians, researchers and systems analysts regarding this distinction, which occasionally led to a lack of focus in the efforts in automating library services in Brazil, which were concentrated on creating cataloging, indexing and metadata search systems. This does not mean that these systems were not important. On the contrary, I am aware of all of the efforts that existed in the country when systems were adopted for generating bibliographic databases (such as ISIS by Unesco), creating record and exchange formats (such as CALCO - *Catalogação Legível em Computador[1]*, based on MARCII), as well as broad spectrum indexing methods, such as LILACS (created by BIREME - Latin American and Caribbean Center of Information on Health Sciences). However, the main focus of this chapter is on automation systems for academic library services in Brazil.

In general, automation should be thought of as comprehensively as possible: using technology so that machines – equipment and computer programs – carry out human tasks: the same definition that appears in the Webster dictionary (Automation, 2012): "automatically controlled operation of an apparatus, process or system by mechanical or electronic devices that take the place of human labor", which means using technology for basic services (cataloging, catalog search and retrieval, acquisition and circulation), user services (reference interviews, document requests), to retrieve information stored locally and in remote providers (using catalogs, metasearch, discovery tools and any other technology applied in libraries), for internal processes and user services, staff management, access control for physical spaces, security, financial transactions (service payments), interactions between the library and all of its stakeholders (employees, users, coordinators, private service providers, print and electronic information providers, the government,…), in other words, all of the services and processes carried out by the library, inside and outside of its building or room.

The automation of library services started to happen in Brazil only from the 1980's on. According to Ohira (1992), "From 1980 on, automation

began to move out of the embryonic and experimental level towards the operational," (p. 234). The first studies and projects for automation, on a national scale, began with the National Institute for Space Research (INPE), between 1960 and 1980, and with the National Library, in 1973, when Manoel Adolpho Wanderley carried out a preliminary and general survey of the possibilities of partial or total automation of the services of the Brazilian National Library, in which the pros and cons of each case are shown (Wanderley, 1973).

On the other hand, in the scope of higher education institutions, it was in the beginning of the 1980's that the first initiatives emerged, the large majority of them using software created by the education institutions themselves. Mr. Jaime Robredo carried out an important research in 1981, the result of which was presented at the Symposium on Library Service Automation (during the second National Seminar of University Libraries, held in Brasília - Federal District). His research was on the *Panorama of the automation plans and projects in Brazilian university libraries*, demonstrated that of the 578 Brazilian university libraries registered in the Coordination for the Improvement of Higher Educational Personnel (CAPES), only 10 had automation systems in operation, 13 in development and 21 in the project phase or with interest in their use. An analytic summary of his research results can be seen in Table 1, which shows the overall data of the data survey, grouped by state:

In his study, Robredo (1981) highlighted that there was "moderate use of terminals in the systems and projects considered, and the predominant tendency to develop one's own software instead of using existing ones" (p.157): of the 10 academic libraries that already had automation systems in 1981, 8 were developed by the institutions themselves and only 2 were acquired from other institutions.

I would like to make a note in this book of the universities that were pioneers in implementing automation systems in their libraries in Brazil, according to research by Robredo (1981):

Table 1. Panorama of the automation plans and projects in Brazilian university libraries (Adapted from Robredo (1981, p.160)).

State	Institutions Contacted	Answers Received	Libraries with Automation or Project	Plans/Projects (Among Respondents)		
				In Operation (% of Respondents)	In Development (% of Respondents)	Project Phase / Interest (% of Inst. Contacted)
Acre	1	-	-	-	-	-
Amazonas	8	2	2	1 (12.5%)	-	1 (12.5%)
Pará	6	2	2	-	2 (33.5%)	-
Maranhão	8	1	1	-	-	1 (12.5%)
Piauí	4	1	-	-	-	-
Ceará	22	3	-	-	-	-
Rio Grande do	9	2				
Norte	19	5	2	1 (5.3%)	1 (5.3%)	-
Paraíba	38	9	-	-	-	-
Pernambuco	3	-	-	-	-	-
Alagoas	5	1	-	-	-	-
Sergipe	32	13	-	-	-	-
Bahia	72	16	11	4 (5.5%)	4 (5.5%)	3 (3.1%)
Rio de Janeiro	3	1	-	-	-	-
Espírito Santo	124	25	7	1 (0.8%)	2 (1.6%)	4 (4.2%)
Minas Gerais	111	46	13	2 (1.8%)	3 (2.7%)	9 (8.1%)
São Paulo	21	2	-	-	-	-
Paraná	27	2	-	-	-	-
Santa Catarina	**46**	15	3	1 (2.2%)	1 (2.2%)	1 (2.2%)
Rio Grande do Sul	4	2	-	-	-	-
Goiás	3	2	-	-	-	-
Mato Grosso	5	2	1	-	-	1 (33.3%)
Mato Grosso do Sul	7	2	1	1 (33.3%)	-	-
Federal District	-	6	-	-	-	-
Others – non-identified						
TOTAL	578	160	43 (7.4% from 578)	10 (1.7%)	13 (2.2%)	21 (3.5%)

- University of Amazonas (UFAM)
- Federal University of Paraíba (UFPB)
- Federal University of Rio Grande do Sul (UFRGS)
- Federal University of Juiz de Fora (UFJF)
- University of Brasília (UnB)
- Federal University of Rio de Janeiro (UFRJ)
- Getúlio Vargas Foundation (FGV)
- University of São Paulo (USP)
- São Paulo State University Júlio de Mesquita Filho (UNESP)

(The number of institutions is not the same as the number of libraries because in some of them, more than one department library was automated and answered the questionnaire individually).

After evaluating the results of his research, Robredo believed that in 1981:

- The panorama of automation in Brazilian academic libraries was not very encouraging, when compared to the situation in the majority of industrialized countries.
- Despite the majority of universities having computers, automating their libraries did not seem to be a priority in applying the techniques of electronic data processing.
- The low automation of libraries was explained by the low representativeness in the number of projects submitted to the authorities.

- There could have been a greater number of systems in operation at that time. They just did not exist because there were not enough projects that deserved to be approved.

From the beginning of the discussions on the use of computers in libraries in Brazil, between 1960 and 1970, higher education institutions always aimed to monitor the development of technologies based on the experiences of institutions in the United States and in European countries (mainly England), by reading bibliographies that began to be published. From the end of the 1970's, the first library automation systems began to be used, and at the end of the 1980's, the first experiences with the development of commercial – and also some free – software for libraries emerged in Brazil. Today, from 2011 on, Brazilian universities began to use discovery tools for their libraries, integrating all of their bibliographic records with electronic resources that can be accessed through the internet.

The development of library automation in Brazil ran a very difficult course in the beginning: there was little experience with information technology and other barriers that needed to be overcome, mainly the restraint of trade on computers and software, which restricted technology imports between the years of 1976 and 1992. The government proposal with this restraint was to stimulate the hardware and software industry, but this development did not happen according to plan, delaying the use of automation in academic libraries in the country.

Despite the barriers and difficulties, there were always many supporters and many actions undertaken for academic libraries in the sense of creating an automation culture and bringing together librarians and computers. One important highlight was the high number of incentives to use bibliographic management applications from the CDS/ISIS family (Computerized Documentation System / Information Storage and Retrieval System; also known as Computerized Documentation

System / Integrated Set for Information System) in Brazil, which culminated in the availability of the ABCD automation system in 2008. This chapter intends to demonstrate how ISIS began to be used in Brazil, how it evolved to this day and the influence it had on automation and libraries.

In the beginning of this automation process in Brazil, there was an important diffusion of knowledge by Brazilian researchers in articles, books and events, mainly the symposium of 1981 and the specific seminars on library automation held between 1984 and 1997. Important works were published between 1972 and the 2000's, enabling the creation in Brazil of a culture of using computers for information retrieval services and the automation of technical services.

During the period of the restraint of trade, inspired by the experiences disseminated by the international literature, some Brazilian public universities that had created their own library automation systems in the 1980's began to substitute them for imported systems, from the 1990's on. After 1993, Côrte et al. (2002) reported that "with the changes in the policies on Information Technology, which provided access to increasingly advanced equipment and software, national computerized systems started to be developed and, today, the large majority of libraries – of all types and sizes – rely on computerized management systems for their services."

The objective of this chapter is to present how the automation of academic libraries in Brazil developed, chronologically, presenting the evolutions that took place over the years, and how it evolved in higher education in Brazil, contemplating the actions carried out by the academic libraries and the accomplishments that have taken place since the 1970's. At the end, an overview is presented of the use of discovery systems, beginning in the country in 2011, new developments made possible with the advent of discovery tools and new human-computer interaction technologies, expectations with respect to next generation systems and research that still needs to be done with Brazilian academic libraries.

BACKGROUND

The early initiatives of computer use in Brazilian universities began at the end of the 1960's, though to a more limited extent, at the same time in which library automation began in other countries (Teixeira, Oliveira, Lapa, & Assunção, 2011). In Brazil, the early initiatives in library automation were based on software developed in the 1970's by the institutions to which they were connected, or based on the Microisis software (Russo, 2010).

Only some federal universities had trained technical personnel and the financial capabilities to import American computers (IBM mainframes, Burroughs, Unisys, among others), since the Information Technology Department was developed in the country under a regime of incentives focused on the substitution of imports and on the local development and production of information technology goods. Federal Law n° 7232 of 1984 established the "National Information Technology Policy" (PNI) in Brazil and made the restraint official for some segments of the market, including software. Since they were still in the development phase, national companies were not able to meet the demands of all sectors, affecting the higher education institutions as well.

During the period in which the National Information Technology Policy was in effect, and even in the following years, between the years of 1981 and 2004, much research was carried out in Brazil with the objective of diagnosing the use of information technologies and automation systems in libraries, such as those of Robredo (1981), McCarthy (1982), McCarthy (1983), McCarthy (1990), Figueiredo (1986), Lage (1989), Sayão, Marcondes, Fernandes, and Medeiros (1989), McCarthy and Neves (1990), Ohira (1992), Ohira (1994), Carvalho (1997), Lima (1998), Lima (1999), Balby (2002), Côrte et al. (2002),

Prado and Abreu (2002), and Burin, Lucas, and Hoffmann (2004).

Brazil has a tradition in teaching Librarianship. Librarianship programs are offered by various federal, state and private universities. The first Librarianship program created in Brazil was by the National Library of Brazil, in July of 1911, by Federal Decree n° 8.835. The country currently has 42 active Librarianship schools, 30 of them in public state and federal institutions, in which the program is free; and 12 in private institutions, in which the program is paid (Brasil, 2012b). Colleges have always had access to literature produced in and out of the country reporting experiences in other countries and possible paths, mainly in Europe and North America. Despite this, the ever-present difficulty for Brazilian academics with the English language did not allow for the automation culture to be as widespread as necessary. The first papers on computer use in libraries were only produced in Brazil from 1972 on. Since then, many other papers were produced or translated, through books, journal articles, theses, dissertations and papers presented at events. In addition to the literature cited above, the following are also worth mentioning: Sambaquy (1972), Eyre (1979), Barsotti (1990), Cunha (1985), Robredo and Cunha (1994), Rowley (1994), Côrte et al. (1999), and, more recently, Lima (1999).

An important inspiration for computer use in Brazilian libraries comes from librarian Lydia de Queiroz Sambaquy, who was a very active supporter of the development and progress of libraries and scientific documentation in Brazil. She studied Librarianship at the National Library of Rio de Janeiro, the first in Brazil, and also at the University of Columbia in 1941 and 1942. She fought for the recognition of the importance of libraries for Brazil, often in charge of librarianship activities and scientific documentation in the country since 1945 (Lydia, 1960). In her article entitled "The Library of the Future" (1972), Sambaquy reports that she would like to have a time machine to visit a library in the year 2000 and have a glimpse of how they would be: libraries with better facilities, in larger environments, in more central locations, with simpler and colorful modern furniture, with a better use of lighting and interior decorating, easier to clean and maintain. In these libraries, she said, there would be books from the 1970's and those that were produced afterwards, in kilometers and kilometers of shelves full of books. These shelves would be controlled mechanically with the use of trays (like existing systems in the United States).

Sambaquy also predicted that computers, in models especially adapted for working with documentation, would definitely be the librarian's great and beneficial solution in the 2000's. With respect to the use of technologies, she also claims that in 2000 "there will be much more and wonderful electronic equipment used for communicating ideas and exchanging information. Closed circuit television and telecommunication, for verbal contacts and for transferring texts and figures, would constitute the wonderful world of tomorrow for librarians and documents" (Sambaquy, 1972, p. 67). She believed that texts transmitted through modern forms of telecommunication would appear on a screen and would be automatically reproduced in the intended location and the problem of automated translations would be completely solved. She concluded the article by saying that "perhaps thirty years from now, sitting in our chairs, comfortably settled in our homes, we can mutually tune our *telephonevision* devices, and we can begin Brazilian congresses on Librarianship and Documentation or worldwide congresses on Documentation Systems and intercontinental Scientific Information" (Sambaquy, 1972, p. 68).

More recently, in 1990, Roberto Barsotti published the book "Computers in Librarianship and Documentation." In addition to presenting a little about the Italian experience with library automation, this was one of the first books published in the Portuguese language in Brazil on the use of computers in libraries and their practical uses, widely used in schools of librarianship. In his book, Barsotti (1990) also made a prediction about the

future of libraries after the 1990's: the placement of complete texts in computer memories with special equipment and their placement by special programs into desired formats; increase in the storage capacity of backup memories; enormous databases with complete texts, complemented by images, graphics and sounds; access to these databases by user-friendly software; availability of differentiated communication terminals, substituting the old cathode ray tubes; increase of the possibilities of communication between computers, thanks to the spread of telephones, satellites and other networks, which will make it very simple to connect to various computer sources in a single network; electronic journals; massive dissemination of videotext (or something similar); cheap, efficient and portable terminals; self-education via computers.

A few years later, another important book was translated and edited in Brazil by librarian and editor Antonio Agenor Briquet de Lemos. Written by Englishwoman Jennifer Rowley, "Computers for Libraries" (1994, 3rd edition, London, Library Association Publishing, 1993), is a very systematic and educational book that was and continues to be one of the most used books in librarianship schools and one of the most cited publications on library automation in the country. It was translated at an opportune moment, in which libraries were beginning to massively adopt automation systems in their libraries and needed to be prepared and more knowledgeable in the subject.

Another important collaboration for spreading the culture of library automation was the creation and publication in 1994 of *Biblioinfo*, the database about library automation (information technology for documents): 1986-1994. It had 390 references with abstracts from Brazilian journals on librarianship and information science. Its objective was to make available to researchers, students and other users, the literature that was published in Brazil from 1986 to 1994. To record the data, a database developed in MicroISIS was created, which enabled the retrieval of references by author, title and subject, in addition to printing research reports. The database was also planned to work as an educational tool for teaching librarianship in Brazilian colleges (Ohira, 1994).

Finally, still between the 1980's and 1990's, Brazil held eight important events, six of them focused exclusively on library automation: the Seminars on Library Automation and Documentation Centers, organized mainly by the National Institute for Space Research (INPE), which was one of the first Brazilian institution to automate a library's services. "The automation of Inpe's technical library procedures contributed to their recognition and dates back from the 1960's to the 1980's, with the use of large computers" (Marcelino, 2009, p. 89). The events were as follows:

- Symposium on the Automation of Library Services, held during the second National Seminar on University Libraries, held in Brasília-FD, in 1981.
- National Meeting on Librarianship and information technology. October 1984. Brasília-FD.
- Seminar on Automation in Libraries and Documentation Centers, 1. December 4-7 of 1984, INPE, São José do Campos-SP.
- Seminar on Automation in Libraries and Documentation Centers, 2. 1986, São José dos Campos-SP.
- Seminar on Automation in Libraries and Documentation Centers, 3. March 14 to 16 of 1989, Vacance Hotel, Águas de Lindóia-SP.
- Seminar on Automation in Libraries and Documentation Centers, 4. December 3 to 6 of 1990, INPE, São Paulo-SP.
- Seminar on Automation in Libraries and Documentation Centers, 5. July 19 to 22 of 1994, UNIVAP, São José dos Campos-SP.
- Seminar on Automation in Libraries and Documentation Centers, 6. September 9 to 11 of 1997, Águas de Lindóia-SP.

AUTOMATION OF ACADEMIC LIBRARIES IN BRAZIL

Higher Education in Brazil and the Evaluation of Library Quality

From a historical-conceptual point of view, the trajectory of the academic library on the national stage reflects the history of education in the country, which was strongly influenced by Portugal during the colonial period. Brazil was the last country in Spanish America to have a university, which occurred only in 1920, 31 years after the Proclamation of the Republic, despite previous attempts and isolated programs created over the course of the 19th century (Vieira & Souza, 2010).

The expansion of higher education in Brazil began in 1808, with higher education programs created by Dom João VI, the King of Portugal. By official initiative, these programs continued in the Portuguese Empire with the creation of the Schools of Law. During the first republic, between 1889 and 1930, the "free institutions" began to be created, by private initiative. After 1930, public participation resumed, with a marked increase in the 1940's, 1950's and the beginning of the 1960's, through the federalization of state and private institutions and the creation of new federal universities.

From 1960 to the Constitution of 1988, the Napoleonic model prevailed in the organization and expansion of higher education in Brazil, characterized by isolated vocational schools, with teaching and research dissociated and a large state centralization. From the 1990's on, in a process that is still going on today, new changes that emerge are characterized by the diversification of the forms of organization of higher education institutions altering the university model towards the Anglo-Saxon model in the North American version (Saviani, 2010). Today, Brazil has 2,653 higher education institutions, 294 public and 2,359 private, divided into the following groups: 193 universities, 2,289 colleges, 171 University

Centers and 40 Federal Centers for Technological Education (Brasil, 2012a).

Due to the large increase in the number of higher education institutions and the competition that began, many private universities lost many students to the recently created institutions. In the mid-2000's, the country suffered a moment of crisis in the private universities (Carvalho, 2004), in which the reduction of revenue caused mass layoffs of professors and staff, as an alternative found for reducing fixed expenses. There was a decline in investments, and academic libraries were also affected. For the sake of illustration, there is a case of the University of Santa Cruz do Sul (Unisc), which stopped using the Aleph™ system to start using the Pergamum™ system, which has a lower acquisition and maintenance cost when compared to the Aleph™. Silva (2005) reports that "as a justification for the change of Unisc library systems, one can say that it was an institutional decision based on technological and financial criteria" (p. 107). Still according to Silva, "It is fundamental to understand the modern concept of information management and have a true understanding of the institution as a whole, to present a system that meets the items that are obligatory, essential and desirable to the managers" (p. 107).

The educational and cultural sectors live in the country simultaneously with disparities: there are institutions that are still outdated, with little or no applied technology; and there are others that have already made significant scientific and technological advances, truly established in the 21st century. This disparity, and also a lack of librarians – there are 33,446 registered in the Federal Council of Librarianship (CFB) and only 19,214 working in the whole country, prevents libraries and library systems in the country from becoming more consolidated, efficient and effective. One of the symptoms of this frailty is the omission that the university library suffered from the University Reform Law of 1968-1969 (Law n° 5.540, of 28/11/68), put forth as an attempt to adapt the educational system to a system resulting from the military coup of 1964 (Vieira & Souza, 2010).

The libraries of Brazilian higher education institutions, in general, remained with the classical structure of print archive and catalog storage and loan cards until the end of the 1990's. It was only with the expansion of higher education in the beginning of the 1990's and the consequent need for improvement in the quality of service management and services for the academic community that libraries started to become more modern. Parallel to this change, the Brazilian Federal Government, through the Ministry of Education, has been evaluating Higher Education Institutions in Brazil since 1995 (Law 9.131/1995), establishing criteria and standards for their operation, providing a range of information to society and to educational administrators (Barcelos & Gomes, 2004). In 1996, the Federal Decree 2.026/96 established additional measures for evaluating higher education. In 2001, the Brazilian government launched the National Education Plan (PNE), enacted by Law n° 10.172, on January 9th of 2001, which defined a total of twenty-three objectives and goals for higher education in Brazil.

The PNE was a great stimulus for the improvement of academic libraries, since "based on the minimum standards established by public authorities, the government began to demand progressive improvements in the infrastructure of laboratories, equipment and libraries, as a condition for the reaccreditation of higher education institutions and the renewal of program recognition" (Brasil, 2004, p. 32).

Today, in Brazil, the Anísio Teixeira National Institute for Educational Studies and Research (INEP) manages the entire evaluation system of higher education programs in the country, producing indicators and an information system, which sanctions the regulatory process, exercised by the Ministry of Education (MEC), and also guarantees the transparency of the data on the quality of higher education for society as a whole. The instruments that sanction the production of quality indicators and the evaluation processes of the programs developed by Inep are the *National Exam of Student Performance* (Enade) and the evaluations carried out on-site by committees of specialists for the Authorization and Recognition of programs and the Accreditation and Reaccreditation of HEIs, anticipated in Laws 9,131/95 and 9,394/96 (National Education Guidelines and Framework Law - LDB) and in Decree 3 860/01, which consolidated Decrees 2.026/96 and 2.306/97.

The instruments used by the Brazilian government to evaluate academic libraries are:

- An evaluation instrument for the accreditation of Higher Education Institutions (Brasil, 2010a).
- An external institutional evaluation instrument (Brasil, 2010b).
- An evaluation instrument for on-campus and distance Undergraduate Programs (Brasil, 2012b).

These three instruments are all used together, and contain criteria for evaluating libraries, including the status of their service automation. Table 2 presents the criteria relative to library automation, which are used in each of the instruments.

An important step in the evaluation of libraries in Higher Education Institutions occurred in February of 2012, with the publication of the new edition of the Evaluation Instrument for on-campus and distance Undergraduate Programs (Brasil, 2012b), in which MEC began to consider as part of the availability criteria of the Library electronic book collections and journals, using the expression "virtual." Until the release of the mentioned instrument, publications subscribed to and acquired in electronic media were not considered as the basic bibliography of programs, only as additional bibliography.

Initial Barriers Faced between the 1970's and 1980's

Between 1960 and 1970, all computers used in Brazil were imported. It was only in 1974 that the first computers produced in the country began to be commercialized, by the Cobra company. According to Azolin (1999), in the following period, between 1976 and 1985, engineering production in an emerging industrial sector took place in Brazil, which had never taken place before in the country. Languages, compilers, operating systems and communication programs were written; memory boards and other circuits were projected and manufactured; video terminals, keyboards and other peripheral units were conceived, projected and manufactured; entire CPUs were built. Large computers were developed to interconnect banking systems, capable of communicating from one end of Brazil to another. This happened due to Federal Law n° 7232 of 1984, in which the country established the National Information Technology Policy (PNI) and made the restraint official for some segments of the market, including software.

This restraint of trade generated considerable development in the national information technology industry with the objective of increasing computer use in the country and having a national industry that was independent of the external market. The idea was excellent. However, the high prices provoked by the restraint prevented the expected growth. Moreover, the Brazilian technological evolution did not accompany that of other countries and complaints began to emerge from many sectors, mainly from the industry. The literature points out that this was the wrong path. The restraint was already collapsing and its consequent discontinuation in 1992 during the administration of President Fernando Collor de Mello was a predictable result of the combination of offering technically outdated products to the market at high prices with North American pressure for Brazil to open its computer market (Marques, 2003).

Table 2. Criteria from the Federal Government used to evaluate academic libraries in Brazil (Sources:Brasil(2010a), Brasil (2010b), Brasil (2012b)).

Ministry of Education Instrument	Year Published	Purpose of the Instrument	Indicators	Library Automation Criteria
Evaluation instrument for the accreditation of Higher Education Institutions	2010	This instrument is aimed at evaluating the initial conditions required for the accreditation of the higher education institution (college) whose result will serve as a basic reference for the decision of regulatory bodies. It was elaborated by the Secretary of Higher Education and the Anísio Teixeira National Institute for Educational Studies and Research according to the guidelines and foundations for national education (Law n° 9.394/1996), the principles of evaluation (Law of Sinaes n° 10.861/2004, Decree n° 5.773/2007 and Regulatory Ordinance n° 40/2007, among others), in addition to taking into account the expansion policy with quality specified by the Ministry of Education.	Dimension 3: Physical facilities. Criterion 3.7 Library: computerization.	The maximum score (5) is obtained when library automation is such that the computers, programs and applications use current technology and in quantities that fully meet the demands foreseen for the use of archives, enabling different forms of research, online book reservations, access via the internet.
External institutional evaluation instrument	2010	Elaborated jointly by the National Higher Education Evaluation Commission (CONAES) and by the Board of Higher Education Evaluation (DAES), of the Anísio Teixeira National Institute for Educational Studies and Research (INEP), the formulation of which was based on the SINAES principles and guidelines and the higher education quality standards. The Institutional Evaluation Instrument will be used by evaluators in on-campus and distance modalities. In this sense, it is necessarily comprehensive and flexible enough to ensure a reliable evaluation of the institutions, to highlight the specificities that distinguish each one of them, and make it possible for their use to be associated to diagnostic indicators that will contribute to a more substantial analysis of reality. With the new External Institutional Evaluation Instrument, INEP is apt to implement this stage of SINAES' integrated evaluation process, ensuring a higher education with academic quality and a social commitment to the development of the country.	DIMENSION 7: Physical infrastructure, especially for teaching and research, library, information and communication resources. Criterion 7.4. Library: archive, services and physical space.	Minimum referential concept of quality: When the adequate actions of updating and broadening bibliographic archives and library* (-ies) services can be verified. (*) Library services: Included among the library services are: computerization system, book reservations through the internet, take-home loans, and internal document copies in the HEI, offer of bibliography exchange services in the country and abroad.
Evaluation instrument for on-campus and distance Undergraduate Programs	2012	This instrument sanctions the authorization of programs – authorization, recognition and renewal of recognition – in technology degrees, teaching degrees and bachelor's degrees for on-campus and distance modalities. According to art. 1 of the Regulatory Ordinance 40/2007, consolidated on December 29 of 2010, the application of the indicators of this instrument will occur exclusively electronically in the e-MEC system.	Dimension 3: Infrastructure Indicator: 3.6. Basic bibliography (For purposes of authorization, consider the archive of the basic bibliography available for the first year of the program, if a Higher Program of Technology (CST), or first two years, if bachelor's/teaching degrees) In the programs that have a virtual archive (at least 1 virtual title per curricular unit), the proportion of students per physical copy are counted in the following way for the scores 3, 4 and 5: Score 3 – 13 to 19 annual places Score 4 – 6 to 13 annual places Score 5 – less than 6 annual places	The maximum score (5) is obtained when the basic bibliography archive, with at least three titles per curricular unit, is available in the average proportion of one copy for less than 5 annual intended/authorized places, of each curricular unit, of all the programs that effectively use the archive, in addition to being computerized and declared as part of the patrimony of the HEI.

With the new university reform, which took place during the same period, when the Brazilian university model stopped being Napoleonic and started to adopt the Anglo-Saxon model in the North American version, hundreds of new colleges, university centers and private universities and their libraries emerged in the country. Consequently, the market of automation systems began to have many new clients, and new needs, which demanded from the companies improvements in their systems, as well as the creation of new companies and new systems.

1980's and 1990's: The ISIS Platform and the First Homemade Systems

Between the 1980's and 1999, six factors were very important for Brazilian academic institutions to begin a major movement towards the automation of their libraries:

1. **Knowledge in Information Technology:** Acquired by recent librarian graduates who specialized in information technology and began working as professors of Librarianship in Brazilian colleges, publishing works and presenting reports at national events, such as Murilo Bastos da Cunha, José Fernando Modesto, Ligia Maria Arruda Café, Cristina Dotta Ortega, Maria Lourdes Blatt Ohira, Elisabeth Márcia Martucci, Janise Silva Borges da Costa, Gercina Ângela Borém de Oliveira Lima, among others;

2. **Opening of the Hardware and Software Market in 1992:** Which in addition to enabling the acquisition of more modern equipment and systems, also offered the opportunity for national software companies for libraries to emerge and compete with the foreign ones;

3. **Increase in the Amount of Literature:** In the Portuguese language produced in Brazil or translated from foreign publications, related to computers for libraries and also automation for libraries;

4. **The Events Held from 1984 to 1997:** With a specific focus on library automation, in which many experiences were shared;

5. **The Availability of CDS/ISIS Software in 1985:** By Unesco, free and capable of running on microcomputers, which was widely used in the country;

6. **Criteria and Standards:** The Brazilian Ministry of Education, with the publication of guidelines and foundations for national education (Law n° 9.394/1996), started to evaluate Higher Education Institutions in Brazil since 1996, establishing criteria and standards for their operation, including the computerization of their libraries.

Professor Fernando Modesto (2006) summarizes the importance that ISIS had in Brazil: "The history of library automation systems cannot be told without opening a broad space for CDS/ISIS, popularly known as Microisis (MS/DOS version) and Winisis (MS/Windows version). The system, promoted by UNESCO (United Nations Educational, Scientific and Cultural Organization) to support library computerization projects in developing countries, certainly has had significant impacts on these information systems."

In 1976, UNESCO created the General Information Program (PGI), which included a plan supporting library automation, especially in developing countries, without financial resources. The Brazilian Institute of Information in Science and Technology IBICT, of the Brazilian Ministry of Science and Technology, was the national distributor of the CDS/ISIS software in Brazil, in its MS-DOS version, until 1998. From 1999 on, it shared the responsibility of providing licenses with BIREME, which was responsible for the MS-Windows version. By 2006, according to data from UNESCO (www.unesco.org/webworld/isis), the national distributors of ISIS added up to 95 countries: Africa (16); Latin America and the Caribbean (25); Asia and the Pacific (18); the Arab States (4); Europe (32) (Modesto, 1996).

The ISIS, as it is commonly called, was launched by Unesco in 1985, initially for IBM 360-30 mainframes, and had a version for local networks (LAN) and another for Linux (both launched in 1993) and also a version for MS-Windows (1997/1998), capable of running on a computer or even on a local network (Modesto, 2006). It represented a great advance ever since it began to be distributed in Brazil. It is, to this day, considered a very potent bibliographic management system, because it has, from its first version, characteristics and functionalities that distinguish it from other software: flexibility to define databases with the desired number of required fields; compatibility with the MARC format; possibility of including, modifying and excluding records in their database online; an ample capacity to personalize the innumerous output formats, short or long, for the screen as well as for print; generating indexes and catalogs of good technical quality; the capacity to index any fields in any indexes in inverted files, for word searches and for browsing headings indexes, exchanging, by export and import, its records by means of the ISO 2709 norm. In retrieving records, it enables the use of simple forms and forms with multiple fields, Boolean logic, stemming and verifying the presence/absence of fields.

Moreover, special modules and programs are always being created to create applications based on ISIS, in large part created by BIREME, such as *GENISIS*: interface development software for searching CDS/ISIS databases, available in a format for the Web environment or in CD-ROM; the *JAVAISIS*: the multilingual TCP/IP client-server suite for CDS/ISIS databases; the *CDS/ISIS Pascal*: a programming language that enables the development of applications integrated according to user needs; the Isis DLL: a function library for Ms-Windows, *WWWISIS*: application to make Isis databases available on the Web, the *WXIS/WXIS-php* modules, for retrieving Isis data

in XML format, the *CISIS*: a function library developed by BIREME in C language to enable the manipulation of ISIS databases without the need to install or use ISIS (MX, MXCP, MXTB, MSRT, etc.), the *IAH* (Interface for Access on Health Information) which enables integrated information retrieval from ISIS databases on the internet or intranet, via www, its successor, *IAHx*, created to improve the presentation mechanism of the search results, enabling to see them in an integrated way, individualized, typified and ordered by different criteria.

BIREME has always been a great supporter of ISIS in Brazil and deserves this recognition. It is a specialized center of the Pan American Health Organization (PAHO)/World Health Organization (WHO), created to be a specialized center in scientific and technical information in health for the Latin American and Caribbean region. Established in Brazil in 1967, with the name Regional Library of Medicine (*Biblioteca Regional de Medicina*, which originated the acronym BIREME), it has from the beginning met the growing demand for scientific literature updated by the national health systems and the communities of researchers, professionals and students. In 1982, it started to be called the Latin American and Caribbean Center of Information on Health Sciences to better express its functions aimed at strengthening and broadening the flow of scientific and technical information in health in the entire region, but maintained its known acronym.

Why have I addressed this software so much? ISIS was created by librarians for librarians as a tool for creating robust bibliographic databases. Even though it was never capable on its own of creating a library automation system for the functions of cataloging, acquisition, serials and online catalog via www, ISIS was and continues to be used in the classroom by practically all librarianship schools in the country and by libraries of all kinds, including academic ones, ever since it was created, perhaps because it was free (Castro & Barboza, 2011; Lamas, 2007; Modesto,

2006; Lima, 1998; Ohira, 1992; Silveira, Knoll, & Araújo, 1990). Various groups of ISIS users were also created to hold meetings and exchange experiences in the states of Bahia, Minas Gerais, Rio de Janeiro, Rio Grande do Sul, Santa Catarina, São Paulo, Paraná and Pará.

It was precisely for having been very widespread in various countries, and for allowing libraries to create software for integration with their databases, that innumerous additional systems emerged in Brazil, to work jointly with ISIS and this way fulfilling the functions it did not have, such as for example: *Zeus, Zeus Web, Winbusca, Kami, Olho de Isis and Oráculo* (from Control Consulting in Information and Documentation), *WinisEMP, WinisETIQ and WebISIS* (from BiblioShop Library and Software Automation), *Masterisis* (from Modalnetworks information technology services).

Other Brazilian initiatives went a step further: they created complete automation systems based on the "ISIS platform," also taking advantage of librarians' knowledge of the ISIS working model and the growing need to adopt "automation" systems to fulfill the need for circulating literature (loans, returns, reservations and charging fines) and also for the availability of an online web catalog, such as the *Suíte Saber* (Control Consulting in Information and Documentation), *PHL* (created by Elysio M. S. Oliveira), *BiblioBase* (Salvato Consulting and Technological Innovations Ltda.) and *Gnuteca* (free software with of MARC format support, developed by Solis – Free Solutions Cooperative).

Finally, according to Guilherme (2009), in 2006, the "ISIS concept" evolved drastically: in the month of May of that year, UNESCO received a group of specialists from BIREME (Brazil) to hear their proposal for a new ISIS software programmed with contemporary languages, with an open code compatible with the old databases from the time of microISIS, capable of running on the MS-Windows and Linux platforms. The proposal was approved and from that moment on, construc-

tion of the ABCD system began: *Automação de Bibliotecas e Centros de Documentação*, which means Library Automation and Documentation Centers, a www-based integrated library management software, comprising the main functions of the basic library.

During the III World Congress of ISIS Users, known as ISIS III, held in September of 2008 in the city of Rio de Janeiro, Brazil, ABCD was presented to the world in its alpha version for evaluation and tests by Guilda Ascencio (Venezuela), Egbert de Smet (Information and Library Science, University of Antwerp - Belgium) and by Ernesto Spinak (BIREME/OPAS/OMS - Uruguay), so that various users from various countries could collaborate with critiques and suggestions about the system.

It is also worth mentioning that at this same event, Paulo Cattelan, an experienced ISIS specialist and applications developer, presented *IsisHome* – the first webhosting service totally dedicated to CDS/ISIS, provided by the company Control Information and Documentation Ltd (Brazil). One year later, on December 3rd 2009, the official launch of the ABCD system version 1.0 (beta) took place, and its release to the community of libraries, available at http://reddes.bvsaude.org/projects/abcd. ABCD is available in Spanish, English, French and Portuguese and can be translated into other languages the same way that the CDS/ISIS applications always could.

1990's: The Use of Foreign Systems and the Creation of a National Industry of Library Automation Systems

At the end of the 1980's, the opening of the information technology market in the country allowed for the import and acquisition of foreign equipment as well as foreign automation systems, enabling a significant advance in academic libraries, a position that was consolidated in the 1990's, precisely at the time in which companies and library automation software in Brazil proliferated, created as an alternative to the market for foreign systems, which were more expensive. The early systems based on free software also emerged at this same time. Côrte (2002), in his book "Evaluation of software for libraries and archives" made an exhaustive survey of software for libraries in 2002. In this year, there were already 60 different software available in the Brazilian market (some of them were not integrated automation systems), from national as well as foreign companies.

Some aspects surrounding the evolution of library systems pointed out by Côrte et al. (2002):

- The majority of the fourth generation systems predicts their customization and expansion or inclusion of new modules;
- In the 80's, applications were developed for their management, guaranteeing the librarian greater speed in information processing and retrieval and command of technology, making it more user friendly;
- The 90's were characterized as a period in which the most companies emerged in the area, and new versions with updates and improvements were made available with some frequency.

In this decade, some Brazilian academic libraries started to create automation systems in-house, and others began to acquire foreign systems.

One of the first foreign automation systems to be used in the country was Aleph™, in its version 300 (based on Cobol, on a Terminal Server platform). The first institution to use it in Brazil was the Pontifical Catholic University of Rio Grande do Sul (PUCRS), of Porto Alegre-RS (in which I have worked as a systems librarian since 1999). Its choice took place by the initiative of the then-President Brother Norberto Francisco Rauch, who after getting to know it in the Vatican Library on a trip to Europe in 1992, brought the proposal to the technical team at the University. The team then evaluated the Aleph™ and two

other solutions (the SAB-II™, developed by IBM, University Foundation of Rio Grande and Getúlio Vargas Foundation; and the Bibliodata/CALCO of the Getúlio Vargas Foundation) and opted to acquire it from the Ex Libris company. The Aleph™ has been used at PUCRS since 1993, when it bought version 300. Currently, PUCRS uses version 500.20, including the SIP2 protocol used together with its self-service equipments; as well as the Aleph Digital Asset Module (ADAM) for local document storage in electronic format. Between 1998 and 2001, through a cooperative agreement with Ex Libris Israel, PUCRS was also responsible for supporting and installing Aleph™ in the country.

In 2001, *Ex Libris Brasil* was created in the city of São Paulo-SP, which began to be the official sales and support representative of the company in the country. The Aleph™ had a group of users in the State of Rio Grande do Sul (between 2001 and 2006), in addition to attempts to create a national group, one in 1998 (GUAL-BR) and another more recent one in 2011, in a meeting held in the city of Maceió, during the XXIV Brazilian Congress on Librarianship, Documentation and Information Science (Group of Brazilian Ex Libris product users), which is still under discussion.

Another foreign system used in the country was Virtua™, from the Virginia Tech Library System (VTLS), from the United States. In 1996, the Federal University of Minas Gerais was the first Brazilian institution to start using it. Between 1996 and 1998 the Virtua™ was commercialized in Brazil by the Getúlio Vargas Foundation (FGV), which was able to secure sales contracts with the Federal University of Minas Gerais, the University of the State of Santa Catarina, the State University of Campinas, the Pontifical Catholic University of Campinas, the Federal University of Goiás, the Federal University of Uberlândia, the Federal University of Ouro Preto and the University of Viçosa. In 1999, *VTLS Americas* was created, subsidiary of VTLS Inc. in Brazil. Virtua™ had a group of users, the Americas Users Group (AUG) – Virtua™ software users in Brazil and Latin America, in addition to holding periodic national meetings for users.

During this same period – at the end of the 1980's and the beginning of the 1990's – various national companies were created to produce and commercialize Brazilian systems, applying the knowledge acquired during the time in which the National Information Technology Policy (1984 to 1992) was in effect, to fulfill libraries' needs, which inspired by experiences of other countries had already begun to want automation systems. They emerged as a national alternative to the foreign software used in the country, such as Aleph™ (Ex Libris) and Virtua™ (Virgínia Tech Library System), which were already used in the large academic libraries of the country and were beginning their expansion phase.

Among the commercial Brazilian software, the following are used by academic libraries in the country: Ortodocs™ (1986), Pergamum™ (1988), SAB-II™ - Sistema de Automação de Bibliotecas II (1989), Sysbibli™ (1990), Gestor de Bibliotecas™ (1990), GVDASA Biblioteca™ (1992), Sagres Acervo™ (1993), Multiacervo™ (1993), Arches Lib™ (1995), Thesaurus™ (1995), CARIBE Biblioteca™ (1995), RM Biblios™ (1996), Sábio™ - Sistema de Automação de Bibliotecas (1997), Informa™ (1997), Biblioteca Argonauta™ (1997), SophiA Biblioteca™ (1997), Alexandria™ (1999), Librarium™ (1999), DINS™ (1999), PHL™ (2001), SIAB™ (2004), and Bibliosys™ (2008). Other smaller commercial software were created in the country during this period and there are no reports on their use by academic libraries, such as Biblioteca Fácil™ (1993) and Minibiblio™ (1996).

A success story in the country without a doubt is Pergamum™ - Integrated Library System, which was developed in a Brazilian higher education institution. It was created in 1988, by the then student of the Computer Science program of the Pontifical Catholic University of Paraná, Marcos Rogério de Souza, current coordinator of the Pergamum™ department at the University. It started to be commercialized in 1996 and is currently the most used library automation system in Brazilian academic libraries. According to the data from its website (www.pergamum.pucpr.br), it is currently installed in approximately 2,500 libraries all over the country. "The Paranaense Association of Culture, parent organization of the Pontifical Catholic University of Parana (PUCPR), holds the rights and produces this software. It was developed with the objective of managing all the services of a small, medium or large library and implemented in the client/server architecture, with a graphic interface, using an SQL relational database" (Anzolin, 2009, p. 496).

Anzolin also reports that in 1999, the Pergamum Network was formed, constituted by the institutions using the Pergamum™ software, with the purpose of improving the global quality of the user services, to promote cooperation in information processing and sharing information resources. The Network initially brought together 13 higher education institutions. Over the years, the number of institutions has increased and become diverse: in the year 2000, there were 36, 2001 (47), 2002 (58), 2003 (97), 2004-2005 (106), 2006 (159), 2007 (180), in 2008 (216). In March of 2012, there were 350 institutions.

The 2000's: Free Software and the First RFID Equipment and Self-Service

In the 2000's, the first free software appeared in Brazil: Gnuteca™ (www.solis.org.br/projetos/gnuteca), created in 2002 by Solis – Free Solutions Cooperative with the support of UNIVATES – University Center of Lajeado, RS; and Biblivre™ (www.biblivre.org.br), created in 2005 through the project called "Free Library" by the Society of Friends of the National Library (SABIN), with the support of the Alberto Luiz Coimbra Institute for Graduate Studies and Research in Engineering at the Federal University of Rio de Janeiro (COPPE/

UFRJ), in developing the project in the 1.0 and 2.0 versions. The project was initially sponsored by IBM-Brazil and since 2007 its exclusive sponsor has been the Itaú Cultural Institute. Currently, Biblivre is in its 3rd version.

The beginning of the 2000's was also the period during which some universities began to use self-service systems for loans and returns and to equip its libraries with radio-frequency identification (RFID). It is worth noting the case of Pontifical Catholic University of Rio Grande do Sul (PUCRS), which was a pioneer in the use of self check-in equipment (acquired from ID Systems in 2004) and the only one which currently uses self check-out and sorting equipment (acquired from 3M in 2008). The installation carried out at PUCRS is the largest installation of RFID equipment in Latin America, with 4 self check-out machines, 2 self check-in and sorting machines, 10 integrated workstations (EM+RFID), 330,000 RFID tags, 2 conversion stations (barcode to RFID), 2 portable RFID readers and 1 conversion workstation (RFID barcode). The initiative for using new technologies happened at the same time of the reform for broadening and modernizing PUCRS' Library. The old three-floor building with 10,000 square meters became 14 floors and 21,000 square meters. This modernization was based on a comprehensive study of experiences at other universities, many of them American, with the purpose of creating an alternative for the increase in user demand for speed in internal processes and in those provided by the library to its users, in addition to delivering a better user experience by offering them the autonomy to loan and return materials without the intervention of the library team.

Radio-frequency identification (RFID) technology began to be used in 2008, for managing the collection, circulation and autonomous user services. The same is applied to inventory processes, locating materials in the collection and organizing items on the bookshelves, through portable RFID readers. In the circulation area, *there is integrated equipment for loans, returns and a security system. For the self-service, there is self check-in and self check-out and sorting equipment, with a color liquid crystal flat screen and interactive software with step-by-step orientation. The self check-out and sorting equipment also has a conveyor belt that separates the returned items according to their location in the library collection. This equipment is totally integrated into the Aleph automation system for recording transactions, issuing receipts and clearance in the security system. (Mazzillo, Araujo, Viana, Crespo, & Naumann, 2011, p. 234-235)*

Currently, very few higher education institutions already use RFID technology and self-service equipment. It is a market that still needs to be more fully tapped by suppliers. According to the data from the Brazilian subsidiary of the 3M Company, the current number of Brazilian higher education institutions using their self check-in equipment is as follows in Table 3:

The 2010's and the Discovery of Discovery Tools

Providing access to all aspects of the library collections, not only those that are available in the traditional library catalog, which is limited to content managed by the integrated library system, and helping users discover the content available through the library in all formats, independently of whether they are in the physical library or among its electronic collections, comprising both on-site materials and those accessed remotely through subscriptions. This concept of *discovery* (Breeding, 2011) began to be used in Brazil in 2011, when the first Brazilian institutions started to use discovery systems for libraries, for an integrated document search in their catalogs, digital libraries and information repositories.

Using information "discovery" has various benefits, such as: the integration between different catalogs and sources of information, creating

Table 3. Number of 3M self check-out equipments in Brazilian higher education institutions. (Source: Brazilian subsidiary of the 3M Company, April 2012).

Higher Education Institution	Amount of Self Check-Out Equipments
ENIAC College (Guarulhos-SP)	1
Federal University of Espírito Santo – UFES (Vitória-ES)	4
Federal University of Ouro Preto – UFOP (Ouro Preto-MG)	1
Federal University of Minas Gerais – UFMG (Belo Horizonte-MG)	1
Federal University of Santa Catarina – UFSC (Florianópolis-SC)	11
Pontifical Catholic University of Rio Grande do Sul – PUCRS (Porto Alegre-RS)	4
University of Brasília – UnB (Brasília-DF)	2
University Center of Brasília – UNICEUB (Brasília-DF)	3
Catholic University of Brasília – UCB (Brasília-DF)	1

a single search point and access for other users, speed in obtaining texts, ordering results by relevance, filtering results, among others.

Of the four main discovery systems for libraries available on the Brazilian market – WorldCat Local™ (OCLC), Ebsco Discovery Service™ (Ebsco), Primo™ (Ex Libris) and Summon™ (Serials Solution) – three of them are already being used in the country, but only at a small number of institutions. I believe that soon, the number of Brazilian higher education institutions that use discovery tools will grow in Brazil, but for this it is necessary that, from now on, librarians, professors of librarianship, researchers and the suppliers themselves start to further promote their potential for integration and their benefits for academic libraries. If the suppliers do not make more of an effort to sell in Brazil, soon the institutions will start to use free discovery tools (such as VUFind) or even develop their own tools,

which was the case in the 1980's with the library automation systems.

WorldCat Local™ (OCLC) is being evaluated by libraries for their possible acquisition in the near future, but it still has not been implemented by any Brazilian libraries. The Ebsco Discovery Service™ (EBSCO) has been in operation since 2011 only in the oil company Petróleo Brasileiro S.A, Petrobrás. The other two (Primo™ and Summon™) are already in operation in academic libraries, according to Table 4:

In August of 2011, the São Paulo State University Júlio de Mesquita Filho (UNESP) was the first to implement a discovery system in the country's higher education institutions, with the objective of allowing its users to explore archive collections of the Library Network – Athena (Bibliographic Databases) and its digital collections such as C@thedra (Digital Library of Theses and Dissertations) and C@pelo (Digital Library of Undergraduate Theses) – and, at the same time, the resources indexed to this new system. In the future, the intention is to include journals from Unesp and other publishers. Flávia Maria Bastos, coordinator of the General Coordination of Libraries (CGB), declared that she intends to develop a discovery tool for mobile equipment, such as cellular telephones, as well as use web 2.0 resources (Universidade Estadual Paulista, 2011).

FUTURE PERSPECTIVES FOR BRAZILIAN ACADEMIC LIBRARIES

Some concerns that I bring to the debate in this chapter refer to the format of the interaction interfaces of automation systems, and the current focus of control over systems. The former is with respect to the fact that we have the legacy of systems created in the last century, made exclusively for desktop computer screens. In terms of format, interfaces have evolved little with respect to the "dumb terminals" used with mainframes. Today,

Table 4. Use of Library Discovery Systems in Brazilian Universities.

Higher Education Institution	Discovery System	Supplier	Status	Web address
São Paulo State University Júlio de Mesquita Filho (UNESP)	Primo™	Ex Libris	In operation since August/2011	www.parthenon.biblioteca.unesp.br
Pontifical Catholic University of Rio de Janeiro (PUC-Rio)	Summon™	Serials Solution	In operation since April/2012	pucrio.summon.serialssolutions.com
University of São Paulo	Primo™	Ex Libris	In operation since 12/March/2012	sibi.usp.br/buscaintegrada
Catholic University of Brasília (UCB)	Summon™	Serials Solution	In operation since 15/March /2012	ucbbr.summon.serialssolutions.com
University of Brasília (UnB)	Summon™	Serials Solution	In operation since 2012	unb.summon.serialssolutions.com
State University of Campinas (UNICAMP)	Summon™	Serials Solution	In operation since June/2012	unicamp.summon.serialssolutions.com

users use the library from anywhere on the planet, with innumerous fixed or mobile devices such as smart televisions, laptops, netbooks, cellular telephones, smartphones, e-book readers, tablets… and the library systems interfaces available in the majority of the systems continue to be made only for PC monitors. It is recommended that companies supplying systems start to offer, out of the box, interfaces for mobile devices with the same usability and responsiveness offered for PCs.

The latter involves the persistence of systems in focusing on the final record of the transaction, that is: create a catalog record, create a loan record, create an acquisition record…instead of focusing on the flow of the libraries' business processes, in attributing tasks and in controlling their execution, so that coordinators and managers can know in real time how staff work performance and user requests are taking place. This new form of control may enable managers of academic libraries to evaluate the execution of services and obstacles, in order to optimize the processes. The final record of the process is no longer sufficient. Along these lines, it is no longer practical for managers to know how many records of books, loans or people are registered in the databases, but how much time it is taking to put a book at the disposal of the user

on the bookshelf, for example. The next generation systems, associated to discovery tools intend to fulfill many of these needs, separating service management from information management.

Teaching Brazilian Librarianship and the discussions held at events continue to disseminate the use of fourth generation automation systems, according to the classification by Rowley (2004). Meanwhile, early discussions have already begun to appear in the literature on the use of online catalogs as a global tool for locating information, regardless of where the information is (Keiser, 2010), and also the advent of discovery tools and "next generation automation systems" (Breeding, 2011). In the opening talk of the 6th Annual Conference of the International Group of Ex Libris Users (IGeLU), held in Haifa, Israel from September 11 to 16 of 2011, entitled "The New Frontier: Libraries seek new technology platforms for end-user discovery, collection management, and preservation," Marshall Breeding, from Vanderbilt University (United States), highlights:

1.   Academic libraries are in transition (print to electronic archive; digitalization of local collections) and new technologies are emerging

to offer more integration between systems and increase use of the library through new devices;

2. It is necessary to reevaluate metadata management and think about sharing metadata, selectively, since knowledge production is quicker than traditional cataloging;

3. Libraries tend not to be able to manage their local systems in the long term which are currently used independently: there is no space for outdated technologies;

4. The priority is to manage workflows, instead of managing bibliographic and administrative records in databases;

5. The next generation of library systems, which will substitute the current automation systems:

   a. Should serve as a platform for connecting external systems (using APIs - Application Programming Interfaces), as well as providing internal functionalities;

   b. Should synchronize internal management (selection, acquisition, cataloging, circulation...) and that which is offered to the user (localization and access) to avoid inefficiency between automation and discovery platforms.

Such next generation systems will start to be commercialized in the country soon. It is necessary for higher education institutions to start evaluating the possible impacts of using new generation systems on academic libraries:

- The financial and academic systems of the higher education institution, and the internal library control and security systems, nowadays integrated with the current automation systems, will need intervention from the team of systems analysts and developers from the institution to integrate them to the next generation system. This integration will probably only be carried

out through the APIs made available by the suppliers;

- The next generation systems will offer a variety of new business indicators to the library managers, based on the accomplishment of results over time, by analyzing the execution of the workflows (not only of the number of transactions and records such as what happens today), enabling the constant adjustment of work tasks;

- The next generation systems, which will be hosted in remote datacenters, according to the concept of "cloud computing" (infrastructure, software, data, tests and support, all offered as services), eliminate the need to maintain local hardware infrastructure, resulting in saving or freeing up the institutional budget for acquiring other resources;

- These systems will need a permanent, stable and high speed Internet connection, since they depend entirely on Internet access with remote servers, based on operation and research terminals;

- The choice of discovery tool needs to be made with strict criteria, since the next generation systems tend to no longer have incorporated online catalogs, and the discovery tool chosen should be able to increasingly improve the user experience with the library, and not the contrary.

Some reasons for higher education institutions to change their current automation systems to next generation systems include:

- **Total Cost of Ownership:** Next generation systems, which are possible thanks to cloud computing technologies, will lower investments in servers and software; there will be no need for local maintenance of hardware/software, database, backup, updates, upgrades; Consequently, there will be less investment in resources and IT personnel, and reduced local management.

- **Return on Investment:** Next generation systems will allow for an improved user experience, optimized use of available print and electronic collections and an increase in personnel productivity, with maximized collaboration.

- **Proven Institutional Value:** Next generation systems will generate more involvement between teaching and learning processes; they will increase productivity of academic research; they will generate more emphasis on strategic initiatives; they will offer new possibilities of adding services for students and researchers.

Today, there are technologies capable of using speech and gesture recognition, eliminating the need to use a physical and virtual alphanumeric keyboard (on computers, smartphones, tablets…), so the user can formulate his questions and the systems present sources of information that meet his information needs or desires, or even guide him regarding the use of library services. Put together multidisciplinary teams of systems librarians, systems analysts, computer scientists, mathematicians, specialists in artificial intelligence, game builders, and linguists to build what I call "intelligent library assistants." The academic libraries are the ideal environment to create them, since higher education institutions have in their staff all these types of researchers. In 2012, the Microsoft Company released the Kinect Software Development Kit version 1.0, which allows for the creation of a practically infinite number of applications for the computer with the "Kinect Sensor," a device originally created for people to control actions in the games from the X-Box™ videogame from Microsoft™ through speech or gestures. Why not use it for information retrieval in catalogs and databases, and to obtain information on the use of resources, spaces and library services? It would be a "machine" similar to "VOX" from the movie "The time machine" (2002, directed by Simon Wells and produced

by Arnold Leibovit), in which the actor Orlando Jones plays a holographic "being" in the New York Public Library in 2030, connected to all of the databases in the world, "a compendium of human knowledge," capable of interacting with people by speech and gesture commands.

Another possible perspective that I observe is with respect to electronic document management in academic libraries. In sources of information as well as in administrative documents, the focus continues to be on paper. We have a lot of information in physical media, which still requires manual work (delivery, receipt, handling, stamping, inclusion of an anti-theft sensor, putting them back on bookshelves etc.). This is changing to the extent that more and more materials are being digitalized. On the other hand, there are various administrative processes carried out in academic libraries that move piles of paper: letters, lists, and printed reports. Hence, it would be very useful for the suppliers of automation systems to include an option to attach digitalized images of printed documents in the systems, for internal use, such as, for example, documents with acquisition orders and vendor invoices.

Many internal library processes still require the use of "office" applications, such as spreadsheets and text editors, to control details of the workflows (technical processing, acquisition, circulation…) which are not contemplated by automation systems in the interactions between all *stakeholders*: colleagues, bosses, suppliers, government sectors, users. Among all of them, communications still take place through printed letters and email messages created manually, however they could adopt electronic protocols or standard messages sent automatically through the system, triggered from certain events or actions performed on the automation system.

There are still services that have little or even no form of management by automation systems, such as, for example, the reference service, which is the main service for users in academic libraries. To date, I am not aware of a system that incorporates the scheduling of reference interviews for users based on the availability of librarians, which creates a calendar of services for the public with the dates and times available, which sends a memo to the librarian with the date and time of the scheduled interview, which records the answers for the users to create a searchable knowledgebase, among many other functions necessary for the appropriate management of this service. I agree that in small libraries this automated control is not very necessary, but when there are thousands of clients to attend to, automating this service makes it much more efficient.

FUTURE RESEARCH DIRECTIONS

It is recommended that institutions as well as companies reevaluate the paths of library automation in Brazil, in general. The software and equipment companies need to create and offer financially viable, efficient and effective solutions to the new Brazilian university library. The demand for equipment and solutions in support of user autonomy is growing in the country, but there are few existing options in the Brazilian market. Among the existing options, many need to be imported and their high cost still makes them inaccessible for the large majority of academic libraries. Brazil has many professionals in the area of control and automation engineering, computer engineering, production engineering, electrical and electronic engineering, but different from the American and European reality, there are few companies that offer specially developed equipment solutions for libraries. Brazil is going through a period in which the economy is booming and with a high capacity for investment, without facing major financial crises such as what is happening in other countries, being an excellent business opportunity for companies that act in the library market as well as higher education institutions.

Some libraries are already advanced in terms of technology, but others still need to advance. But how many academic libraries are there in Brazil? Which automation systems do they use? Which needs are met and which ones are still not met? It is necessary to carry out a new diagnosis or census of the use of automation systems in higher education institutions in the country. I particularly believe that this diagnosis needs to be done as soon as possible, such that institutions have an updated panorama based on these results and are capable of:

- Knowing how many academic libraries there are in Brazil in the more than 2,650 higher education institutions.
- Knowing exactly which needs are being met in these libraries by the systems and equipment and which are still not.
- Evaluating the level at which the use of automation is taking place.
- Knowing which systems and equipment are used.
- Knowing how libraries evaluate the systems and equipment in use: what are the strong points and weak points of each one, from their points of view.
- Knowing how the systems can be improved.
- Knowing whether higher education institutions wish to change their automation system, for what reasons and which systems they intend to start using.
- Creating technical cooperation and information resource networks, to the extent possible, respecting the existing competition in private education, sharing experiences, solutions and resources.

The answer to these and other related questions can contribute to a new advance in the automation of Brazilian university libraries, which will preferably happen sooner than later.

Figure 1. Important facts about the automation of academic libraries in Brazil

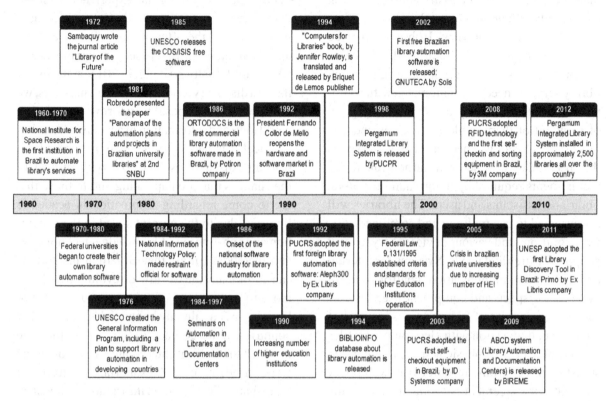

CONCLUSION

The following timeline (Figure 1) illustrates when significant historical facts took place on Brazilian academic libraries automation.

By analyzing the views of Sambaquy and Barsotti, one can observe that practically everything they predicted for the "library of the future" in fact happened, though not exactly as they had imagined. Libraries, mainly academic ones, have become bigger and are currently capable of accommodating a growing number of students and providing spaces for research and knowledge production. Instead of *telephonevision*, we have all benefited from the Internet and the www since the 1990's in Brazil. Not only is the exchange of information from a distance constant, but also the publication and instantaneous diffusion of texts, figures, videos, through fixed and mobile devices, from and to anywhere on the planet. Free

databases and commercial providers inundate the market and have been increasingly integrated with library automation systems, no longer being treated as separate resources.

Despite the advances that have occurred in Brazil, academic libraries still cannot serve the user exclusively through the current automation systems when using the physical space of the library or even when accessing the online resources available for remote use, if we consider that not all of the services and processes have been contemplated by automation systems. A lot of manual labor is still carried out, and this lack of automation generates slow services difficult to be measured, controlled and managed. It is a global situation, but I believe that the situation is worse in Brazil, mainly since we have, in the 21st century, libraries without librarians (public and private institutions) attended by laymen, libraries without any kind of systematic organization of their archives (decimal classification), and even others without any kind of catalog.

From my point of view, I observe that the automation of academic libraries should be as comprehensive as possible, without limits. Which library services need to be automated with equipment and information systems? From my point of view, all of them, middle processes and end processes. Certainly, there is no ready system that meets all of the functional and non-functional requirements required by library administrators, librarians, assistants and users. The libraries will always need to use multiple and combined software solutions and various devices to try to meet people's desires and needs. Systems for internal and external communication control, control systems for space and equipment use, systems for online payment of fees and fines, control of locker use, integration with electronic turnstiles to control access to physical spaces, integrated statistics of electronic resource use, web proxy automatic management, systems for electronic loan of e-books, etc. None of this comes out of the box in automation systems, and higher education institutions need to create or acquire various systems separately and try to "fit" them in the Library system, or they end up using them separately, generating a duplication of efforts and data and the same inconsistencies of the past. Some may think that many of these processes are not the main objectives of the library, but no one can deny that these needs exist, they are real, they happen on a day to day basis in libraries and need to be met somehow, automated or not: the reality has changed since the 1970's.

It has been observed through this process of historic evaluation that the automation of academic libraries has evolved considerably in Brazil, from the time of the use of the first systems based on mainframes, the important experience with the ISIS systems, moving on to the software created in the country, the offer of free software and the use of foreign systems, self-service equipment and RFID, until finally arriving, in 2011, at the use of the first discovery tools. In the following years, we will have at our disposal new forms of managing workflows and information in academic libraries.

I feel it is necessary to open up a broad debate and discussions in Brazil with respect to the changes that are happening and to those that are to come regarding automation in academic libraries, due to the current speed in the generation and substitution of knowledge, technologies and methods. I hope that Brazilian institutions can accompany the worldwide evolution and thus move to a new phase of automation in academic libraries, as has already begun to happen in other countries. If we go back to the past, to the time in which the Callimachus created a catalog of the works in the Library of Alexandria, we will see that more than collect and hold information, he wished that they could be located and used. In the Middle Ages, the monastic scribes were also concerned with this, but they believed that the new media of information created with the printing press could spread non-censurable ideas, contain and spread spelling errors and take power from the hand of who had possessed it for so long. Thankfully, the opposite happened: the technologies used in spreading knowledge have improved every day and bring more and more benefits to those who organize them and use them. I see a day in which the software and equipment used in academic libraries will be capable of managing all of their functions and interactions, and automation will enable librarians to have more time to dedicate to bringing together those who produce information and those who need this information.

REFERENCES

Anzolin, H. H. (2009). Rede Pergamum: história, evolução e perspectivas. *Revista ACB: Biblioteconomia em Santa Catarina, 14*(2), 493-512. Retrieved March 30, 2012, from http://revista.acbsc.org.br/index.php/racb/article/view/640

Automation. (2012). In *Merriam-Webster.com*. Retrieved March 20, 2012, from http://www.merriam-webster.com/dictionary/automation

Azolin, B. R. (1999). O futuro da informática no brasil. Retrieved March 30, 2012, from http://www-usr.inf.ufsm.br/~cacau/elc202/futuro.html

Balby, C. N. (2002). *Estudos de uso de catálogos on-line (OPACs): revisão metodológica e aplicação da técnica de análise de log de transações a um OPAC de biblioteca universitária brasileira.* (Unpublished Doctoral dissertation). Universidade de São Paulo, São Paulo, SP.

Barcelos, M. E., & Gomes, M. L. (2004, October). *Preparando sua biblioteca para avaliação do MEC*. Paper presented at the Seminário Nacional de Bibliotecas Universitárias, 13. Retrieved March 30, 2012, from http://repositorio.cfb.org.br/handle/123456789/495

Barsotti, R. (1990). *A informática na biblioteconomia e na documentação*. São Paulo: Polis, APB.

Brasil. Ministério da Educação. (2004). SINAES – Sistema Nacional de Avaliação da Educação Superior: da concepção à regulamentação. Retrieved March 30, 2012, from http://www.abem-educmed.org.br/pdf/sinaes.pdf

Brasil. Ministério da Educação. (2010a). Instrumento de avaliação para credenciamento de Instituição de Educação Superior (Faculdade). Retrieved March 30, 2012, from http://download.inep.gov.br/download/superior/institucional/2010/instrumento_avaliacao_para_credenciamento_IES.pdf

Brasil. Ministério da Educação. (2010b). *Instrumento de avaliação institucional externa.* Retrieved March 30, 2012, from http://download.inep.gov.br/download/superior/institucional/2010/instrumento_avaliacao_institucional_externa_recredenciamento.pdf

Brasil. Ministério da Educação. (2012a). Sistema e-MEC: instituições de educação superior e cursos cadastrados. Retrieved March 30, 2012, from http://emec.mec.gov.br

Brasil. Ministério da Educação. (2012b). Instrumento de Avaliação de Cursos de Graduação presencial e a distância. Retrieved March 30, 2012, from http://download.inep.gov.br/educacao_superior/avaliacao_cursos_graduacao/instrumentos/2012/instrumento_retificado_fevereiro_2012.pdf

Breeding, M. (2011, September). *The new frontier: Libraries seek new technology platforms for and end-user discovery, collection management, and preservation.* Paper presented at the Conference of the International Group of Ex Libris Users. Retrieved March 30, 2012, from http://igelu.org/conferences/haifa-2011/archive-of-presentations

Burin, C. K., Lucas, E. R. D. O., & Hoffmann, S. G. (2004, October). *Informatizar por quê? a experiência das bibliotecas universitárias informatizadas da região sul.* Paper presented at the Seminário Nacional de Bibliotecas Universitárias, 13. Retrieved March 30, 2012, from http://www.pergamum.pucpr.br/redepergamum/trabs/Camila_K_Burin-Informatizar_por_que.pdf

Carvalho, C. H. A. D. (2004). Agenda neoliberal e a política pública para o ensino superior nos anos 90. Retrieved March 30, 2012, from http://www.anped.org.br/reunioes/27/gt11/t114.pdf

Carvalho, I. C. L. (1997, July). *Bibliotecas universitárias federais: o cenário da informatização*. Paper presented at the Congresso Brasileiro de Biblioteconomia e Documentação.

Castro, A. D., & Barboza, T. L. (2011, August). *Família ISIS: do Microisis ao ABCD*. Paper presented at the Congresso Brasileiro de Biblioteconomia, Documentação e Ciência da Informação, 24. Retrieved March 30, 2012, from http://www.febab.org.br/congressos/index.php/cbbd/xxiv/paper/view/510

Côrte, A. R. E., Almeida, I. M. D., Pellegrini, A. E., Lopes, I. O., Saenger, J. C., Esmeraldo, M. B. P., et al. (1999). Automação de bibliotecas e centros de documentação: o processo de avaliação e seleção de softwares. *Ciência da Informação, 28*(3), 241-256. Retrieved March 30, 2012, from http://dx.doi.org/10.1590/S0100-19651999000300002; doi: 10.1590/S0100-19651999000300002

Côrte, A. R. E., Almeida, I. M. D., Rocha, E. G., & Lago, W. G. D. (2002). Avaliação de softwares para bibliotecas e arquivos: uma visão do cenário nacional (2.ed. rev. e ampl. ed.). São Paulo: Polis.

Cunha, M. B. D. (1985). A informática e a biblioteconomia: uma união de muito futuro. *Revista de Biblioteconomia de Brasília, 13*(1), 1-7. Retrieved March 30, 2012, from http://www.brapci.ufpr.br/documento.php?dd0=0000001822&dd1=fa22d

Figueiredo, N. M. D. (1986). Aplicação de computadores em bibliotecas: estudo comparativo entre países desenvolvidos e o Brasil. *Revista de Biblioteconomia de Brasília, 14*(2), 227–244.

Guilherme, R. C. (2009). Introdução ao ABCD (Automação de Bibliotecas e Centros de Documentação). Retrieved March 4, 2012, from http://abcdisis.files.wordpress.com/2009/03/microsoft-word-apostila-abcd.pdf

Keiser, B. (2010). Library of the future -- Today! *Searcher, 18*(8), 18-54. Retrieved March 30, 2012, from http://www.allbusiness.com/media-telecommunications/information-services-libraries/15180365-1.html

Lage, Â. (1989, June). *Automação de bibliotecas universitárias do Brasil: tendências e perspectivas*. Paper presented at the Seminário Nacional de Bibliotecas Universitárias, 6. Retrieved March 30, 2012, from http://www.dominiopublico.gov.br/download/texto/me001650.pdf

Lamas, S. D. F. T. B. (2007). Automação de bibliotecas: do ISIS à biblioteca do futuro. *Pesquisa Brasileira em Ciência da Informação e Biblioteconomia, 2*(1).

Lima, G. Â. B. (1999). Softwares para automação de bibliotecas e centros de documentação na literatura brasileira até 1998. *Ciência da Informação, 28*(3), 310-321. Retrieved March 30, 2012, from http://dx.doi.org/10.1590/S0100-19651999000300009; doi: 10.1590/S0100-19651999000300009

Lima, G. Â. B. D. O., & Mendonça, A. M. (1998). A Utilização do MicroISIS no Brasil. *Perspectiva em Ciência da Informação, 3*(1), 125-136. Retrieved March 30, 2012, from http://portaldeperiodicos.eci.ufmg.br/index.php/pci/article/view/601

Marcelino, S. C. (2009). A contribuição da biblioteca para a construção e difusão do conhecimento no Instituto Nacional de Pesquisas Espaciais (Inpe). *Ciência da Informação, 38*(2), 80-95. Retrieved March 30, 2012, from http://revista.ibict.br/index.php/ciinf/article/view/1090

Marques, I. D. C. (2003). Minicomputadores brasileiros nos anos 1970: uma reserva de mercado democrática em meio ao autoritarismo. *História, Ciências, Saúde-Manguinhos, 10*(2), 657-681. Retrieved March 30, 2012, from http://www.scielo.br/scielo.php?script=sci_arttext&pid=S0104-59702003000200008

Mazzillo, C. A., Araujo, D. K. D., Viana, M. M. M., Crespo, I. M., & Naumann, P. (2011). Die Zentralbibliothek Irmão José Otão an der Päpstlich-Katholischen Universität von Rio Grande do Sul: Ein Beispiel für Innovation. *BIBLIOTHEK Forschung und Praxis, 35*(2), 231-235. Retrieved March 30, 2012, from http://www.degruyter.com/view/j/bfup.2011.35.issue-2/bfup.2011.032/bfup.2011.032.xml

McCarthy, C. M. (1982). *The automation of libraries and bibliographic information systems in Brazil. Leicestershire.* Loughborough University of Technology.

McCarthy, C. M. (1983). Library automation in Brazil: The state of the art. *Program, 17*(4), 233-240. Retrieved March 30, 2012, from http://www.emeraldinsight.com/journals.htm?articleid=1671037&show=abstract; doi:10.1108/eb046868

McCarthy, C. M., & Neves, F. I. (1990). Levantamento geral da automação de bibliotecas no Brasil. *Revista Biblioteconomia de Brasília, 18*(2), 51-57. Retrieved March 30, 2012, from http://www.brapci.ufpr.br/documento.php?dd0=0000002623&dd1=cd841

Modesto, F. (2006). O CDS/ISIS morreu? Viva o CDS/ISIS livre. Retrieved March 30, 2012, from http://www.ofaj.com.br/colunas_conteudo.php?cod=274

Ohira, M. L. B. (1992). Automação de bibliotecas: Utilização do MicroISIS. *Ciência da Informação, 21*(3). Retrieved March 30, 2012, from http://revista.ibict.br/index.php/ciinf/article/viewArticle/1306

Ohira, M. L. B. (1994). Biblioinfo base de dados sobre automação em bibliotecas (informática documentária): 1986-1994. *Ciência da informação, 23*(3). Retrieved March 30, 2012, from http://revista.ibict.br/cienciadainformacao/index.php/ciinf/article/viewArticle/1159

Paulista, U. E. Assessoria de Comunicação e Imprensa. (2011). Unesp lança novo sistema de busca bibliográfica. *Unesp Informa, 2*(21). Retrieved March 30, 2012, from http://www.unesp.br/unespinforma/21/novo-sistema-de-busca-bibliografica

Prado, N. S., & Abreu, J. D. (2002). Informatização das bibliotecas universitárias do estado de Santa Catarina: cenário. *Revista ACB, 7*(2). Retrieved March 30, 2012, from http://revista.acbsc.org.br/index.php/racb/article/view/394

Robredo, J. (1981). *Panorama dos planos e projetos de automação das bibliotecas universitárias brasileiras.* Paper presented at the Seminário Nacional de Bibliotecas Universitárias, 2.

Rowley, J. (1994). *Informática para bibliotecas.* Brasília: Briquet de Lemos/Livros.

Russo, M. (2010). *Fundamentos de Biblioteconomia e Ciência da Informação.* Rio de Janeiro: E-papers.

Sambaquy, L. (1960). Article. *American Documentation (pre-1986), 11*(3), 205-205. Retrieved March 30, 2012, from http://search.proquest.com/docview/195446789

Sambaquy, L. D. Q. (1972). A Biblioteca do Futuro. *Revista da Escola de Biblioteconomia da UFMG, 1*(1), 62-68. Retrieved March 30, 2012, from http://www.brapci.ufpr.br/documento.php?dd0=0000001890&dd1=1b795

Saviani, D. (2010). A expansão do ensino superior no Brasil: Mudanças e continuidades. *Poíesis Pedagógica, 8*(2), 4-17. Retrieved March 30, 2012, from http://www.revistas.ufg.br/index.php/poiesis/article/view/14035

Sayão, L. F., Marcondes, C. H., Fernandes, C. C., & Medeiros, L. P. M. (1989). Avaliação dos processsos de automação em bibliotecas universitárias. *Trans-in-informação, 1*(2), 233-254. Retrieved March 30, 2012, from http://www.brapci.ufpr.br/documento.php?dd0=0000000139&dd1=d6193

Silva, F. C. C. D., & Favaretto, B. (2005). Uso de softwares para o gerenciamento de bibliotecas: Um estudo de caso da migração do sistema Aleph para o sistema Pergamum na Universidade de Santa Cruz do Sul. *Ciência da Informação, 34*(2), 105-111. Retrieved fromhttp://dx.doi.org/10.1590/S0100-19652005000200011; doi: 10.1590/S0100-19652005000200011

Silveira, A., Knoll, M. M. D. D. C., & Araújo, F. M. B. G. D. (1990). Mini-micro CDS/ISIS: Uma proposta de aplicação no ensino da informática na Biblioteconomia e Ciência da Informação. *Revista de Biblioteconomia de Brasília, 18*(2). Retrieved March 30, 2012, from http://www.brapci.ufpr.br/documento.php?dd0=0000002624&dd1=912cd

Teixeira, L. A., Oliveira, L. R. V., Lapa, R. C., & Assunção, R. V. D. (2011, August). *Pergamum: Serviços web e auto-atendimento na Biblioteca da Universidade Federal de Mato Grosso do Sul (UFMS)*. Paper presented at the Congresso Brasileiro de Biblioteconomia, Documentação e Ciência da Informação (p. 24). Retrieved March 30, 2012, from http://febab.org.br/congressos/index.php/cbbd/xxiv/paper/view/293

Vieira, A. D. S. (1972). A automação no currículo de Biblioteconomia. *Revista da Escola de Biblioteconomia da UFMG, 1*(1), 12-31. Retrieved March 30, 2012, from http://www.brapci.ufpr.br/documento.php?dd0=0000001886&dd1=58ac4

Vieira, D. K., & Souza, A. L. D. (2010). Bibliotecário X Indicadores: Biblioteca universitária nos resultados de avaliação do MEC. Retrieved March 30, 2012, from http://issuu.com/biblioteconomiafatea/docs/indicadores_avaliacaomec

Wanderley, M. A. (1973). Utilização de processos de automação na Biblioteca Nacional. *Ciência da Informação, 2*(1). Retrieved from http://revista.ibict.br/index.php/ciinf/article/view/1631

ADDITIONAL READING

Amorim, A. M., & Damasio, E. (2006). *O Gnuteca e o OpenBiblio avaliação de softwares livres para a automação de bibliotecas*. Paper presented at the Simpósio de Diretores de Bibliotecas Universitárias da América Latina e do Caribe, 4, Salvador. Retrieved March 30, 2012, from http://eprints.rclis.org/handle/10760/8490

Balby, C. N. (1994, July). *Conversão retrospectiva: Para consolidar a automação e a cooperação nas bibliotecas brasileiras*. Paper presented at the Seminário sobre Automação em Bibliotecas e Centros de Documentação, 5, São José dos Campos.

Bireme. (2012). Lilacs: Literatura Latino-Americana e do Caribe em Ciências da Saúde. Retrieved March 30, 2012, from http://lilacs.bvsalud.org

Café, L., Santos, C. D., & Macedo, F. (2001). Proposta de um método para escolha de software de automação de bibliotecas. *Ciência da Informação, 30*(2), 70-79. Retrieved March 30, 2012, from http://revista.ibict.br/cienciadainformacao/index.php/ciinf/article/view/198

Cunha, M. B. D. (1985). Reflexões sobre a informática na Biblioteconomia. *Boletim ABDF. Nova Série, 8*(3), 180-186. Retrieved March 30, 2012, from http://repositorio.bce.unb.br/handle/10482/5726

Cunha, M. B. D. (2000). Construindo o futuro: A biblioteca universitária brasileira em 2010. *Ciência da Informação, 29*(1), 71-89. Retrieved March 30, 2012, from http://www.scielo.br/scielo.php?pid=S0100-19652000000100008&script=sci_arttext

Damasio, E., & Ribeiro, C. E. N. (2006). Software livre para bibliotecas, sua importância e utilização: O caso GNUTECA. *Revista Digital de Biblioteconomia e Ciência da Informação, 4*(1), 70-86. Retrieved March 30, 2012, from http://www.sbu.unicamp.br/seer/ojs/index.php/sbu_rci/article/view/347

Eyre, J. J. (1979). O impacto da automação nas bibliotecas: Uma revisão. *Ciência da Informação,* *8*(1), 51-57. Retrieved March 30, 2012, from http://revista.ibict.br/index.php/ciinf/article/view/1547

Feijó, H. C. (2009). *A participação das bibliotecas universitárias em redes cooperativas no Brasil.* (Bacharel em Biblioteconomia), Universidade Federal de Santa Catarina, Florianópolis. Retrieved March 30, 2012, from http://www.cin.publicacoes.ufsc.br/tccs/cin0011.pdf

Figueiredo, N. M. D. (1992). Serviços de referência & informação. São Paulo: Polis: APB.

Frauches, C. D. C. (2007, November/December). *A livre iniciativa e reforma universitária brasileira.* Paper presented at the Colóquio Internacional sobre Gestão Universitária na América do Sul, 4., Mar del Plata. Retrieved March 30, 2012, from http://www.inpeau.ufsc.br/wp/wp-content/BD_documentos/coloquio4/IV-381.pdf

Gomes, L. C. V. B., & Barbosa, M. L. A. (2003, June). *Impacto da aplicação das tecnologias da informação e comunicação (TICs) no funcionamento das bibliotecas universitárias.* Paper presented at the Encontro Nacional de Ciência da Informação, 4., Salvador.

Gonçalves, E. M. S., Costa, J. S. B. D., Caregnato, L. F., & Fraga, T. M. D. A. (1998). Informatização da informação: A experiência do Sistema de Bibliotecas da Universidade Federal do Rio Grande do Sul. *Ciência da Informação, 27*(1), 99-102. Retrieved March 30, 2012, from http://www.scielo.br/scielo.php?script=sci_arttext&pid=S0100-19651998000100014&nrm=iso

Ikehara, H. C. (1997). A reserva de mercado de Informática no Brasil e seus resultados. *Akrópolis, 5*(18). Retrieved March 30, 2012, from http://revistas.unipar.br/akropolis/article/viewFile/1694/1466

Instituto de Pesquisas e Estudos em Administração Universitária. (2009). *Colóquios Internacionais sobre Gestão Universitária na América do Sul.* Retrieved March 10, 2012, from http://www.inpeau.ufsc.br/wp/?page_id=44

Knoll, M. M. D. C. (1986). Bibliografia brasileira sobre automação em bibliotecas e sistemas de informação: 1980-1986. São José dos Campos.

McCarthy, C. M. (1983). Problems of library and information system automation in Brazil. *Journal of information science, 7*(4-5), 149-158. Retrieved March 30, 2012, from http://jis.sagepub.com/content/7/4-5/149.abstract; doi: 10.1177/016555158300700403

Mêgnigbêto, E. (2012). ISIS2ABCD: A batch program to automatically create a database in ABCD from an existing CDS/ISIS database. *Information Development, in press, Epub ahead of print February 2, 2012.* Retrieved March 30, 2012, from http://idv.sagepub.com/content/early/2012/01/27/0266666911433313.abstract; doi: 10.1177/0266666911433313

Miki, H. (1989). Micro-isis: Uma ferramenta para o gerenciamento de bases de dados bibliográficas. *Ciência da Informação, 18*(1). Retrieved March 30, 2012, from http://revista.ibict.br/index.php/ciinf/article/viewArticle/1299

Nocetti, M. A. (1982). *Bibliografia brasileira sobre automação de serviços bibliotecários: 1968-1981.* Brasília: EMBRASA.

Pontes, A. M. D. (2003). *Percepção e utilização do aplicativo Microisis sob o ponto de vista do bibliotecário.* (Bacharel em Biblioteconomia), Universidade Federal da Paraíba, João Pessoa. Retrieved March 30, 2012, from http://www.biblioteca.sebrae.com.br/bds/bds.nsf/d086c43daf-01071b03256ebe004897a0/ef02c58e3b3bc867 03256fd30066fe10/$FILE/NT000A5FEA.pdf

Robredo, J., & Cunha, M. B. D. (1994). *Documentação de hoje e de amanhã: Uma abordagem informatizada da Biblioteconomia e dos sistemas de informação* (2nd ed.). São Paulo: Global.

Rowley, J. (2002). *A biblioteca eletrônica*. Brasília: Briquet de Lemos/Livros.

Sambaquy, L. D. Q. (1978). Da biblioteconomia à informática. *Ciência da Informação, 7*(1). Retrieved March 30, 2012, from http://revista.ibict. br/index.php/ciinf/article/view/1548

Smet, E. D. (2009). ABCD: A new FOSS library automation solution based on ISIS. *Information Development, 25*(1), 61-67. Retrieved March 30, 2012, from http://idv.sagepub.com/content/25/1/61. short; doi: 10.1177/0266666908101265

Souza, N. J. S. D. (2009). *Avaliação de softwares livres para bibliotecas*. (Bacharelado em Biblioteconomia), UFRN, Natal. Retrieved March 30, 2012, from http://repositorio.ufrn.br:8080/ monografias/handle/1/131

Vargas, F. G. (2003). BIBLIODATA/CALCO. Retrieved March 30, 2012, from http://www8.fgv. br/bibliodata/geral/modelos/historico.htm

Velho, M. E., & Neves, I. C. B. (1987). *Automação em bibliotecas: Produtos e serviços*. Porto Alegre: NID.

Vicili, M. (2010). Biblioteca tecnológica: Equipamentos e softwares facilitam a vida dos usuários. *PUCRS Informação, 33*(151), 33. Retrieved March 30, 2012, from http://www.pucrs.br/revista/ pdf/0151.pdf

KEY TERMS AND DEFINITIONS

Biblioinfo: The database about library automation (information technology for documents): It had 390 references with abstracts from Brazilian journals on librarianship and information science. Its objective was to make available to researchers, students and other users, the literature that was published in Brazil from 1986 to 1994.

Bireme: Latin American and Caribbean Center of Information on Health Sciences. The acronym comes from its original name: Biblioteca Regional de Medicina (Regional Medical Library).

CAPES: Coordination for the Improvement of Higher Educational Personnel, Ministry of Education, Brazil.

CDS/ISIS: Computerized Documentation System/Information Storage and Retrieval System.

LDB: National Education Guidelines and Framework Law (Lei de Diretrizes e Bases da Educação): Brazilian Federal Law n° 9,394 of 1996.

Microisis: MS/DOS version of the CDS/ISIS software.

PNI: National Information Technology Policy (Política Nacional de Informática): Brazilian Federal Law n° 7,232 of 1984. Made the restraint official for some segments of the market, including software.

Winisis: MS/Windows version of the CDS/ISIS software.

ENDNOTE

[1]   Translation: *Machine Readable Cataloging*

Chapter 9
Transforming Technical Services:
Maximizing Technology to Minimize Risk

Lai Ying Hsiung
University of California, Santa Cruz, USA

Wei Wei
University of California, Santa Cruz, USA

ABSTRACT

The current economic downturn has resulted in constantly shrinking budgets and drastic staff reduction at the University of California, Santa Cruz (UCSC) Library. Meanwhile, rapid shifting to digital formats as well as dramatic growth in social networking, mobile applications and cloud computing continues. To face these challenges, the Technical Services (TS) at the university library at UCSC need a transformation. This chapter discusses how the authors have adopted the strategy of maximizing technology in utilizing "robot-like" batch processing tools in house to minimize the risk of becoming ineffective or irrelevant. In aligning human resources to apply those tools to achieve our goals in tandem with the mission of the library, the authors learn to work with the various issues and the barriers that we have encountered during the past decade. The authors are examining the changes brought to the department through the process, highlighting a plan of action, and providing guidance for those interested in bringing about a technological transformation that will continue into the future.

INTRODUCTION

The current sustained downward budget pressure and rapid technological advances have created unprecedented challenges for all academic libraries including the library system at the University of California, Santa Cruz (UCSC). The mechanisms for learning and information dissemination in our society have been shifting from authoritative sources to people's social networks. Social media have grown to be an integral and increasingly important part of the everyday lives of hundreds of millions of users around the world. Explosive growth of mobile devices and applications has driven new services. Next generation fast paced users require next generation solutions in library

DOI: 10.4018/978-1-4666-3938-6.ch009

services to ensure library relevance in an increasing digital, online and networked world. Data is becoming obsolete faster and knowledge built on the data less durable. We are seeing powerful technologies and information systems causing a parallel change in the knowledge base and in learning, communicating and collaborating approaches in libraries. In order to remain relevant, to identify key factors in technology use, as well as to face rapidly changing user behavior, the UCSC library system has continuously reviewed its services and collections.

Technical services are an important creator and manager of information. Their competency in identifying, acquiring, analyzing and delivering appropriate information plays a significant part in the efficiency and effectiveness of the organization's performance. As information is rapidly moving from print to digital and to online, users are flocking to technology to discover and use information in the new formats. Many previous manual operations to handle tangible objects have to take advantage of technology to migrate to new services. Dunlap (2012) gives a good description of the situation. According to him:

The internet turned the whole library-world upside-down. Instead of going to the library as the easiest course of action for students ... Google has replaced that whole paradigm. We now have to go to the students (and also demonstrate to faculty and administrators) to tell them that we have a part to play in the education of students. Technical Services provides the additional human element to organizing information and making it "find-able" quickly. Google's search ranking algorithms remain a trade secret. We can do better, and must.

Fessler (2007) also states that we need:

A hybrid chief information officer, systems engineer, Internet architect, and strategic planner who also happens to manage the selection, cataloging, acquisition, organization, and labeling information packages for a library. Other duties will also include digitization project management and digital archiving. This new librarian has been dubbed "Librarian 2.0" in library blogs ... However, all that said, how does today's library plan for tomorrow['s] move from theory to reality at a time of stringent budgets, aging workforces, and unprecedented technological development? Addressing the competency gaps for present and future librarians is the most obvious and critical need. The technical services librarian/department of tomorrow will need a whole expanded set of competencies to swiftly respond to the vast changes in the bibliographic/information universe ... Tomorrow's technical services librarians will not be able to survive without sound grounding in technologies as they evolve. (p. 139)

There is a sense of great urgency for many technical services departments to change, to think outside the box, to re-organize, and to seek alternatives or aggressive methods of stretching resources as well as equipping library staff with the necessary skills to meet the organizational missions.

BACKGROUND

Many academic libraries have been working with their technology (IT) department to seek a technology solution. The Technical Services (TS) at UCSC is also seeking technological solutions to save human labor and provide prompt services to support the library missions amidst dwindling resources. Developing and implementing automated means to perform repetitive, mechanical or time-consuming tasks, especially in the manipulation of massive bibliographic data, will allow us to respond quickly to users and expand the scope of available resources. During the past decade, we have found it cost-effective for our own Technical Services staff to develop the needed

IT skills, instead of competing for attention from another department for specialized IT support as most academic libraries of our size normally do. These skill set "robots" have kept us nimble and self-sufficient in meeting new service demands when the workforce in our department is on a continuous decline.

How we embarked on this journey of IT self-sufficiency was not by design, but by default. It could be traced to the well-established tradition of shared initiatives at the University of California (UC), a consortium of ten campuses, and the emergence of big deals, aggregator packages and bundled subscriptions. Melvyl and shared regional storage facilities were started in the 1980's. Shared acquisition of electronic resources among campuses under the California Digital Library (CDL Shared Collections and Services, 1998) was initiated in 1998. The Shared Cataloging Program (SCP) stationed at UC San Diego (French, Culbertson & Hsiung, 2002) to provide all campuses with thousands of MARC records and Uniform Resource Locator (URL) maintenance for massive number of electronic titles followed suit in 1999. The successive drastic budget cuts in state funding mandated the Next Generation Technical Services (NGTS) initiative for UC to develop effective communication strategies and shared operations among university campuses. UC Council of University Librarians (2012) describes NGTS as:

An initiative developed by the University Librarians and SOPAG as a consequence of work conducted in other key UC efforts ... the UC Libraries are involved in a strategic partnership with OCLC to develop a "Next-Generation Melvyl" intended to re-architect the systemwide OPAC and transform the user experience of search and retrieval. NGTS will build on and complement that work by redesigning technical services workflows across the full range of library formats in order to take advantage of new systemwide capabilities and tools, minimize redundant activities, improve

efficiency, and foster innovation in collection development and management for the benefit of UC library users.

The global worldcat.org database, i.e. World-Cat Local (WCL), offering services to cataloging, resource sharing, information discovery and delivery became the basic infrastructure of our Melvyl in 2011. Hidden collections have become a high priority to use the "More Product, Less Process" approach (Greene & Meissner, 2005) as revised processing methodologies are developed for their handling. The report submitted by the American library Association (2012) confirms the current prevailing sentiment:

As academic librarians and their colleagues in higher education in the United States continued to navigate the "new normal," characterized by stagnating budgets, unsustainable costs, increased student enrollments, and reduced staff, the pressure on higher education to demonstrate value took on new urgency and importance in 2011–2012 ... Academic libraries are working to transform programs and services by repurposing space, migrating collections, and redeploying staff in the digital resources environment.

In 2002, the UCSC library had 130 FTE to serve 13, 642 students, 4,198 faculty and staff, 9,000+ print serials, and 1.8 million volumes. In 2012, the number of library FTE to serve 17,525 students, 4,548 faculty and staff, 1600+ print serials, 45,000+ online serials, and 2.18 million volumes dwindles to below 90. The TS alone have experienced a staffing reduction from 23 FTE in 2002 to 8 FTE in 2012, with only one librarian in the department since 2004.

One common theme from all such developments for the TS at UCSC is the need to manage the onslaught of digital information with fewer and fewer staff resources. We envision that a sustainable solution in the batch processing of massive bibliographic and/or holdings data will not only bring breakthroughs, but also will launch us into a paradigm shift. In 2004, the organization of our department had been traditional, with three functional areas: Monograph Acquisitions, Cataloging, and Serials/Electronic Resources. The catalogers previously handled only monographs. The serials staff handled the acquisitions, cataloging and management of continuing resources. The only paraprofessional serials cataloger had the sole knowledge of serials physical processing in the Library. The technical services staff had clear work boundaries and each staff was responsible for one's own area. Materials might pass through several hands and be trucked from one person to the other before a task could be completed or a problem could be solved. Bottlenecks, duplication of efforts and overlaps in the problem solving process existed. Copy cataloging and record downloading were done one at a time. One person's absence could delay certain core routines from being performed promptly. The TS staff took pride in their expertise, but some found it difficult to break free of their silo mentality. There were very few batch processing operations in the department.

JOURNEY TO THE TRANSFORMATION

Batch Processing Solution

In 2000 and early 2001, the UCSC library was faced with unloaded files of thousands of MARC records from the CDL Shared Cataloging Program created the year before. The library decided to have two experts work together on the files. One was an IT staff member who had been formally trained in batch loading techniques; the other was a newly recruited serials cataloging librarian, supervisor of the Serials Unit Head and the serials copy cataloging para-professional. After some initial experimentation, the serials librarian also received formal batch loading training to expedite

the development and maintenance process. This was the beginning of UCSC's journey in growing IT skill sets within Technical Services Department.

Within a year, the title of the serials librarian was changed to Electronic Resources Librarian to reflect the real nature of the work being performed by that position. Learning how to batch load, and later, how to create and customize powerful loaders within TS for our rapidly growing record loading activities became our number one priority. This had turned out to be the most impactful technology skill set TS staff had developed during the past decade. In 2001, the batch load solution, together with the adoption of Request Tracker, a web-based ticketing tool, the infrastructure supporting cross-training and team work to maximize technology was seamlessly created. The unique opportunity to expand this infrastructure took place in 2004, when the Electronic Resources Librarian was appointed the Acting Head of Technical Services to take over the responsibilities of three retiring heads, Head of Technical Services, Head of Serials Unit and Head of Cataloging Unit. She became the direct supervisor of twelve library assistants in the cataloging and serials units, comprising 80% of the technical services FTE. TS staff gradually picked up other IT skill sets to complement the work of the loaders in the Request Tracker environment, resulting in an evolving organizational and cultural transformation.

Loaders for SCP record sets

Since 2000, the weekly distribution to UCSC of hundreds and sometimes, thousands of SCP MARC electronic resources records, in both the serials and monograph formats, had driven us from individual downloads to batch loading, and from individual to collective endeavors to strive for timely access to e-resources. Each format had its own challenges. For SCP e-serial records under the "single cataloging record approach" (Princeton University, 2004), we had to ascertain whether we already had a record for our local print version or for another version on a different platform; and whether all versions had the same publication history. Only one MARC record was to be used in our local catalog to describe all versions and platforms for better display to the users. Our print holdings records might require adjustments if discrepancies in publication history were found between our print and SCP serial. Serials records are notorious for the need of regular maintenance due to changes in title, frequency, publishers, etc. On average, over 60% of SCP serials records were updates. Record overlay, using OCLC record number as the match point, had been found to be the most reliable approach to find matches in our local database. We defined what local tags to be protected to preserve our local data and accepted SCP records as is. Unfortunately we did not always have an OCLC number in our local ILS record and even if we had one, it might not be the same one found in the SCP record. Thus, we used to search every incoming serials record in our local ILS to guarantee that our local matching record would have the same OCLC number and publication history as that in the SCP record for accurate overlay.

The less likelihood of having duplicates or updates and the systemwide adoption of "separate record cataloging approach" (UC Cataloging Metadata Group, 2009) for SCP e-monograph records had simplified our loading process for them by removing the need for overlaying the print version. Yet SCP e-monograph records did not always have OCLC record numbers. We still needed an acceptable match point. The challenge created by the mix of records with and without OCLC numbers in the same SCP file was solved through our design of a special loader using two match points for overlay. The loader moved the stripped OCLC number to the match point field (from MARC 035 to 001) if it existed and relocated the vendor record ID, preceded by a locally defined vendor qualifier, from tag 001 to 035 to become two unique match points for accurate matching.

To handle the initial unloaded serials files, the weekly pre-processing, the record loading

and occasional record clean-up projects promptly and efficiently, we had to train three or four staff members to share the big workload. During the first few years, when the manual pre-processing work load was heavy, having several staff members to share the work simultaneously allowed record checking and loading to be completed within a day after the files were received. We were able to accomplish this weekly task efficiently because we had designed our loaders to insert, delete and update data in the loaded records without much manual intervention. Most other libraries dealing with the same files had taken a week or more to do the loading if they did a lot of manual editing. Once our ILS records became more in sync with the SCP records, especially after the completion of the OCLC Reclamation Projects, a rotation system negotiated among team members was implemented, with each person working on the complete process from start to finish within the day. Team members were originally electronic resources staff, but eventually as staff size shrank, they could come from any of the TS units. Cross-training in serials and their associated attached records was a pre-requisite. Most of the time, no team leader was assigned. Everyone took turns and backed up each other when needed. Efforts had been consistently made to train replacements of record loading staff members, when one of them departed from the group. Once record loading skills are mastered, those skills have become very handy for many other purposes. Not only are the loaders used to load other vendor records, new loaders are created or existing loaders modified to work with other IT tools to generate in-house record sets from non MARC metadata and for database maintenance.

Loaders for Other Vendor Record Sets

With the dramatic growth in e-resources, the need to work with non-SCP MARC records in both serials and monograph formats increased, and we have been loading records from different sources. During the past few years, eBooks have rapidly moved into the mainstream of publishing and are often sold in big packages with free record sets. We have to depend on those massive files of eBook records. Several new loaders have been created for their downloading. Some records are to be inserted while others require record overlays. Minor modifications have to be done to these loaders whenever purchasing models or our needs change. The core SCP loader configuration has been applied to all these loaders, such as the protection of local data, including the local URL, retention of the Library of Congress (LC) call number data for collection analysis and other purposes even though call numbers are not used to locate e-resources, and selection of appropriate match points for record overlay.

Loaders to Generate MARC Records from Non-MARC Metadata

We occasionally convert non-MARC metadata obtained from publisher sites into MARC format and create special loaders with constant data for loading them into our ILS. For example, we had conducted such a project for the 1,200 e-serials purchased from Taylor and Francis. We loaded these brief MARC records for prompt user discovery and delivery before SCP had time to distribute MARC records for the set. This gave public service staff the ability to discover and access those e-journals directly in full-text from the web OPAC, without going through the cumbersome process of consulting a paper list while helping users at the reference desk.

Loaders for Data Maintenance

Sometimes we only add title-specific new data to existing ILS records or modify the existing ILS data. This can be done in numerous ways depending on the source of new data and what we wish to achieve. We use loaders to create constant data based on certain coding in the records. We

batch create thousands of attached records by loader scripting. We have batch deleted existing data based on incoming data, and have moved or have modified data at loading time when the record sets need to be edited or enhanced. We use loaders to upgrade record sets requiring subject headings, content summary or table of contents notes. During the OCLC Reclamation projects, we enhanced our ILS records by extracting LC call number data, OCLC record numbers, missing serials title change tags (780/785), ISSN, ISBN or government document numbers from each MARC record returned by OCLC and by loading the extracted OCLC data into our corresponding ILS records. We have created authority record loaders to insert or overlay authority records to streamline the authority control operations.

In other words, many operations once were done by the staff manually one by one have been automated through the use of record loaders. We are able to create or modify effective loaders promptly, because the staff responsible for the loaders has in-depth cataloging, acquisitions and serials background from cross-training, and can already perform the tasks manually, often from start to finish, in the first place. Record loader is among the most sophisticated tools offered by the ILS. Maximizing its potential does require scripting skills and knowledge of natural language processing, however, minor loader customization and modification does not require much scripting knowledge. In-depth knowledge of technical services operations and needs, the willingness to think outside the box, plus an idea about how a machine works are a must to decide on what customization and modification to explore. The trend of using more and more vendor record sets and the frequent need for loader tweaking for various purposes justifies TS performing basic loader maintenance, without relying on another department to make minor changes. Customization and testing done in house mitigate the possible "lost in translation" issue (Leonard, 2011). Inclusion of technology in technical services tasks means that IT is no longer a separate function, but part of the solution components to achieve organizational efficiency and operational cost-effectiveness. More and more tasks in the department adopt this model and more and more staff members apply technology to their work through cross-training teams, resulting in a seamless transformation of the department.

Web-Based Help Desk Ticketing System

As mentioned earlier, our unique adoption of the low-cost Help Desk ticketing system reinforces our team approach, propelling us towards transformation. In 2001, in order to provide prompt responses to electronic resource queries and reports of broken URL links, we introduced the web-based Help-Desk Request Tracker (RT) ticketing system to be our communication tool with our stakeholders, such as users, public services, vendors, publishers and California Digital Library (CDL). All such emails go to one central site. Acquisitions and Serials units has its own unique RT contact email address, while all the rest use the generic techserv@library.ucsc.edu email address within RT. As a side benefit, we have built up the RT site to be our permanent database of correspondences and comments to document acquisitions, serials/ electronic resources and technical services activities and transactions. Requests and notifications coming in through RT are no longer sitting in personal mail boxes. Public services staff members are allowed access to the site to increase the transparency of our problem-solving process. The web-based system provides multiple staff with simultaneous access and we have assigned an average of four staff members to monitor the site several times a day and users often express their amazement at our quick response time. Staff working on the RT site have to be trained on electronic resources to resolve many of the tickets received by the system. The way we use

RT, designed for team collaboration, has become an additional catalyst for cross-training in our department's metamorphosis.

SYSTEMS AND TOOLS FOR TECHNOLOGY MAXIMIZATION

Nowadays most Technical Services use turnkey vendor systems with built-in functions and report features. If technical services departments could use these systems fully, they would already be on their way to maximizing technology in-house. The vendor systems that we use daily at UCSC are the ILS and OCLC.

ILS

The TS at UCSC use existing functionality of their ILS to do various tasks, such as automatic foreign currency conversion for Acquisitions and synchronization of location links between bibliographic and item records. Electronic Data Interchange (EDI) and other automated ordering and invoicing features are also employed. We set up a schedule in our ILS to routinely generate a complete file of all serials records, so that this time-consuming task can run in the middle of the night and be ready for anyone in the library to use during the work hours without paying the vendor for this service. Frequently used searching and exporting strategies saved in the ILS have greatly simplified tasks.

If ILS data to be modified is the same for a group of records, using the ILS global update functionality will be cost-effective. Our ILS has an extremely powerful reporting module to gather records using Boolean commands on data in the bibliographic, order or holdings records. Data in such records gathered can be globally changed, updated, inserted or deleted for record maintenance and manipulation. We batch insert special markers into our bibliographic records to identify which records have been submitted for our weekly batch setting of OCLC holdings. We combine our ILS's automatic authorities processing and global updates to do certain in-house authority control work, saving an average of 25-30 hours per month. We use codes to batch update the PRICE fields in item records in the whole library database weekly to ensure that all new item records have the correct replacement cost. Once efficient procedures are set up, we train students to perform some of them, giving us more cost savings.

OCLC Connexion

OCLC Connexion has ample batch processing functionality to replace one-by-one operations. In our WorldCat Local (WCL) environment, we have to ensure an OCLC number in each local ILS record for automatic holding linking between WCL and our ILS. To obtain such numbers for our ILS records in batch mode or to batch derive record sets from OCLC require several types of record manipulation skills. We batch extract record identification numbers like ISBN's or music numbers from MARC records or even from publisher's list onto Notepad or Excel. We then batch match them in OCLC using various search keys and download them into our local ILS. Several iterations in searching, supplemented by manual review, may be necessary to ensure thorough batch searching. We completed the batch OCLC searching and loading of 2,300+ DRAM (Database of Recorded American Music) records within a couple of days. It would have taken us months to complete the project if we downloaded records one by one.

We employ OCLC macros, constant data, text strings, keymaps and save files to batch update OCLC records. As an example, we use the OCLC macro to create records for electronic resources from records for the print version. This reduces the operation from 30 minutes to a couple of minutes per record. We batch set and delete holdings

weekly in OCLC. As a result, our staff members no longer handle this one by one on a routine basis.

Not all ILS offer the same functionality; and ILS alone certainly cannot cater to all the needs in the manipulation of massive amounts of data. There are IT tools that can work with vendor systems to extend the power of the ILS and to save money without purchasing services from ILS vendors for one-time data maintenance projects, as well as for processing ongoing tasks.

MarcEdit

MarcEdit, a free software program developed and regularly updated by Terry Reese of Oregon State University, is one of those free IT tools designed specifically to work with MARC data structure. It is easy to learn and has a large user group to provide help through an active listserv. It facilitates the batch processing of records that is not possible within many ILS systems and can automate many of the basic edits without necessitating assistance from staff with a programming background. It can convert a MARC or tab-delimited file into human-readable text file, use its various batch edit tools to make changes to all the records in the file and re-convert the file back into MARC for loading into the ILS. One can query other library systems and download records, and join, split, extract and delete MARC records. It allows users to create default template files to reduce the amount of manual repetitive data input. Its Script Maker helps novice users to automate simple maintenance tasks. For those with a scripting background, the Script Maker can be used to simply generate a common script framework, sparing the coder from typing hundreds of lines of code. MarcEdit has saved us many person-hours.

Excel

Excel is a spreadsheet software which has been used by many libraries. Besides using it for data entry, for quickly manipulating rows and columns and for a variety of calculations, we find it to be a great tool as a bridge to transition data between systems. It is often used together with other software tools to receive data extracted from an external source or as data source for input into another IT tool. We have used Excel to translate many types of data into meaningful reports for making technical services decisions and for crunching numbers. Using Excel to paste in data from another source, to collocate data from two different sources, to add different data for every record, or to customize the scope of one's data can be very efficient; as an example, we re-purposed non-MARC metadata in Excel from the publisher site monthly by machine generating brief MARC records for the constantly changing 6,000+ LexisNexis resources. This monthly maintenance process took only one to two hours to batch create all the needed records and to batch remove obsolete ones.

Miscellaneous

There are word processing and database software that we have used for our TS operations. For example, we use FileMaker Pro to cross match serials publisher names from a vendor site with our ILS data and then load these names into our ILS to collocate serials by publishers or by vendor packages. This technique has reduced our processing time from months to a matter of days. We have locally developed a replacement database using FileMaker Pro. This makes it easier for us to share evaluation information about lost items and to determine our user needs. This replacement database can also generate statistics and information that was not previously available. The savings in ordering time has now translated into labor savings; as an example, a paper order that costs thirty-five to fifty dollars in labor is now only costing us about thirteen cents.

Other IT software like AutoIt, TeraTerm, MarcXGen, and Xenu's Link Sleuth has helped us achieve labor savings. We are adopting web

2.0 technology such as Google docs, Google calendar, Drupal, PBworks and PBwiki. Perl is great for extracting information from text files and we will describe how we have used it in different projects later.

PROCESS OPTIMIZATION TECHNIQUES FOR TECHNOLOGY MAXIMIZATION

Lean Library Management

Huber, a management consultant, has devised a "lean management" approach to help organizations and libraries improve their customer service by identifying and eliminating "all waste in a process" (Huber, 2011). We find a similar theme between his approach and our process optimization technique that has "dramatically reduced the lead-time required to deliver products and services". Our cross-training and team culture has allowed us to "define, measure, analyze, improve and control the performance of our service delivery chain," crossing unit boundaries. Our "lean" approach has contributed to our transformation in many ways.

Our most recent example is to use a special loader to batch insert the needed data to identify all the eBooks that are in the patron driven acquisitions program (Anderson, 2011). After purchases have been triggered by users, the staff loading the records will perform the acquisitions, cataloguing and invoicing functions for those eBooks all in one setting by using multiple tools as needed. We have indeed designed many processes employing systems and tools that cross unit boundaries.

Interoperable, Modular Set-up

Often a task cannot be easily achieved by using a single system or tool. Collaborative work often requires systems that can talk to each other. Integrating multiple tools and systems that are interoperable provides the ability to seamlessly reuse or move data from one platform to another, eliminating re-keying of data, reducing errors and time. Lang, Gerz, Meyer, and Sim (2008) note that:

The development of an interoperability solution should follow a well-documented and well-understood methodology ... The methodology should be based on a modular approach to provide the necessary flexibility, enabling individual modules to be changed independently ... there is still the need for harmonization between the modules to ensure that the information exchanged between the modules ... can be orchestrated. (p. 1)

Can ILS accept pre-order or selection from any book vendor's databases? What sort of order, invoice and barcode data can be embedded in the MARC record and how do we balance the use of several systems? Our selectors and acquisitions staff directly place orders in the fulfillment vendor database (Gobi) with the specific order information. We batch download them to the local ILS for export to generate the necessary information for use further down the acquisitions/cataloging cycle. Batch movement and massaging of bibliographic records between WorldCat and our local ILS, using MarcEdit, Excel and the ILS loaders, has revolutionized our bibliographic production and maintenance. Our experience shows that a model, making use of modular systems or tools that can talk to each other, will provide much greater flexibility than what a single turnkey system or tool can offer.

Stable Technology, Robust, Secure and Low-Cost

Library technology for general adoption does not need to use the latest or the best systems or tools, but they have to be decent, robust and stable, to minimize unnecessary migration or constant changes for staff. We are dealing with

more and more applications in the cloud. Privacy and data security are two issues that have to be considered when using web as infrastructure for creating contents using web 2.0 tools. Our tools and techniques can be described as "low threshold applications" (Gilbert, 2002) which are free, inexpensive, and easy to implement or may have only "sunk cost." With severe budget and staff constraints, such applications will go a long way to help maximizing technology.

Manual and Automated Hybrid Process

Adoption of technology is not an end, but a means to an end. Not all processes can be or should be automated. Proper identification of projects or parts of the projects identified for automation is essential. We frequently mix manual and automated steps, since with certain activity, human intervention could be more appropriate or indispensable. One of the most important reasons for automation is not to eliminate manual operations, but to manage the whole project in a cost-effective and reliable fashion through systems analysis ahead of automation. According to International Records Management Trust (1999), "Automation projects cannot be planned for in isolation from the organisation's mandate, mission, functions, resources and other systems" (p. 16). How to select tools or systems, which steps should be manual or which ones should be automated, and how to align them well would be topics of interest for technology maximization. Setting goals with an in-depth knowledge of the operations as well as with an understanding of the available technology options, issues, and functionality will lay a good foundation for automation. Bulk and consistent data in electronic formats that can make use of default settings, templates and algorithms for decision-making to eliminate labor intensive and tedious clerical duties will be the most adaptable to technology; however, decisions requiring human visual investigation or judgment will, perhaps, be difficult to automate. Automation should reduce redundancy, errors and labor cost, and enable us to generate reports across multiple modules and automatically track transactions with consistent results. Automation should incorporate or integrate information into a seamless flow. Sometimes unintentional consequences from automation can be avoided when human review is incorporated. Manual handling to verify intermediate automated results at appropriate break-points would help staff trust the procedure and secure stakeholder buy-in. Manual handling allows staff to learn whether proper actions have been applied to the data. It also helps staff to figure out how to upgrade when changes are made in an organization.

Facilitating and Sustaining Technology Maximization

Some automation projects are implemented mainly for getting staff to become familiar with technology, and those projects usually are not transformational. To really reap the benefits from automation at UCSC, we recently dedicated one FTE in the department to work on "Process Optimization Research and Development," to meet one of nine stated goals in our Mission and Services Statement (UCSC Library Technical Services, 2012). This FTE had already been cross-trained from the entry level in all core areas for a couple of years. She was not assigned specific routines, but was directed to work on new initiatives, and to identify low hanging fruit, by discovering, learning, and applying the appropriate new skill sets. Even though the person holding this position was promoted outside the department after a year, transformational results were already evident during the year of her appointment.

Coming up with the best process cannot depend on IT skills alone. Keeping up with trends and technology related to library developments and possibilities of tools is a must. There are many

resources that can be helpful, such as vendor support, services, documentation, tutorials, and webinars. Professional associations and their services (such as ALA Connect), conferences, workshops, etc. as well as colleagues from consortium and other libraries are also increasing their roles in networking and professional development. Professional literature and information databases provide good overviews, insights and shared information. Resorting to free web tools would make things happen faster. Google has become a convenient starting point for locating and verifying information.

Cross-training experience in many different areas, with an in-depth understanding of what needs to be accomplished is fundamental to success. Training has to be geared to an individual, considering a person's skills, aptitude and interests. Breaking up a process into segments, with incremental training, can facilitate understanding. Clear documentation with an outline summary, plus detailed steps, illustrated by visual aids or recording can be helpful. Complicated and detailed, but ancillary, procedural information can be moved to appendices or footnotes. Collaborative documentation, using Wiki, allows staff to easily write down each step performed. Incorporating new tasks into staff routines and allowing them to experience new changes on a regular basis can help staff retain their skills and sustain any new changes. Easy access to IT consultations will facilitate technology maximization.

There are certain techniques that can help to start or to speed up a development process. Building on other libraries' work, or even borrowing the entire process, if appropriate, can prevent re-invention of the wheel--an important strategy for the have-not departments. NOBLE SwapShop, maintained by Thomsen (2008), is an excellent example of members of a library community sharing scripts, graphs, training material, presentation and other expertise. Involving staff at an early stage to take advantage of group input and perspective and using appropriate outside resources can reduce unnecessary mistakes and introduce no surprises. Willingness to take risks, to weigh returns of resource investments, to set priorities and to have project management training will be helpful when looking for workflows to streamline. Periodic evaluation of workflows for bottlenecks and redundancy, for an ongoing review of established procedures and for applications of batch processing will keep processes relevant to current needs. Timelines for tasks could be shortened dramatically. As an example, our SCP record loading has been reduced from 8 hours to half an hour per week within a decade. Our SCP batch loading experience, our gradual cultural and organizational transformation, and our in-depth understanding of what a record loader can do have been instrumental in technology maximization in the TS department at UCSC. With flexible and knowledgeable staff members who are enthusiastic to learn new approaches, the TS have become a fertile ground for change.

USAGE BEYOND ORIGINAL INTENT

As a matter of fact, the tools and techniques described above may have already been known or available to many; however, they alone may not create the transformation. It is the way we expand their usage beyond their original intent that has made us unique. We question the status quo. We apply unconventional thinking that facilitates the cultural and organizational transformation which in turn catapults our technological transformation. Below are presented three of our recent projects as illustration.

Shelf-Ready

The basic level of vendor shelf ready service requires visual checking of every book and MARC record at receipt to guarantee that a correct and preferred MARC record has been received. To guarantee correct record receipt, the TS at UCSC has developed a Perl script to generate cross-matching Excel reports to expose discrepancies between the MARC data from the vendor's invoice file (Gobi) and its corresponding bibliographic data found in the MARC records from OCLC's WorldCat Cataloging Partners (WCP, i.e. Prompt-Cat) Service. All items with no discrepancies found in the reports can be pulled for shelving within a couple of hours after the shipment is received. Vendor duplicate reports and ILS duplicate headings reports from the record loading process are used to identify unwanted duplicates. Global updates and the local ILS location software are used to insert location codes for those items not using the default location. Our shelf ready loader inserts, among other things, a donor note and a web link to the donor page into the MARC record based on the fund code supplied when the order is processed. Catalogers normally only need to replace wrong records or fix incorrect labels for books identified as requiring human review from the reports. Such problems average about 10% of the titles in a shipment. The time taken to fix those problems is usually short since the majority of the processing has been taken care of by global updates or the loader or by the shelf-ready vendor. These problem books are normally ready for the shelf within 24 hours. This process is a good illustration of how we efficiently integrate manual and automated processing in a modular model using Gobi, OCLC WCP, Perl, Excel and ILS features. The cross-matching reports allows the elimination of one-by-one manual record review by having the machine compare GOBI invoice MARC and WCP MARC on data elements specified by us. This usage of MARC data is unconventional. Our willingness to re-define record acceptance level, to depend on subsequent batch processing for record upgrade and to accept "good enough" records as is at receipt has enabled us to de-deploy a couple of FTEs from routine copy cataloging to higher level or other needed tasks.

Using WCP Service with YBP e-Invoicing for eBooks Ordered in GOBI

In OCLC, both the "provider neutral cataloging approach" (Provider-Neutral, 2009) and the "separate record cataloging approach" are allowed to describe eBooks. Normally, it is impossible to use WCP service to secure appropriate MARC records for eBooks ordered in GOBI without record-by-record manual scrutiny (Wu & Mitchell, 2010). Yet we have devised a simple process to make use of both WCP service and vendor record sets to obtain YBP's e-invoicing service, while ensuring that every e-book MARC record has the vendor-specific URL and the correct OCLC number through record loading. We do this by first exporting select data from previously loaded WCP or Gobi eBook records into Excel, including the ILS record number as the overlay match point. Each vendor record should have the correct URL, but not every record has an OCLC number. If there are missing OCLC numbers, batch searching OCLC is required as an intermediate step. Using MarcEdit to add OCLC numbers and other related data to the vendor records follows. Then we load the enriched vendor record sets into the ILS to overlay those WCP or Gobi records. The process can save one or more hours per load, depending on the number of vendor records missing the OCLC number. This process is another illustration of integrating manual and automated steps using OCLC WCP, MarcEdit, Excel, various ILS features, and eBook Vendor data. This is an utterly unconventional way of using the WCP service. What really gets retained is the vendor record set which completely overlays every WCP record received to achieve 100% record accuracy.

UCSC Collection Statistics e-Count

We generate collection titles and volume statistics monthly entirely from Boolean searches of the entire ILS database for all online, print and non-print library resources to eliminate manual tallies by different departments. To develop this process, we conducted careful reviews of statistics instructions from the Association of Research Libraries, the Association of College and Research Libraries and systemwide UC specifications. We decided on the criteria and methodology to be used. We did comprehensive data clean-ups, ILS record enhancements and coding convention re-alignments along the way. To overcome the challenge from incomplete, inaccurate or inconsistent holdings data and coding, we use IT tools like Perl to harvest data from the descriptive and coded information in both the bibliographic and attached records in our local ILS. To generate counts for our collections in the science/engineering building, the social science/humanities building and the Special Collections area, call number data and location codes are employed. The number of item records is used to calculate monograph items. Existence of government document numbers in the bibliographic records is the criterion to identify government documents. More than half of our serials records do not have item records, necessitating the use of Perl to estimate the number of serials volumes from the summary of holdings data. Benchmark standards and assumptions based on instructions from the national statistical agencies are used to decide on a formula for the Perl script. To calculate the number of free continuing resources, local free text coding such as "open access," "free," "depository," and "CalDocs" are used to gather records. Local coding for item type, material type, item status, etc. is also employed extensively. It took one FTE three months to develop the procedure. This process, utilizing many of the ILS functionality such as Saved Searches, Saved Export, Global Update, has to be carried out by a staff member following meticulous documentation every month. Results are manually recorded in the Excel Spreadsheet which has built-in formulas and functions to compile the recorded data. The integration of manual and automated operations allows results to be easily verified at each step and facilitates process update when coding conventions, statistical definitions and organization needs change. The execution of the monthly process takes two days since it involves many Boolean searches on the whole database. Applying machine methodology on existing ILS data by using ILS built in features and functions has generated consistent and sustainable statistics which approximate very closely our established manual count figures. The elimination of daily manual tallies and compilation is a welcome change for all departments. This is an illustration of a hybrid solution using Perl and Excel to support our ILS functionality. The use of Perl script to estimate missing serials item records from a formula based on in-house benchmarking is borrowing from the basic concept of fuzzy methods (Workshop on Fuzzy is Scalable, 2008). The extensive use of descriptive and coded bibliographic and holdings data for gathering collection statistics is definitely beyond what the data is intended for. Systematically deriving comprehensive and complex non-precise and precise data from detail ILS record coding to yield collection e-counts on a routine basis is unconventional.

BARRIERS TO TECHNOLOGICAL MAXIMIZATION

TS at libraries take great pride in an orderly application of standards and a systematic implementation of details to bibliographical control. Its work has been conducted mostly in a manual way and backlogs have been accepted as unavoidable. With users and library services rapidly embracing technology, profound technological changes with their associated data and human barriers are inevitable in TS.

Data Barriers and Strategies

As described previously, inconsistent, incorrect or missing data require continuous maintenance efforts and are not amenable to automatic processing.

Unusable data due to lack of standards and lack of appropriate tags for storing information would make it more difficult to transmit data, jeopardize the integrity of data during migration or use, and also would make operation management more difficult. The power of batch processing could be easily compromised by mistakes and typos in data and in coding. Inconsistency in word division/spacing, in coding format, and in mismatch in scope could corrupt data. Duplicate records in OCLC have been perceived by many as a big problem. Varying cataloging treatments have added to the complexity. A book within a multi-volume set, an issue within a serial, or an accompanying item could be cataloged separately or as one entity. Different interpretation or opinions of serial title changes will create multiple records for the same title. In general, de-bugging, back-tracking and continuous monitoring of data integrity could be time-consuming or require higher level staff.

The consistent application of logical, but simple coding standards; removing obsolete, inaccurate and misleading data; provision of useful administrative data; as well as application of techniques such as fuzzy methods will facilitate removal of data barriers. For example, if a value has been coded consistently for active subscriptions, a simple Boolean search can pull up all the active subscriptions. Some reliable codes allow the machine to identify related incorrect codes in other fields and clean up inaccuracies on thousands of records very efficiently. Identifiers are very important in batch record management (Coyle, 2006). The TS at the UCSC library have defined a simple convention for record control numbers in the e-monograph records created by the CDL Shared Cataloging Program (SCP). Once it is implemented, batch clean-up of all those control numbers facilitates proper record display, indexing, overlays, linking, and their use by other applications interacting with the data. This enables efficient global updates and record extraction for projects like collection moves, digitization projects, and Boolean searches. Organizing related contents within the same record or in the same string (with fields and subfields) will facilitate data management.

Vendors could be a source of help to acquire the data needed. They can either generate the data for us or provide us with the data to generate the information we are looking for. We have established an optimal procedure to export Local Holdings Records (LHR) from our ILS to World-Cat by working with the vendor using the system's output tables and then asking OCLC to programmatically massage data in our LHRs, allowing us to eliminate the MarcEdit export /global update /re-import steps. We may submit enhancement requests to vendors to improve their functionality and indexing practices. We review our indexing scheme used for batch record loading, automatic overlay, and linking among systems. We batch load package title "hooks", such as EBL eBook library, to collocate titles for each collection. We use various technologies to provide relevant system generated reports by extracting data from our ILS and/or vendor management databases for staff to use in projects or routines, securing significant time savings. This has reduced the time to perform a process from weeks to a matter of hours. We export order numbers from Ebsconet and cross-match them with order record numbers from our local ILS to facilitate our updating of subscription formats in our ILS order records. We design appropriate processes to global update codes for items that are moved. Hybrid manual and automation solutions allowing human review also help to reduce the data barriers.

As explained above, quality data are important for reliable user access, management decision-making and number crunching. Keeping data entry as simple as possible can preserve data consistency. Batch acquisitions, creation, correction and updates help keep data up-to-date. Finding ways to use machine text matching for data comparison between two data sources will save time, ensure analysis consistency, and quality for subsequent

manual data evaluation that can detect duplicates, identify errors as well as point out inconsistent coding or discrepancies. Eventually knowledge base and artificial intelligence, and using concepts such as fuzzy methods, can help technical services maximize technology to combat data barriers. However, to arrive there will require in-depth collaboration with IT departments. No matter how much technology can be applied, there is still a significant place for human intuition and judgment in meeting the challenges of data barriers.

Human Barriers and Strategies

Human barriers are other factors that could slow down or even cripple the use of technology in TS. We find that territorial ownership can be stifling to change. Some staff members might not be willing to let others intrude into their areas of responsibilities. There could be fear of loss of expertise status and fear of exposing weaknesses or lack of understanding. Some might rightfully resent the implication that their work of past years or decades has been inadequate or incompetent. We also hear remarks such as "We have enough staff to do the work. There is no need to batch." Job loss has often been attributed to the adoption of technology.

On the other hand, as the economy declines, people are more worried about losing their jobs. Consequently staff might be more willing to accept change. Yet, lack of technology orientation could be another big barrier. Automation might entail redesigning work processes, and thus, changing people's jobs or the way services are offered. Fear of technology makes non-tech savvy staff unwilling to learn new skills or change their familiar preferences. Fear of failures and lack of incentive for change can make some people unwilling to adapt to new methods, making compliance difficult. We also hear remarks that "if it is not broken, why do we have to fix it?" Some staff claim that they can do their work just as fast; and there is no need to automate. They might not see the real benefits from automation and perceive that the new process is more cumbersome.

Those who are unfamiliar with technology grow suspicious of automation. They think that global updates are risky. They believe that unintentional changes can be made to data, and such changes cannot easily be reversible. There is fear of losing data in an unpredictable way. If errors are discovered, they would blame them on some mysterious automated process. When tasks are done by machines or when several automatic processes interact, they cannot figure out the why and the how. They regard automation as being too complicated. To them, the loss of control of familiar processes is unwelcome.

Some people might expect too much of automation and be disappointed if it is not 'perfect'. Poorly planned systems will not serve organizational requirements adequately and so be considered poor substitutes for previous manual systems. Insufficient training of personnel using the system could result in morale problems and in data that is not reliable. Vague or imprecise systems will not serve needs well and might become obsolete quickly. Then there is a possible problem of human physical limitations. Some staff still prefer a paper trail. Looking at a computer screen for a long period of time might hurt their eyes; too many mouse movements by hands could cause carpal tunnel syndrome. Finally, lack of library administration support and leadership could be detrimental to adopting technology in TS if the expertise of higher level staff to set up an automated process and in de-bugging is not available.

It is crucial to assess people's capacity for change in their work environment and to prepare them for change. Whether an automation program succeeds or fails often depends on the support received from stakeholders. Introducing technology is the easiest when starting new initiatives or projects that no one has done before. Recognizing individual preferences whenever possible will reduce obstacles. Similarly, users may have to

be given guidance about how automated systems work, if they are to be expected to use technology themselves.

Cross-training and team work have played big roles in overcoming the human barriers at UCSC. They have enhanced the communications among our staff members, and helped to gradually overcome their territorial silos. "People of different age groups have varying work ethics, communication habits, adaptability, and technology skills. Soliciting ideas and opinions from employees of all ages and encouraging questions so the different generational groups can learn from one another can minimize workplace problems" (Smith, 2011). We have reduced redundant steps to allow automation to cross boundaries for optimization. We have trained all TS staff members to read MARC. Our staff have been cross-trained to work with our bibliographic, order and holdings records in both the monograph and serials formats. Every staff member can perform basic copy cataloging or verify URL when needed. We have enough Acquisitions staff to switch to copy catalog when the ordering is down. Similarly, we can have Electronic Resources staff work in the Acquisitions area when the ordering is up. One staff member can be on several teams. Since our work no longer depends on one person, our staff absences and turnover will not affect our work flow; and we are prompt and reliable in addressing issues that come our way. Our "train the trainer" approach raises the number of staff who can train and document procedures and provides mutual support when assuming new work. Informal problem solving discussions are common among the team members to identify the methods and the needs for change, re-enforcing the team spirit. Our integrated processes have reduced passing work back and forth from one staff to the next. Every staff member has a job specialization, but each is encouraged to assume work in new areas. The supervisors are relieved from being the only person to handle the continuous training workload. This informal training from peers fosters mutual working relationships. Our collective memory contributes to sustainability. The monopoly of work knowledge is kept to a minimum. This democratic approach increases the comfort zone for the staff. This expands the number of staff who can perform each core function. A high level of collaboration and coordination has developed throughout the TS department. The same staff can place orders, receive, batch load records, process invoice and even copy catalog. Efficiency is evident wherever it is applied. This environment gives every staff an opportunity to nurture IT skills with each other at their own pace. The staff does not need to compete with each other. All these changes have further facilitated the acceptance of automated processes that can cut across the functional lines vertically and horizontally, reduce the duplicated steps, increase the built-in backup redundancy and streamline the workflow.

We target technical skills each time we recruit for a new position. For the existing staff, we have assessed the staff abilities and orientations and are looking for the best fit on an ongoing basis. We acknowledge different learning styles and satisfy the staff needs for self esteem and social actualization by spending time with each individual; also, we try to understand their work preferences and receive feedback about how to provide the best support as each person approaches a new task. Behind all this is a strong commitment from the TS department leadership. To address ergonomics issues, we seek campus assistance in meeting the need for special equipment and furniture. We communicate and incorporate the performance expectations into the TS mission and vision statements. In doing so, staff can set personal goals in the context of the department's direction. Those who cannot pick up an automated task can perform alternative complementary work. We emphasize the equal value of automated and non-automated work, and instill pride in all types of work; at the same time, we also convince the staff the benefits of using automation. Once more and more repetitive and monotonous tasks are automated or outsourced,

we then re-classify the staff to a higher level of job classification. During the past decade, we have not devised a one-time roadmap or a written comprehensive plan for our transformation. We can afford this approach because we have started our journey to transformation long before this current economic crisis and rapid technological development hit us. We only have a clear vision that TS need to maximize technology to achieve an ongoing transformation to stay prepared. We set priorities. This allows us to target willing participants to soothe the way of implementation so that changes do not come to everyone all at the same time. We definitely allow plenty of time for staff to get used to the new ways of doing things. We introduce improved processes, not as a response to staff reduction, but as our long standing goal for streamlining and efficiency. When a staff vacancy occurs, we conduct a careful analysis of our staff needs. We do not automatically request for a replacement unless there is a pressing need. Original cataloging expertise is a good example of how we evaluate priorities. With the system-wide direction towards collaboration under the Next Generation Technical Services, the limited amount of original cataloging that we have to do, and the many new responsibilities we have to assume amidst greatly dwindling resources, we decide to postpone the training on Resource Description and Access (RDA), the new cataloging rules, until we are ready to use them in January 2013. Another example is web 2.0 technology for technical services. In general, most Web 2.0 tools are less stable and we have not strived for its extensive adoption. TS staff will no longer embark on projects unless we can devise optimal and sustainable strategies to guarantee worthwhile returns and to retain workflow efficiency.

Major Breakthroughs and Outcomes

Major breakthroughs in maximizing technology in TS during the past decade have saved thousands of person-hours per year while maintaining quality service. The first breakthrough was defining a local tag indicator when we learned how to load records in early 2001, an innovative idea in record loading at the time of introduction. This has proven to be a simple and yet reliable way to preserve local data, even when one tag, such as the 856 URL, is used for both universal and local access. Our record loading process has been very efficient using this technique. This exciting solution has encouraged us to always think outside the box right from the beginning. It paved the way for our learning to design loaders in house that can accomplish many tasks with as little human intervention as possible. It also laid a foundation for our growing IT skill sets within TS. The second milestone was adapting Request Tracker (RT) to manage communications with our users, publishers and vendors. This facilitated work sharing and documentation, another innovation for the TS department. The third breakthrough was the move towards unusually extensive cross-training which provided the cultural and organizational infrastructure for our departmental transformation. The fourth was the adoption of the "good enough" concept in copy-cataloging and other operations which allowed us to avoid wasting time on unnecessary details. The fifth breakthrough was the extensive adoption of batch processing which helped us to streamline many of our TS processes. The creation of cross-matching reports, the use of WCP service for eBook e-invoicing as well as the collection e-counts could be labeled as our most recent major breakthroughs.

Many of our routine batch processing tasks have been developed and designed for the technologically challenged staff members, in order to help them feel that they are moving forward with the rest of the group. The size of our ILS internal processing file was expanded from 200,000 to 3 million records and transaction files to 2.5 million transactions. Our technological breakthroughs have brought numerous FTE savings, and have improved consistency, accuracy and speed in our operations. We can do more with less. We assume new activities, such as taking over the entire end processing operation, running the weekly batch item price update, the weekly batch OCLC holdings updates, the maintenance of EZProxy domain names for off campus users, the weekly Local Holdings Records submission to OCLC, the monthly global update of local note tags to protect them from overwrites, as well as the routine verification of item types from status codes in item records. Our prompt and smooth implementation of the Reclamation Project radically increased UCSC's local holdings in WorldCat from 38% to 100%. Technology was employed to improve the bibliographical control of the UCSC collections by synchronizing holdings among multiple networks, enhancing access, linking and database quality, and greatly increasing efficiency in both backend management and in bibliographic production and maintenance.

Aggressive strategies have been employed to stretch our resources, as well as to optimize and to equip the department with the necessary technological skill sets. We have undergone organizational and cultural transformations, characterized by many technological breakthroughs. We have emphasized good returns on investments (ROI) by generating low cost but high impact transformation by automating manual processes. The number of staff members being trained to perform each core function has been expanded. It has also raised the number of staff who could participate in productive group work, as well as train and document procedures. Each staff member has been given a diversified range of responsibilities and tasks that are drastically optimized, both through cross-training and automation. We all have learned to delegate and to let go of unnecessary details so that we can focus on setting expectations, directions and priorities. Furthermore, we are able to allocate resources and to monitor progress and decision-making. At the same time, we can quickly address the issues that demand direct attention.

As a result, we can handle the core TS functions without backlogs even when our staff size has been reduced from twenty-three to eight FTE during the past decade. By now, our staff has become so small that dividing the department into three or four official units no longer makes sense. We stress on functional areas instead. The pioneering application of EDI had propelled Georgia State University to recognition as being on the leading edge of technology (Stephens & Presley, 1996). This could be another positive outcome.

Looking Ahead

Transformation of Technical Services will continue to unfold and will occur at an even much faster pace than in the last decade. It will be an endless journey of renovation, remodeling, expanding, and rearranging from creative minds that will inspire new thoughts and new ideas. Systems and tools may come and go, but technology and massive data are here to stay. Users' behavior, needs, and expectations are drastically changing. The demise of the integrated library system has been predicted ever since early 1990's (Breeding, 2010; Lynch, 1991; Silvis, 2010). OCLC's WorldShare Management Services is now beckoning libraries as well as third parties to contribute "apps" to the WorldShare Platform in the cloud. There will be more tech-savvy staff managing massive data at different institutions or organizations, such as libraries, museums, vendors and publishers. The realization that all such data need to talk to each other to allow users seamless and instant access to information will promote networking, self-help and collaboration among all parties. Metadata is taking on many new formats such as Dublin Core (Dublin Core Metadata Initiative, 2011), MODS and METS (Cundiff & Trail, 2007). Even Library of Congress is working on Resource Description Framework (RDF) to replace MARC (Library of Congress, 2011). Coyle, a linked data and semantic web expert (2012), states that:

We need to be creating data, not records, and that we need to create the data first, then build records with it for those applications where records are needed. Those records will operate internally to library systems, while the data has the potential to make connections in linked data space. I would also suggest that we cease creating silo'd RDF record formats, as these will not move us forward. Instead, we should concentrate on discovering and defining the elements of our data, and begin looking outward at all of the data we want to link to in the vast information universe. (p. 39)

Murray (2011) talks about local, digital material being the third wave of new contents:

It takes the form of article pre-prints/post-prints, working papers, technical reports, datasets from experiments, slide collections, lecture notes and recordings, blogs, wikis, and corporate publications. To manage this new wave of content, a new suite of tools are emerging: content management systems, institutional repositories, e-print software, and collaborative writing applications ... applying library expertise to new views of corporate intellectual assets, such as the long term management and 'exposure' of both research and undergraduate outputs, in a multimedia and collaborative world ... A focus on curating local, digital content, however, means that libraries can more directly insert their services at the point where content is being created.

We have moved from centralized processing in the mid 20th century to distributed processing at the end of the century. It seems that we are now on the verge of a new round of centralized processing again, this time in the cloud, but surrounded by many available distributed local customizable options. Are we ready to be avant-garde in the exploration of yet another set of strategies and tools designed to do things differently, to gain efficiency and to venture into new services and to better serve our users? The Ministry of Justice in the government of Catalonia in Spain (As cited in Mann & Chan, 2011) proclaims,

Nowadays the Internet is about sharing, co-producing, transforming and personalizing to create new products and services. To create, it is necessary to be able to make use of knowledge that already exists, without limits, and to share it afterwards. This is the philosophy of innovation that is now all-pervasive thanks to the democratization of technology. (p. 129)

CONCLUSION

The successful adoption of technology is a matter of survival for libraries. The TS at UCSC have set clear goals to align with institutional priorities, introduced services and strategies to support its vision, manipulated systems and tools with dramatic outcomes and cleared barriers to facilitate technology maximization. We manage all processes in a synchronized way to achieve optimal cost and service performance through cross-training, team culture and extensive documentation. We take baby steps in all directions, making it simple for the staff to embrace technology. Growing IT skill sets within the department is the secret to success.

Technology is here to stay. Frey (2010) and Lynn, FitzSimmons, and Robinson (2011) all support the view that users will have even greater needs for libraries in the future, but the roles of librarians have to change. We should expect more changes in Technical Services staffing, training and prioritization of workloads as libraries continue to turn their focus increasingly to electronic and digital resources. What are the new skills needed by technical services staff in the next few years? How can libraries provide further training for their staff members? How do managers develop necessary mindsets to lead changes as libraries move in new directions? The Australian Finance

Minister, Tanner (as cited in Ricketts, 2008) states that "We need to recognize that not only must we adapt to a world moving online, but it will likely have to do so at an ever increasing pace. As a huge creator and manager of information … we have little choice." These remarks also apply to library technical services in this country as well, lest we become the new human barriers for the next generation of innovators. Matthews' (2012) statement that "Facing the future, we don't just need change, we need breakthrough, paradigm-shifting, transformative, disruptive ideas" is intended to "inspire transformative thinking using insight into startup culture and innovation methodologies … to stir the entrepreneurial spirit in library leaders at every level." As libraries are immersed in these thought-provoking discussions, technical services leaders should be active participants and partners. During the entire process of transformation, technology is definitely a major solution component.

REFERENCES

American Library Association. (2012). Academic libraries. *State of America's Libraries Report 2012*. Retrieved April 21, 2012, from http://www.ala.org/news/mediapresscenter/americaslibraries/soal2012/academic-libraries

Anderson, R. (2011, November 16). *It is about us, or is it about them? Libraries and collections in a patron-driven world*. Retrieved April 21, 2012 from http://www.alaeditions.org/blog/categories/acquisitions-and-collection-development

Breeding, M. (2010). *Next-gen library catalogs*. New York, NY: Neal-Schuman Publishers.

Cataloging, U. C., & the Metadata Common Interest Group. (2009). Change in cataloging policy for government document monographs. Retrieved April 21, 2012, from http://libraries.universityofcalifornia.edu/hots/camcig/GovDoc-MonoCatChangePolicy.pdf

CDL Shared Collections and Services sets up Acquisitions at UC San Diego. (1998, June 10). *CDLINFO* News. Retrieved April 21, 2012, from http://www.cdlib.org/cdlinfo/1998/06/10/cdl-shared-collections-and-services-sets-up-acquisitions-at-uc-san-diego/

Coyle, K. (2006). Identifiers: Unique, persistent, global. *Journal of Academic Librarianship, 32*(4), 428–431. doi:10.1016/j.acalib.2006.04.004.

Coyle, K. (2012, January 11). *Bibliographic framework: RDF and linked data.* Retrieved April 21, 2012 from http://kcoyle.blogspot.com/search/label/linked%20data

Cundiff, M., & Trail, N. (2007, June 25). *Using METS and MODS to create XML standards-based digital library applications.* Retrieved April 21, 2012, from http://www.loc.gov/standards/mods/presentations/mets-mods-morgan-ala07/

Dublin Core Metatdata Initiative. (2011). *User guide.* Retrieved April 17, 2012, from http://wiki.dublincore.org/index.php/User_Guide

Dunlap, S. (2012, March 25). *The philosophy of technical services.* Retrieved March 25, 2012, from http://www.linkedin.com/groups/Philosophy-Technical-Services-3265114.S.102889148

Fessler, V. (2007). The future of technical services (it's not the Technical Services it was). *Library leadership & management, 21*(3), 139-144, 155.

French, P. S., Culbertson, R., & Hsiung, L.-Y. (2002). One for nine: The shared cataloging program of the California Digital Library. *Serials Review, 28*(1), 4–12. doi:10.1016/S0098-7913(01)00169-1.

Frey, T. (2010). *The future of libraries: Beginning the great transformation.* Louisville, CO: DaVinci Institute. Retrieved April 16, 2012 from http://www.davinciinstitute.com/papers/the-future-of-libraries/

Gilbert, S. W. (2002, February, 12). *The beauty of low threshold applications.* Retrieved April 17, 2012, from http://campustechnology.com/articles/2002/02/the-beauty-of-low-threshold-applications.aspx

Greene, M. A., & Meissner, D. E. (2005). More product, less process: Revamping traditional archival processing. *The American Archivist, 68,* 208–263.

Huber, J. (2011). *Lean library management: Eleven strategies for reducing costs and improving services.* New York, NY: Neal-Schuman.

International Records Management Trust & International Council on Archives. (1999). *Automating records services.* Retrieved April 22, 2012, from http://www.irmt.org/documents/educ_training/public_sector_rec/IRMT_automating_rec_serv.doc

Lang, B., Gerz, M., Meyer, O., & Sim, D. (2008). *An enterprise architecture for the delivery of a modular interoperability solution.* Retrieved April 21, 2012, from http://ftp.rta.nato.int/public//PubFullText/RTO/MP%5CRTO-MP-IST-101///MP-IST-101-08.doc

Leonard, E. (2011). *Lost in translation: The emerging technology librarian and the new technology.* Retrieved April 21, 2012, from http://www.slideshare.net/eleonard/lost-in-translation-8440125

Library of Congress. (2011, October 31). *A bibliographic framework for the digital age.* Retrieved April 21, 2012, from http://www.loc.gov/marc/transition/news/framework-103111.html

Lynch, C. A. (1991). Evolution in action: The demise of the integrated library system and the rise of networked information resources. *Library Software Review, 10*(5), 336–337.

Lynn, V. A., FitzSimmons, M., & Robinson, C. K. (2011). Special report: Symposium on transoformational change in health sciences libraries: Space, collections, and roles. *Journal of Medical Library Association, 99*(1), 82-87. Retrieved April 16, 2012, from http://www.ncbi.nlm.nih.gov/pmc/articles/PMC3016656/

Mann, L., & Chan, J. (Eds.). (2011). *Creativity and innovation in business and beyond: Social science perspectives and policy implications.* New York, NY: Taylor and Francis.

Matthews, B. (2012). *Think like a startup: A white paper to inspire library entrepreneurialism.* Retrieved April 21, 2012, from http://vtechworks.lib.vt.edu/handle/10919/18649

Murray, P. (2008, June 20). *Riding the waves of content and change.* Retrieved April 16, 2012, from http://dltj.org/article/riding-the-waves/

Princeton University Cataloging. (2004, February 24). *Single-record approach for e-journals –Princeton practice, mid-1999+.* Retrieved April 21, 2012, from http://library.princeton.edu/departments/tsd/katmandu/electronic/single.html

Provider-Neutral E-Monograph Record Task Group. (2009, July 30). *Provider-neutral e-monograph record task group report.* Retrieved April 21, 2012, from http://www.loc.gov/catdir/pcc/bibco/PN-Final-Report.pdf

Rathemacher, A. J., Cerbo, M. A., II, & Li, Y. (2011). New England technical services librarians Spring 2011 Conference: 2020 vision: A new decade for technical services. *Technical Services Department Faculty Publications. Paper 42.* Retrieved April 16, 2012, from http://digitalcommons.uri.edu/lib_ts_pubs/42

Ricketts, P. (2008, November 5). *Australian federal government and Web 2.0.* Retrieved April 21, 2012, from http://oracle-gtmi-anz.blogspot.com/search?q=little+choice

Santa Cruz, U. C. University Library. (2012). *Technical services mission statement.* Retrieved April 21, 2012, from http://library.ucsc.edu/content/technical-services-mission-statement

Silvis, G. A. (2012). *The impending demise of the local OPAC.* Retrieved April 21, 2012, from www.wils.wisc.edu/events/opac06/impending_demise.pdf

Smith, S. (2011). The technical services-public services connection: Tips for managing change. *American Assoication of Law Libraries Spectrum, 16*(2). Retrieved April 16, 2012, from http://www.aallnet.org/main-menu/Publications/spectrum/Vol-16/No-2/tech.html

Stephens, J., & Presley, R. (1996). *EDI: Slow walk to fast forward.* Retrieved April 21, 2012, from http://www.ala.org/acrl/sites/ala.org.acrl/files/content/conferences/pdf/stephens99.pdf

Swierczek, J. (2010). Using Web 2.0 applications in technical services: An ALCTS webcast. Retrieved April 16, 2012, from http://www.ala.org/alcts/confevents/upcoming/webinar/092910web

Thomsen, E. (2008). Welcome to the NOBLE SwapShop. Retrieved April 20, 2012, from http://www.noblenet.org/swapshop/

UC Council of University Librarians. (2012). *Next-generation technical services (NGTS).* Retrieved April 21, 2012, from http://libraries.universityof-california.edu/about/uls/ngts/index.html

Verzosa, F. A. (2010). *The changing library environment of technical services.* Retrieved April 16, 2012, from http://www.slideshare.net/verzosaf/the-changing-library-environment-of-technical-services

Workshop on "Fuzzy is Scalable: Managing Huge Databases Using Fuzzy Methods" at the International Conference on Soft Computing as Transdisciplinary Science and Technology CSTST08. (2008). *Summary.* Retrieved April 21, 2012, from http://www.lirmm.fr/~laurent/FiS/

Wu, A., & Mitchell, A. M. (2010, July 1). Mass management of e-book catalog records: Approaches, challenges, and solutions. *Library Resources & Technical Services*, *54*(3), 164–174.

ADDITIONAL READING

American Library Association. (2010). *Academic library trends and statistics*. Chicago, IL: American Library Association.

Bazinjian, R., & Mugridge, R. (2006). *Teams in library technical services*. Lanham, MD: Scarecrow Press.

Calhoun, K. (2003). Technology, productivity and change in library technical services. *Library Collections, Acquisitions & Technical Services*, *27*(3), 281–289. doi:10.1016/S1464-9055(03)00068-X.

Canepi, K. (2007). Work analysis in library technical services. *Technical Services Quarterly*, *25*(2), 19–30. doi:10.1300/J124v25n02_02.

Carter, R. C., & Smith, L. (Eds.). (1995). *Technical services management, 1965-1990: A quarter century of change and a look to the future: Festschrift for Kathryn Luther Henderson (Haworth Series in Cataloging & Classification.)*. London: Routledge.

Gibbons, S. (2010). *Time horizon 2020: Library renaissance*. Retrieved April 16, 2012, from http://hdl.handle.net/1802/10051

Greever, K. E., & Andreadis, D. K. (2006). Technical services work redesign across two college libraries. *Technical Services Quarterly*, *24*(2), 45–54. doi:10.1300/J124v24n02_05.

Medeiros, N. (2005). Factors influencing competency perceptions and expectations of technical services administrators. *Library Resources & Technical Services*, *49*(3), 167–174.

Medeiros, N. (2011, February 15). Transformation: Next generation technical services at the University of California Libraries. *OCLC Systems & Services*, *27*(1), 6–9. doi:10.1108/10650751111106500.

Mitchell, M. (Ed.). (2007). *Library workflow redesign: Six case studies*. Washington, DC: Council on Library and Information Resources. Retrieved April 16, 2012, from http://www.clir.org/pubs/reports/pub139/sum139.html

RT. Request Tracker. (2012). *Somerville, Massachusetts: Best practical solutions*. Retrieved April 21, 2012, from http://bestpractical.com/rt/

Senge, P. M. (2006). *The fifth discipline: The art and practice of the learning organization*. New York, NY: Doubleday/Currency.

Stalbberg, E., & Cronin, C. (2011). Assessing the cost and value of bibliographic control. *Library Resources & Technical Services*, *55*(3), 124–137.

Williams, D. E. (2001). Developing libraries as nimble organizations. *Technical Services Quarterly*, *18*(4), 35–46. doi:10.1300/J124v18n04_04.

Wiser, J. (2005). Kaizen meets Dewey: Applying the principles of the Toyota Way in your library. *Information Outlook*, *9*(6), 27–37.

Zhu, L. (2011). Use of teams in technical services in academic libraries. *Library Collections, Acquisitions, and Technical Services*, *35*(2-3), 69-82. Retrieved April 21, 2012, from http://dx.doi.org/10.1016/j.lcats.2011.03.013

Chapter 10
Empires of the Future:
Libraries, Technology, and the Academic Environment

Denise A. Garofalo
Mount Saint Mary College, USA

ABSTRACT

Exploring technology and academic libraries concerns more than just machines, functions or processes; the human factor is as important as the equipment. Implementing successful technology changes requires attention to the people involved, and academia is no exception. Technology can be divided into either disruptive or sustaining technologies, and these technological changes impact students, faculty, and staff. In higher education, technology changes are shifting knowledge transfer to a more participatory environment and a more synergistic experience. The academic library is in a transitive state of change, evolving from a warehouse of things to a collaborative learning destination for resources. Both the library and the academic environment must adapt to survive. Overcoming the challenge of changes to the delivery of instruction may lead to extensive restructuring of courses and curriculum. The academic library can serve as a collaborative partner with faculty, leading by example to incorporate technological changes.

INTRODUCTION

The empires of the future are the empires of the mind.
—*Winston Churchill, September 5, 1943, Harvard University, Cambridge, Massachusetts*

Technology is more than just machines or functions or processes; the human factor is as important as the equipment (Rizzuto & Reeves, 2007).The prevalence of technology in teaching and learning

and the changes brought through technological innovations in education may seem at odds with any observation about reluctance to embrace technology, but as Clark (1983) observed, although academia has been a source of social change it continues to remain entrenched in tradition, and reactions to technology and change can fall into this traditional perception. Comfort with technology is not universal, and technology is impacted by "an organization's culture and people [that] can significantly impact the success or failure of technology implementation" (Averett, 2001, p. 34).

DOI: 10.4018/978-1-4666-3938-6.ch010

Academic Environment Changing

The college and university environment is in the midst of changing from the comfortable standard of the preceding decades to a brave new world of learning. The paradigm shift is here--the model of the lecturer sitting at a desk or standing in front of the blackboard, imparting his knowledge to the quiet, listening students, is changing into a more synergistic experience (Adelsberger, Collis, & Pawlowski, 2008, p. 253), where the instructor may be in a classroom with interactive whiteboards and response systems, or in a room all alone with students "attending" from around the world, or a mix of both. These technological changes in the higher education environment have the promise to disrupt as well as to transform the learning milieu, impacting the delivery, location, and form of learning and requiring the development of new literacies.

These changes in learning are not always embraced by all (Bower, 2001; McBride, 2010), and human barriers to technology and the resulting resistance to change can have significant, detrimental impact on efforts to integrate new technologies in learning. Knowledge about learning technology innovation is not extending across academia as fast as expected (Hannafin & Kim, 2003), resulting in a failure of teaching practices keeping up with student expectations (Oblinger, 2005). Resourceful use of technology in the classroom occurs only after any resistance to change from faculty, staff, administrators, and students is surmounted and technology is integrated and incorporated effectively into the delivery of learning, and can support "the philosophy of integrative learning, a view that encompasses student integration of the various disciplines into a way of understanding the world and ideas rather than simply acquiring pieces of knowledge through a college education" (Hinchliffe & Wong, 2010, p. 215).

Academic Library and Response to Technology Change

The academic library, too, can be seen as being in a transitive state of change. The traditional but outmoded service model, where the library serves solely as the solitary gateway to knowledge and information for scholars, students, and faculty, is no longer applicable today. "Library space has moved from an emphasis on storage to an emphasis on creating learning environments" (Stoffle & Cuillier, 2011, p. 147). Just as technology plays a role in changes to access in information and knowledge, the library is transforming, assisting in the effort to "prepare students for a society in which information took center stage" (Owusu-Ansah, 2004, p. 4). Library users no longer are required to visit the library to access information and resources; they can bypass the library and access databases of journal articles and e-books and streaming videos from any location where they can connect to the Internet.

One means to remain relevant and viable is for libraries to redefine how they advance learning and scholarly activities and promote how they are still essential and support their institutions' missions and visions. Stoffle and Cuillier note, "To thrive, libraries will need to use sound business-management practices, align themselves with campus learning and research goals, [and] nimbly apply new technologies" (2011, p. 130). Technology will certainly remain a constant force for change in libraries as they grapple with how to adapt to survive and then thrive. Online services continue to proliferate, as libraries constantly identify the needs of our users and revise how these needs can be met. Rethinking the role of the library as a service provider can assist in processing these revisions. "Library as service focuses on the customer. It results in everywhere-you-are access—pushing the library into the research and learning environment at the desktop, in the lab, and in the classroom" (Stoffle & Cuillier, 2011, p. 146).

Signs of such service adjustments include the implementation of the various flavors of virtual reference services (IM, chat, Skype) and the installation of self-checkout systems to address users' needs for speedy transactions outside the parameters of a physical building's open hours or a physically staffed service desk.

Technology as Catalyst for Campus Changes

Technology is the overall underpinning structure driving these transformations, and the fact that "college students are avid users of technology" (Junco & Cole-Avent, 2009, p. 4) aid in influencing changes on campus. The best indicator of the changes is the ubiquitous computer, found all across the academic campus, whether in classrooms, labs, libraries, dorm rooms, lounges, or dining halls. With the increased accessibility to computer technology, the academic environment has transformed to allow for the synchronous and asynchronous delivery of instruction, and thus distance education grows, prospers, and thrives. The software tools for distance education, the learning management systems, are also thriving not only for remote learning but also for traditional learning delivery in the classroom as well as for blended delivery where learners attend some sessions remotely and some in traditional classrooms on campus (Osguthorpe & Graham, 2003; Vaughan, 2007).

Web, Resources, Mobile Computing, and Distance Education

The World Wide Web goes hand-in-hand with the computer in transforming the academic environment and this partnership has improved information and communication technology. Students no longer are limited to what resources are available on their own campus, but rather have access to a surplus of knowledge from across their region and around the world (Lin, Chen, & Chang, 2010).

The vast expanse of information now available and the need to evaluate its quality is obvious (Smith, 1997). But the reality is that any barrier that may have existed to access information has been eroded with the omnipresent Web.

The transition to digital information is an obvious change on college campuses. The communication and fostering of knowledge at colleges and universities has had to transform as the academic campus has embraced, to varying degrees, the use of electronic information in achieving this integral vocation. The library is, "becoming a people place, providing the tools to support learning and scholarship and the environment for social interaction" (Duderstadt, 2009, p. 220).

Academic libraries are still grappling with the consequences of embracing digital resources. From e-books to e-journals to e-textbooks, digital information is combining with print resources to provide the means of communicating knowledge to the learner. Databases are a bastion of academic research, with a decreasing number of print-only serials available to scholars. Although many predict that electronic information will supplant print entirely as scholarly communication transforms to an all-digital model, those stating that by some date in the near future we will see the total elimination of print material from the academic milieu may be accepting a stance that is too radical to comfortably adopt right now. Suffice it to say that electronic information will occupy an ever-growing portion of scholarly communication as we drive towards the future (Dempsey, 2006).

Another noticeable change is the prevalence of mobile computing. Students have embraced and incorporated mobile computing devices into all aspects of their daily lives (Johnson, Levine, Smith, & Stone, 2010; Smith, Rainie, & Zickuhr, 2011), and colleges and universities must establish and adopt pedagogic techniques that take advantage of mobile computing. A merging of electronic learning and technology can provide endless opportunities to enhance teaching and learning. Effectively utilizing mobile computing will mean delineating ways to offer students the delivery of learning that are convenient for the student and not location specific. Then add into the mix the competition traditional colleges and universities now have with a for-profit business model, and a critical aspect of survival will be their reaction to the ascendancy of distance education and ever-present mobile computing.

Distance education is impacting different facets of the academic environment. Faculty may perceive a diminished role in the transmission of knowledge to the students as their engagement in classroom teaching is reduced through distance education. Meeting the challenge of delivery of instruction to students outside of a traditional classroom may lead to extensive restructuring of courses and curriculum, and not all faculty may support these necessary adjustments (Bower, 2001).

DISRUPTIVE AND SUSTAINING TECHNOLOGIES

Definitions

Technology bombards the academic environment with changes. These changes have come from both sustaining technologies and disruptive technologies. Christensen described these technologies in 1997, but the concepts are still relevant today. Sustaining technologies are generally considered those technologies that, for most traditional users, replace traditional activities while improving the performance of an established technology (Christensen, 1997, p. xv). An example of a sustaining technology is word processing. Typewriters were an established technology for composing letters, papers, and other documents. Word processing replaced the typewriter for most users, and these users abandoned the typewriter for word processing with few regrets. Other examples of sustaining technologies are color television as a replacement for black and white television, or digital telephones

as a replacement for analog telephones. In the academic realm an interactive white board is a sustaining technology that replaced the previous methods to present learning and information to students in a classroom, via a chalkboard and an overhead projector.

Disruptive technologies are generally defined as technologies that do new things for new users (Christensen, 1997, p. xv). The digital camera is considered a disruptive technology; it displaced film photography in a relatively short time span and encouraged new users to embrace digital photography. Another example of a disruptive technology is desktop publishing, which allows anyone with a computer and software to produce a professional-looking document and distribute such without the use of a traditional publisher. In the academic environment, distance education is a disruptive technology, in that the target student body is remotely located and the pedagogy to impart learning remotely is different than that of the traditional student seated in a classroom. Cloud computing is another disruptive technology, as it removes the need for vast computer processing and storage to be housed and tended on campus and replaces it with access that provides those services offsite.

With sustaining technologies, most people make the transition without too much resistance, primarily because the change is not seen as really a change but rather is an improvement of an existing function. The move towards disruptive technologies is inherently less welcomed. Of course, there are always those who readily embrace the latest and greatest technological innovation, but when a disruptive technology is involved, the majority of users resist rather than accept (Lockwood, 2001; Oakey, 2007).

Simply displacing one technology with another is disruptive but is not an implementation of a disruptive technology. Meyer states that "disruptive technology must interrupt our usual policies, practices, and assumptions" (2010, p. 6). Further, if academia is to advance education and teaching, then further introduction of disruptive technologies will continue, since disruptive technologies have not yet met their potential to transform education and teaching (Carmody, 2009). In regards to online teaching, "the simple act of teaching online improves faculty opinions of online learning" (Meyer, 2010, p. 7), yet people are generally "unwilling to use a disruptive product" (Disruptive technologies, 1996, p. 45). Overcoming this dichotomy will be a challenge to academic institutions in the coming years, especially with the idea bandied about that "tomorrow's best learning environments will be located outside universities" (Hanna & Johnson, 2006, p. 1). The academic campus as we know it must choose to integrate technologies, both sustaining and disruptive, or be faced with the specter of extinction (Scott, 2003, p. 65).

Successful Integration is Key

With the current financial scarcity and burgeoning technology proliferation, the critical challenge will be to successfully integrate sustaining and disruptive technologies into the academic environment while overcoming any resistance to these changes (Marshall, 2010, p. 181). Many academic institutions may not be proficient at planning for technology adoption and may not consider technology critical to the success of the mission of the institution. Technology changes rapidly, but people generally do not. The ever-changing nature of technology can be perceived as hectic and distressing, and planning for such can be seen as time-consuming and too labor-intensive (Legris, Ingham, & Collerette, 2003). Academic institutions will need to anticipate this perception and be aware of resistance whenever a new technology is considered for implementation, and take advantage of eager participants to aid in communicating the benefits of technology changes and thereby reducing overall resistance. Meyer (2010) provides a sound perspective on technology in academia--"It is not computer use

but how computers are used that makes disruptive innovation possible in higher education" (Enabling Disruptive Innovation in Online Learning section, para. 1). Recognizing the necessity to plan, integrate, implement, and diffuse issues with emerging disruptive and sustaining technologies is vital not only to achieve their successful adoption and integration but to the continued viability of academic institutions.

STUDENT RESPONSE SYSTEMS (AKA 'CLICKERS')

Background

Academic institutions should consider integration of technologies that encourage and support interaction and collaboration. Interactive student-response keypad systems (aka clickers, audience response systems, personal response systems, group response systems, class performance systems, student response systems) are one such technology that can maximize student participation and impact student achievement (Mayer et al., 2008). Essentially these student response systems are a combination of hardware and software that allow the instructor to administer questions and conduct polls and surveys to students and immediately display the feedback. The tests or surveys can be integrated into a graphical presentation software (PowerPoint, Impress) or word processing software (Word, Writer) that may include lecture notes as well as multiple choice or true-or-false questions and surveys. Students use a handset (clicker) to reply to the questions, and that information is transmitted, received, collected, and analyzed by the response system software. The answers can be instantly displayed and the data can be stored for later use in grading.

Clickers provide instructors with a quick and efficient means to increase engagement and improve student motivation and active learning, especially in courses with large enrollments. "Active learning is a diverse learning process in which students are engaged in a guided activity instead of being silent spectators and passive learners to whom the education is only the transfer of information by the instructor" (Mahmood, Tariq, & Javed, 2011, p. 193). Students do meaningful learning activities and think about what they are doing while engaged in active learning.

The use of clickers can indicate to instructors when student comprehension of a concept or topic is clear or uncertain, supporting better learning and increasing student satisfaction. The immediate feedback an instructor receives through the use of student response systems aids in ascertaining the best means to revise their teaching. With clickers all students, those that regularly participate in class discussions and those that are usually content to be on the sidelines, join in the interactions and attend to the discussions more closely. When incorporated effectively into the design of a course, clickers support active involvement in the learning process and initial studies have also shown that effective use of student response systems can increase student learning (Mayer et al., 2009).

Increasing the level of student-instructor interaction is a means of engaging learners in the classroom. The use of student response systems can "stimulate student-instructor interaction" (Mayer et al., 2009, p. 56) and foster generative learning via a questioning method of instruction. One key aspect of generative learning is the student's need to control the learning process (Lee, Lim, & Grabowski, 2010, p. 630). The use of these response systems in a classroom supports active processing of the material, concepts, and topics covered in the lecture. In order to successfully understand the content and perform well on the questions, students must better focus on the lecture and accomplish organizing the new material as well as integrating the content into their knowledge-base. This instruction process, where the learner interaction via the response systems assists in understanding why the answer is correct through a discussion and explanation of the think-

ing that led to the correct answer, is an effective instructional questioning method (Campbell & Mayer, 2009, p. 748). The questioning via the student response system assists the cognitive processes during learning, resulting in a meaningful student-centered learning experience.

Clickers in Information Literacy Instruction, an Opportunity for Library Leadership

Academic libraries can support the integration of this pedagogical tool into instruction through incorporating them into information literacy sessions and other emerging learning opportunities (Buhay, Best, & McGuire, 2010). Such use would both provide a useful technological addition to information literacy instruction as well as a demonstration platform for faculty who may be reluctant to integrate response systems into their instruction. As Chan and Knight note,

Librarians typically welcome pedagogical enhancements that address troubling issues in library user education: faulty assumptions as to students' base level of understanding, the difficulty of creating an active learning environment in a one-shot class, and overcoming perceptions that library instruction is dull. (2010, p. 194)

These negative perceptions are difficult to overcome in a single instruction session, but student response systems may help modify these perceptions. Hudson, McGowan, and Smith state that these response systems can be, "effective in arousing learner attention, demonstrating relevance to student learning goals, inspiring confidence in learners' ability to master the material, and creating learner satisfaction with the library instruction experience" (2011, p. 20).

The use of student response systems by librarians for information literacy sessions provides an opportunity to demonstrate this technology to any faculty who may not have experience with it. If academic libraries can demonstrate that student engagement and participation can be increased through the use of student response systems in these single session information literacy sessions, instruction faculty may be encouraged to attempt to integrate these response systems into their regular classroom teaching. Academic libraries can provide leadership in adopting this technology and demonstrating its effectiveness, but this effort will involve learning and incorporating clickers into the regular semester load of instruction sessions. Student response systems typically are fairly simple to implement in these one-shot instruction sessions, and their participatory nature addresses librarians' concern that students perceive these information literacy sessions as either boring or too complex (Chan & Knight, 2010).

Information literacy instruction sessions typically have several goals: improved research skills, critical evaluation of websites, familiarity with citation styles, awareness of discipline-specific journals and databases. Because of the participatory nature of response systems, students in information literacy sessions that utilize clickers tend to be more involved and participate more than those in sessions without clickers. Incorporating student response systems into information literacy instruction enhances these sessions via active learning approaches, making them more interactive (Hoffman & Goodwin, 2006). Using student response systems in information literacy instruction provides support to those active learners who tend to absorb and retain knowledge more efficiently than passive learners. Student response systems can promote academic performance and stimulate student-to-student and student-to-instructor interaction (Boyle & Nicol, 2003). Well-integrated use of student response systems in instruction promotes participatory learning and encourages student engagement, and the use of these systems in information literacy instruction can demonstrate that they are an overall useful pedagogical tool (Collins, Tedford, & Womack, 2008).

Pedagogical Issues with Student Response Systems

Pedagogical issues involved with integrating clickers in the classroom can be a concern. Implementing student response systems in the classroom requires faculty flexibility, responsiveness, and creativity (Blazer, 2010). If faculty have a basic underlying resistance to implementing the use of a technology-based response system in the classroom, and if that resistance is combined with a commonly held belief that technology interferes with or impedes instruction, then these perceptions must be overcome for a successful implementation of classroom response systems to occur (Blazer, 2010).

Response systems can provide a positive contribution to instruction, and acquiring the knowledge of how best to integrate a technology such as student response systems into instruction supports the aspect of the scholarship of teaching as "both ongoing learning about teaching and the demonstration of teaching knowledge" (Kreber & Cranton, 2000, p. 478). As Dangel and Wang note,

...feedback that an instructor receives about student misconceptions and error patterns in reasoning provide a potentially rich source of information about how one might need to restructure readings, lectures, and course activities to address student difficulties. Without the frequent interactions and systematic display of students' responses, many of the patterns of students' misinterpretation, lack of prior knowledge, or incomplete logic would go unnoticed. (2008, p. 101)

Clickers are useful in keeping the students more engaged in learning the material being covered in class (Shaffer & Collura, 2009). Clickers are a part of the emerging participatory culture of contemporary life, a useful teaching tool in developing the digital learning environments that support content mastery by students with increased engagement, interaction, and satisfaction (Trees &

Jackson, 2007). As noted by Deleo, Eichenholtz, and Sosin, these response systems are

...a pedagogical tool that furnishes on the spot information from student responses, enabling the teacher to evaluate the accuracy of the response and have the opportunity to immediately correct misunderstandings and reinforce conceptual knowledge. (2009, p. 438)

Using the student response systems in information literacy instruction sessions allows academic libraries the opportunity to encourage the adoption of these response systems, whereby the successful demonstration and achievement of integrating clickers into the campus classroom provides the academic library with both a supportive and progressive role in aiding the shift in the way the institution's students and instructors interact, hopefully paving the way for more technology-based innovations being incorporated into instruction. "As new technologies impact higher education, librarians must seize the opportunity to investigate and evaluate the most appropriate ones for delivery of information in order to enhance the teaching and learning process" (Collins, Tedford, & Womack, 2008, p. 24).

Impact of Mobile Technology

The ease with which students utilize technology to maintain connections emphasizes how they rely on social and mobile technology to share and contribute questions, problems, experiences, and so on. The speed with which mobile technology has been embraced underscores the importance of integrating mobile technology into the distance education setting, and the new generations of students expect information access anytime, anywhere (Ally, 2007; Gibbons, 2007). Joining the internet and portable computing devices, the mobile environment is simply another platform where learning can occur. The term 'mlearning' is short for mobile learning, described by John

Traxler as "any educational provision where the sole or dominant technologies are handheld or palmtop devices" (2005, p. 262). Allowing for the provision of learning through a medium effortlessly used by the Internet Generation, those born from the 1990s through today, presents academic institutions with the chance to engage these students via the devices they choose to be tethered to daily. Mobile devices allow users to have the world of information right in their hands. The logical next step for higher education it to utilize this platform for the delivery of learning (Ally, 2009; Shih & Mills, 2007).

However, Koszalka and Ntloedibe-Kuswant (2010) caution that, "Although there is merit in starting the effort, planning and making careful research or evaluation methodology choices are critical to obtaining results that clarify the relationships among teaching, learning, and technology in m-learning situations" (p. 153). Yet connecting with these Internet Generation students through m-learning is another means to encourage participation in a student-centered learning environment. M-learning is distance education delivered through the ever-present mobile device, and where m-learning is available it supports a perception that the academic institution is prepared to provide knowledge distribution and learning via innovative means that are portable, personalized, and motivate the students (Traxler, 2005, p. 262-264).

DISTANCE EDUCATION

Background

Distance education is another area of instruction where a disruptive technology has made an impact, although faculty may be resistant to its integration (Selim, 2007, p. 397). Seemingly at odds with the perceived moribund tradition of the lecturer standing or sitting in front of a classroom of students, distance education physically separates the deliverer of knowledge, the instructor, from the learners but connects them through the use of technology, integrating student-centered instruction with an emphasis on interactive communication, dialogue, and reflective thought. For students seeking alternatives to conventional face-to-face classes, distance education provides the convenience of asynchronous delivery of educational content; since students may never have to physically visit the campus "online learning has thus rapidly become a popular method for traditional and non-traditional students" (Schrum & Hong, 2002, p. 57). The provision of distance education classes allows traditional higher education institutions a channel to supply courses and programs to compete in this changing education delivery setting. The colleges and universities that adapt to student needs through offering distance education suggest they have chosen to change to "serve the needs of customers…by providing them with the products, services" they require (Christensen, 1997, p. 101). These institutions hope to generate additional revenue streams, retain students, reach out and attract new students, and improve the learning environment for students.

Some may view distance education as a panacea for higher education in the 21st century, but perhaps it's more an evolution in the delivery of knowledge. Distance education relies heavily on technologies to deliver the courses to the learners (Beldarrain, 2006, p. 139). The Internet and computers provide the connectivity and the interface, but they are only one facet of the various technology components that support distance learning and provide the means for interaction and content distribution. Students seek more than just convenience and gadgets; they are looking for interactivity as Palloff and Pratt (2001) noted, "the more typical online student is seeking an active approach to learning….[T]he online student seeks to engage with faculty in a more collaborative learning partnership" (p. 2).

Educational tools such as course management systems and virtual learning environments provide essential components for successful distance education course, such as discussion forum areas, study materials, and resources such as calendars, wikis, readings, videos, podcasts, RSS feeds, and web sites, all vital enhancements of the online learning experience (Beldarrain, 2006). Distance education can provide a valuable education experience as long as the college or university has addressed the need for training and support for the faculty teaching these courses (Palloff & Pratt, 2001). These technologies are not difficult to implement; the challenge to a successful distance education program is in getting instructors to use them and to use them effectively, and Selim (2007) notes that the "most critical indicators were instructor's attitude towards interactive learning and teaching via e-learning technologies" (p. 409).

Distance education course content is not solely from the instructor but contains student-generated content. Students are encouraged to collaborate and share with classmates (Belarrain, 2006). Distance education employs a learner-centered approach to instruction, learning that is delivered in such a way that the students are encouraged to participate, think, read, and discuss using the technologies that allow virtual learning to occur but also using these tools and technologies to enhance their understanding of course content and evaluate their knowledge. Collaboration among distance education students creates an "effective learning environment" (Agostinho, 1997, p. 2). Instructors become more of a guide through the concepts rather than the font which distributes the knowledge, assisting those who are unsure, assessing the learners, and evaluating their course experience and effort.

Impediments to Distance Education Integration

Supporting a cultural shift to distance education within higher education takes resources. Academic institutions need to build awareness that distance education is not a fad but is an integral part of a balanced course offering palate in the 21st century. Faculty and staff will need assistance in providing course content through distance education technologies. "IT [Information Technology] is the engine that drives the e-learning revolution. The efficient and effective use of IT in delivering e-learning based components of a course is of critical importance to the success and student acceptance of e-learning" (Selim, 2007, p. 399). The collaboration and connectedness that functions at the heart of distance education can aid in shifting the academic environment to this new learning-delivery model. Coaching, modeling, and other interventions should be made available to ease the transition to the student-centered, demand-driven personalized learning environment that distance education can provide.

There may be a reluctance to embrace distance education due to the greater demands placed on faculty, and "program development costs, concerns for faculty workload, and lack of faculty rewards are significant barriers that prevent institutions from offering distance education" (Chen, 2009, p. 337). The time requirements for the constant monitoring, communication, and interactivity of teaching an online course are additional time challenges. Perceptions and tenure procedures may need adjustment in light of distance education needs, but the bottom line requires that in order for distance education to be successful any resistance must be surmounted. It is essential that faculty new to teaching online courses receive pedagogical and technical training and assistance to become familiar with using these tools and technologies. Hand-in-hand with training is effective time management, because faculty will need time for developing, managing, and teaching online courses as well as for learning the technologies required for distance education (Dempsey, Fisher, Wright, & Anderton, 2008).

Role for QR Codes

As emphasis shifts to mobile access, another relatively recent innovation related to the mobile device proliferation is the QR code. The use of QR or quick response (*OED,* 2011) codes or mobile tagging allows companies and organizations to deliver information to mobile users who photograph the code and retrieve whatever content the organization wants mobile users to know about the product. QR codes can direct students to services, such as research tutorials or to information related to a research project. "Academic libraries are poised to benefit from the momentum created by the uptake of QR codes in the corporate world and popular culture" (Ashford, 2010, p. 530). Libraries can place QR codes in the stacks to direct students to online content on similar subjects in the libraries' research databases, or to the libraries' web forms to request interlibrary loan materials, or to access help via chat or text. Students seeking an open study room might appreciate a QR code posted by the room that leads to an online system showing the room's availability. The potential of using QR codes is vast, and because QR codes provide instant access to content and they are easily portable, they are an easy and low-threat opportunity to provide services and information to the mobile device user.

Ally states that "Mobile learning has growing significance and visibility in higher education" (Ally, 2009, p. 11). QR codes seamlessly integrate into the mobile information interaction framework, and m-learning has great potential for higher education. Integrating m-learning is simply allowing another platform where knowledge transfer can occur; the result is still the provision of distance education. "Convenient availability of information and resources are strong bargaining chips in championing M-Learning" (McConatha, Praul, & Lynch, 2008, p. 20).

Libraries and Distance Education

There are a variety of ways academic libraries can become involved with distance education efforts at their institutions. Students in distance education courses are diverse, and their experience with library and digital resources is just as varied, and "the greatest inequality concerned the need for library orientation for distance learning" (Dempsey, Fisher, Wright, & Anderton, 2008, p. 633). Academic libraries need to be a keystone for the distance education or virtual classroom as much as they are for the physical classroom, and keep in mind the "integration of resources and services that have facilitated learning for distance learners" (Bower & Mee, 2010, p. 477). The nature of online learning means that academic libraries can provide a more user-centered experience where the information literacy instruction, the digital resources, and the library services are relevant to each student. One means to achieve such is through an embedded librarian program.

Embedded Librarians as a Distance Education Scaffold

The phrase "embedded librarian" evolves from the Pentagon's use of a similar phrase, embedded reporters, "journalists assigned to specific military units during armed conflicts" (Chandler & Munday, 2011). In the case of distance education, an embedded librarian is a librarian assigned to a specific user environment, in online courses where they provide information literacy instruction and research support throughout the duration of the course rather than for just for a single session. The use of embedded librarians is one means for libraries to revamp their core services in response to the changes wrought by technology to the academic environment, to develop "a virtual collaboration … interacting with dispersed students in a computer-based distance learning environment" (Shumaker, 2009, p. 240). In addition, an embedded librarian program provides library services and resources to the distance education students that correspond to the services traditional classroom students receive but which are of longer duration and are more personalized. Embedded librarians provide essential support to online and distance education programs through easy, albeit virtual, access to reference and research support and information literacy instruction (Shumaker & Makins, 2012).

Indeed, embedded librarians can provide more than the traditional "one-shot" information literacy instruction sessions that are focused on teaching-centered instruction and offer little continuity; embedded librarians are an integral support for a learning-centered experience that emphasizes the learning process and learning outcomes and enhances student learning. "Embedding librarians brings the learning process in closer proximity to the scholarship on which the disciplines are based and to those that service it–librarians" (Dewey, 2004, p. 8).

Embedded librarians are accessible to the students during the course sessions, and they can offer online tutorials and provide research assistance at the point-of-need throughout the entirety of the online course. Students can more easily develop a personal relationship with their embedded librarian, who can mentor the students through their research and aid in contextualizing and applying the knowledge they gain through the research activities. Embedded librarians can become a meaningful instructional partner to the distance education faculty, further demonstrating the value the academic library can bring to the distance education slice of the higher education venue (Hoffman, 2011).

Challenges and Benefits of Embedded Librarians

Developing an embedded librarian service model may be viewed as a service that is both too broad and requires a great investment in time from the librarian. "Defining an embedded librarian's role in an online course is challenging" (Hoffman, 2011, p. 446). Collaborating with faculty to create the content and then investing the time to take an active part in a course are two labor-intensive aspects of an embedded librarian program that may appear too challenging for libraries to surmount. However, with higher education embracing distance education and academic libraries needing to demonstrate they provide services that are essential to all students, the value of a librarian's presence in a distance education program is intrinsic, and provides students with more opportunities to seek assistance from 'their' embedded librarian. In addition, an embedded librarian brings a sense of personal engagement to the students, which can facilitate inquiry, individualize learning, and increase student involvement (Shumaker, 2009).

As librarians and faculty know, students are not always aware of the resources available through the academic library. Hoffman noted the "overwhelmingly positive response of the distance students to the embedded librarians' presence" and that "embedded librarian services may prove valuable not only for students who actively interact with the librarian, but also for the rest of the class." (2011, p. 453). A librarian embedded into a distance education course can be an invaluable support service, assisting the students in navigating databases and in searching, evaluating, and citing the best sources for papers and projects. "Embedding as an effective mode of collaboration," (Dewey, 2004, p. 6) and embedded librarians support engaged and meaningful learning because they are a fundamental part of the distance education courses in which they participate, a contributing member of the virtual interaction learning space that is the distance education course, a co-teacher of sorts. Their activities and responses sustain collaborative learning, and they contribute to the team effort to create a purposeful education experience. Embedded librarians are of necessity knowledgeable of the course requirements and assignments, and they can provide proactive assistance with resources and ascertain points of

need regarding assignment snags to one, many, or all the students in the course (Hoffman, 2011).

Other Library Support Services for Distance Education

The evolving nature of learning on the academic campus and technology's impact creates an opportunity for the academic librarian to forge into new territory and create different instruction opportunities. Besides implementing an embedded librarian program, there are other initiatives academic libraries can develop to provide learning support to distance education programs. Librarians can provide Web tutorials on using digital resources, so that the distance education student can obtain the information on not only how to access these resources but on how to use them effectively and interact with them meaningfully (Viggiano, 2004). As the intermediary between the virtual learner and the electronic resources, libraries have an essential role to integrate the technology and the learning, providing the learner with the skills to efficiently navigate the resources and successfully interpret the results (Brumfield, 2008). Web tutorials on topics such as composing a useful search string, evaluating the search results, and revising the search string when necessary insert the librarian into the role of virtual facilitator in the distance education students' active learning. Su and Kuo, in their discussion of PRIMO (peer-reviewed information literacy tutorial database), highlight the essential aspects for web-based tutorials, aspects that are essential for tutorials aimed at distance education students: clear objectives, a sense of 'community,' and reference support (2010). These aspects are necessary components for any tutorial intended for use in conjunction with a distance education course.

Tutorial Collaboration

Collaboration with faculty on tutorial and resource development and deployment for distance education programs is another means to showcase the academic library's capabilities to support virtual learning and its inherent changes. Such a collaborative effort can result in the development of flexible learning resources targeted to course content and integrated into the course, learning resources designed to address specific research techniques or course information needs (Arp, Woodard, Lindstrom, & Shonrock, 2006). By building engaging and accessible content-relevant tutorials that are fully integrated into the distance education course, students have the opportunity to obtain necessary information handling skills through step-by-step assistance. Courses can be enhanced through the incorporation of focused, content-relative, collaboratively developed tutorials, because these instructional resources uniformly deliver the relevant content. A simple, but effective collaboration endeavor can focus on reading lists. The library can work with the faculty to make library resources available in digital format, provide links from the course unit reading lists to the digital resources, and then integrate the reading list into each unit. These reading lists can make a significant and positive impact with the students and with the faculty.

Library Resource Guides

Another initiative that libraries can present is the development of web-based library resource guides specific to each distance education course, with resources limited to each particular course. Library resource guides can be provided either in house or through a library content management system hosted by the library, the college, or a vendor. These resource guides can connect the learner with content, and will require successful collaboration between librarian and instructor to ensure that the resource guides contain relevant information. Gonzalez and Westbrock observe that, "Creating contextual and remotely accessible research help is crucial to the relevance of the library in distance education" (2010, p. 638).

These guides might contain links to books and other materials housed in the library's collection, to online resources such as databases and scholarly web sites, to library-created documents on citation style or resources evaluation, and more. Such efforts are time-consuming, but they result in a resource that is intrinsic to a specific course and therefore more useful to the students.

Academic libraries can support the learning and research needs of the distance education student, but to do so will take innovative approaches. Libraries may have to modify their structure and reorganize duties to support distance education needs. The learning community is changing, and academic libraries will need to transform services to respond to the changes and succeed.

ACADEMIC LIBRARIES, TECHNOLOGY, AND THE FUTURE

Difficult to Predict Future Technology

The future is difficult to predict, especially when it comes to technology. Predicting which new trends, devices, platforms, innovations, and applications could be in play and affecting the academic environment 5, 10, or 20 years from now may be nigh unto impossible given the current rapid rate of technological change and development. Throw in sustaining and disruptive technology evolution and the forecast grows even murkier. As Staley and Malenfant state, "Because the future is inherently not open to direct experience, there are no facts established and no evidence to measure" (2010, p. 59). Planning on how to manage technology better and integrate new technologies more efficiently seems intractable if not unattainable. Instead, recognizing the need to manage how faculty and staff learn about, interact with, and implement new technologies is too important to leave to chance. There are some foundational concepts to aid in such planning, thoughts that can be applied to effectively handle future technological trends.

Colleges and universities will need to change from the "world of monolithic, teacher-led content delivery, where the key skills are in holding students' attention" (Christensen, Horn, & Johnson, 2011, p. 228-229) and the instructor is "a provider and evaluator of knowledge" to an environment of learner-centered teaching where the instructor is "a facilitator and guide (Ehlers & Schneckenberg, 2010, p. 21). Such change is systematic, and the pace of technological innovation and change must accelerate if colleges and universities are to survive. "Academic institutions also have to adjust themselves and develop strategies to respond rapidly to the changes in technologies" (Ranjan, 2008, p. 443-444). Successful integration of learning innovation technologies into instruction is a key factor required for their continued success; the academic library can assist in this implementation and provide a means to support the integration campus-wide.

Future of Libraries

Those that forecast the demise of the library in general and the academic library in particular are overly reactionary. Libraries, along with higher education institutions in general, will need to change and adapt in order to survive. The academic library continues as an integral part of the academic environment and will remain such in the future. "The quality of an academic institution is measured by the resources for learning on the campus" (Kargbo, 2009, p. 43). But the library's role will change. The library as a warehouse of physical things (books, journals, DVDs, etc.) is evolving into the library as a place, a community location where students and faculty can go to study, collaborate, work, learn, relax, research, and connect. Changes in technology have influenced this transition with access to digital resources, e-books, online databases, and so on. The library should integrate all available technology to support the

library's contribution to the academic campus as a ubiquitous learning environment as well as a refuge for intellectual pursuits, a place where learning is pervasive and new behaviors among learners, instructor, and technologies are expected.

Academic libraries are challenged by the rapid and relentless rate new technologies are developed, introduced, implemented, become prevalent, and then supplanted seemingly overnight. As we stand at what many call the apparent end of the age of print, libraries must redefine their role in the scheme of the information environment or risk being left to molder with the books they used to warehouse. Libraries should be defined not by the building and what it houses but by the services it provides and the content it curates. The twenty-first century should find academic libraries as more of an informal learning space with tools, resources, skills, and knowledge that needs to be formally meshed into the established teaching, learning, and research of a higher education institution. Libraries as a destination, a location for learning, a collaborative space where students and faculty can come together and meet, work, and learn. Libraries need to embrace the core of their existence, making knowledge accessible and available to organizations and communities, and stop allowing the uncertainty of the future to immobilize them into oblivion.

Build Competencies

Librarians and library staff need to develop competencies in emerging technologies. Some formal professional development will be required to ensure success in the understanding and use of these technologies, but this variety of knowledge and skills is required for the future. To ensure that the library maintains links to teaching, learning, and research, staff should develop an understanding of pedagogical methods relevant to distance learning and to collaborative and active learning, along with a familiarity with any widespread consumer technologies that could enhance education. With a higher technology comfort level comes changes in attitude, which aids in building effective integration of technology into library services, transforming technology hesitation into a level of technology confidence. The successful sustainability of digital learning technologies in academic institutions relies on developing competent users of these teaching and learning technologies. Libraries can support this mindshift on campus through the active support and observable implementation of e-learning technologies in library services. The development of competent digital learning technology users will require planning and a strategic effort to create opportunities and incentives for learning, recognizing that different staff and faculty have different learning requirements and that in order to be successful with technology implementation there will need to be a spectrum of technology learning interventions, and a focus on developing an effective learning design support for faculty.

Distance education or virtual e-learning students tend to be 'invisible' and their needs must be kept as a focal point when developing and implementing services to support these learners. Ensuring that academic librarians receive the proper technology training to successfully support distance and e-learning programs is imperative. Although providing curriculum support services such as information literacy instruction is a foundational aspect of academic libraries, transitioning this instruction to different instructional platforms requires attention and planning to core elements, pedagogy, and methodologies. Collaboration among librarians, faculty, distance education and technology specialists in the development and implementation of strategies to integrate the teaching of information literacy and research skills in a virtual, mobile, or hybrid setting will result in a more unified and successful strategy.

One thing to keep in mind when attempting to peer into the future of technology in the academic environment is the transformational changes learning technologies will undergo over time. The scope of possible applications of new technologies seems almost limitless, as current technologies evolve and morph and extend their capabilities and user expectations. The nature of learning means concentrating on what the learner is doing rather than what the teacher is doing. That focus must be kept in mind when trying to ascertain what the learner is doing with technology so that the academic community can continue to support the innovation and change the new technologies require and the students demand. New skills, perceptions, and training to deal with implementing the new technologies can lead to new services as well as to innovations in providing existing services, and these changes may be due to sustaining or disruptive technologies, or a combination of both. The future will see new models for digital learning and enhanced academic learning institutions, with an academic library continuing to serve as a stanchion for the learning and scholarly activities of higher education, an essential support to an academic institution's mission and vision.

Library, Future, and Disruptive Technologies

What might one see at an academic library of the future? Disruptive technologies, no doubt. There will be issues to overcome, but the explosion of mobile computing logically leads to all-pervading wireless and to such breakthroughs as wearable computing. Gesture-based computing is an outgrowth of computer gaming (Barley, 2012). Technology developments will allow computers to interact with faculty and students in richer ways, utilizing environmental conditions to customize the user experience for increased access and ease-of-use. Devices and rooms will respond to voice, motion, and other subtle signals, acting directly without the users' conscious knowledge to set the lighting, start an Internet search, launch an interactive whiteboard, or begin a movie. Ordinary objects will be given the ability to recognize their physical location and respond accordingly. Gesture-based computing will be widespread, allowing devices to be controlled by natural movements rather than via an input device. Gesture-based computing will create an environment where seamless integration of comments and highlighting into presentations and demonstrations allows collaboration to flow, unfettered by the need for physical devices to support the shared effort, yet continuing to enhance the perception of library as space for learning, sharing, and collaborating.

Geo-Social Networking

Social networking's impact on the knowledge web will continue to forge a more personal Web experience. Tools collect personalized content in customizable ways. The focus will continue to move towards community, creativity, and collaboration. Users will configure, administer, and organize their content in various methods that complement the ways they view and use the Internet and Web. Location-based social networking, also called geo-social services, may become more integral to the social networking fabric (Ahn, Shehab, & Squicciarini, 2011). Geo-aware information environments will go beyond the "find-a-friend" apps, creating place-based interactions with students and faculty. Academic libraries may utilize such social networking outlets as a part of the new student orientation to the library and its services, providing context-specific information about specific locations and service points in the library. Academic institutions and libraries will need to address transmedia navigation, where students and faculty move from channel to channel (music to course content to entertainment) and across platforms (TV to Web to Facebook) and media (word, voice, video), and incorporate social reading, where via an ereader or other device passages read can be highlighted and then shared via Twitter. These evolving social networking facets will need to be addressed by higher education and academic libraries in order to remain relevant to students.

Mobile Apps for Libraries

The development of library-specific apps (applications) is an arena where libraries may discover a useful resource-rich niche. Libraries already have apps to provide displays of library web pages, the library catalog, and digital resources that are better tailored for smartphones and other handheld devices, so there is a familiarity with the necessity and usefulness of apps specific for library uses. Any development of such apps should integrate with whatever efforts are going on campus-wide regarding app development so as not to be counterproductive. In addition, app development should be considered a participatory service, involving the students in the development to ensure that not only is the considered app useful but it will be used by the students (Connolly, Cosgrave, & Krkoska, 2011). Libraries have developed library orientation-type apps as well as other basic guidance and navigation apps. Will future app development find apps for bibliography creation, search query formation, resource assessment, and other basic literacy information skills? Only time will tell.

Cloud Computing

Cloud computing has the potential to serve as a disruptive technology in higher education. But in reality it may be more of a sustaining technology, swapping environment-controlled server rooms on campus with access over the Internet to software and storage and servers located remotely. A move to cloud computing can replace a complex data center configuration and (hopefully) free up technology staff to work more with users rather than supporting equipment and servers. There are risks with moving to the cloud, given that personal and institutional data is housed with a third-party. When navigating among the clouds, academic institutions must pay particular attention to information security and risk assessment methodology as well as a data classification scheme to guide the cloud service provider in case of data breaches or other situations. The redundancy and availability characteristic of cloud services make the switch particularly appealing to institutions that may not be able to provide such on their own, but institutions need to demonstrate due diligence when considering and configuring for the switch, especially in the area of data security (Yan, 2010).

Academic libraries are looking to the cloud as a solution and service provider. Switching to a cloud computing model for an integrated library management system, a discovery service, or any other of the myriad library technology service applications is a viable alternative to either maintaining a library server with library staff or relying on the academic IT department to support and maintain the servers for vital library applications. A move to the cloud provides redundancy that most libraries do not have, and can free up resources to apply towards the mission critical library service of getting the information and resources into the hands of the faculty and students as quickly and easily as possible. Academic libraries just need to ensure that cloud computing will maintain or improve library services and not lead to a deterioration in services. Cloud computing is "a new way of managing critical computing resources in an academic library setting" (Yan, 2010, p. 90).

Personalized Learning

The continuing advancements in technology may impede developing a steady and clear-cut view of how to get to the future, but one factor remains apparent, students want a more customized and personalized education experience. Such personalization enhances course content and the entire student experience, sparking more engagement and, hopefully, leading to higher student retention. The idea of developing a more individualized "student-centric" focus for higher education may be less arduous because of technology changes. Customizable web pages and social networking are just two methods of creating an individualized, student-centered aspect to higher education.

With the current focus on student-centered learning and with technological advances happening seemingly daily, the future academic environment will incorporate personalized learning, an educational model that is individualized, flexible, and multifaceted to accommodate the needs, interests, and range of abilities of each student (Andersen, 2011). Personal learning networks (PLNs) encompass the broad aspects of knowledge interactions today—Twitter, friends, family, co-workers, Facebook, college, and e-mail. Personal learning environments (PLEs) are the higher education facet of PLNs, and they focus on student-driven learning, allowing for differences in learning styles to develop an individualized approach to learning. PLEs will be customized for each student, and follow an instructional approach that includes differentiation and individualization to leverage the differences in learning styles, thereby providing a student-centered learning model where the learner can connect to learning communities, educational tools for communication and information management, personal learning networks, expertise and authoritative sources, online tutoring and guided sources, as well as peer interactions (Hricko, 2010). PLEs differ from the more institution-centered, traditional virtual learning environments through integrating formal and informal learning, and supporting student self-regulated learning so that the learner has control of their learning environment.

Each student can modify the PLE to his or her needs, but as Dabbagh and Kristanas note, students may not "possess the knowledge management and the self-regulatory skills to effectively use social media to customize a PLE to provide the learning experience they desire" (2012, p. 7). Implementing PLEs involves a cultural change at odds with the traditional industrial educational model where the instructor controls the learning environment and students must adapt. Instead, with PLEs each student can make differentiations in how they interact with the learning community. Instructors will need to assist students to regulate and control their learning environment so they can become effective self-regulated learners, and instruct students on how to integrate formal and informal learning to achieve an enriched and effective PLE. Academic librarians can assist in this integration, utilizing information literacy sessions, tutorials, and customized research guides as aspects of the PLEs the students each develop and employ.

Future Library Innovations

What will the future bring, besides more technology and innovations? And how will academic campus and library fare in the higher education environment of the future? No one really knows, but there are suppositions that can be made based on the past and the present. Technology will continue to evolve and to change, and failing to accept the inevitable will be disastrous. Dealing with sustaining, disruptive, and intrusive technologies is a fact of life (Ardner, 2002). Learning how to use these technologies efficiently in learning and education will take not only resources and time but patience. Academic libraries and campuses must find a way to navigate and accept these technologies to survive; ignoring them will only make these institutions slowly cease to exist. For those of us involved in the academic environment, we must discover a means to successfully appropriate and integrate present and future technologies in order to flourish. Or, as Godwin notes, "the question then becomes, how might instructors appropriate the force of disruptive technologies to renew their own teaching, and what response might be formulated for both instructors and students to manage the unwelcome intrusion of technology?" (2010, p. 1). Academic libraries need to remain aware of new technologies as they appear over the horizon and be willing to learn and experiment with such. The opportunity to maintain a vigilant view of new

trends and technologies will be advantageous to the library and the academic community at large.

Library: Physical vs. Virtual Destination

The academic library as solely a virtual destination without a physical location will not happen overnight (Bower & Mee, 2010). The transitional situation may consist of a hybrid, where the library has both a virtual face and a physical reality. The academic library needs to change its image from a warehouse of things to a storehouse of learning spaces. Providing adequate and quality learning spaces, whether they are study rooms on campus or augmented reality spaces on the internet, allows for an enriching knowledge-interaction experience that is really the heart of libraries' missions nonetheless. Shifting the focus of the library from a space that houses physical resources to a service space where there are librarians acting as scaffolds to facilitate learning, research, and resource discovery and support scholarship, knowledge gains, and collaboration in this increasingly more virtual world is key to the future. Will the idea of augmented reality, defined as a "real-time direct or indirect view of a physical realworld environment that has been enhanced/augmented by adding virtual computer-generated information" (Carmigniani, 2011, p. 342), become a part of the future academic library?

Augmented Reality

Augmented reality has seen a fair amount of use in the medical field as a means to aid in diagnosis and surgery, but has not been integrated widely in education beyond tour guide-like information. This technology could have viable applications in higher education. With MIT's SixthSense WUW (Wear Ur World) technology in place in an elearning environment, an instructor could have specific information about students available via his interface to the learning management system, providing details about the students without having to access or consult a document or another system (Carmigniani, 2011, p. 362, 365, 369). Libraries are moving towards augmented reality through Second Life® and its Community Virtual Library (formerly the Second Life Library), where librarians have avatars, graphical representations similar to an icon. Typical interactions in the Community Virtual Library are reference encounters, where questions involve directional issues, learning how to navigate through the Second Life® virtual world, or more library-related inquires about special collections. Tang's article presents basic background information on the library's place in Second Life®, and provides a glimpse into how librarians may interact with users in an exclusively virtual library of the future (2010, p. 513, 516-519).

There is technology in use today with museums that contain aspects of augmented reality (Dede, 2005, p. 9), and the concepts can easily translate to the library world. Academic libraries may use augmented reality to present on online experience relative to browsing the available digital resources or electronic databases, thereby merely reusing content and presenting it in a manner and format more conducive to mobile technologies. Creating an orientation guide to the library and its resources through the use of augmented reality may be another way to utilize this technology in the future, allowing for a customized tour of the library's resources and content, with various aspects of more interest to a particular user highlighted to reflect relevance, i.e. business majors might be made aware that the library has full-text access to the *Wall Street Journal* online while English literature scholars would be led to resources in JSTOR (Forsyth, 2011).

Adjust to Technology for Success

Taking advantage of new technology trends will be essential in the future, necessitating a strategy to handle evolving learner expectations and the breakdown of the distinction between traditional on-campus learning and elearning. Adjusting to these transformations is an important first step; today students "access knowledge and build social communities through 3-D virtual reality environments such as Second Life, the World of Warcraft, and Croquet in which all of the senses are faithfully replicated to enable human interaction at a distance" (Duderstadt, 2009, p. 222). Learning to view these changes as opportunities to utilize rather than challenges to overcome will be key to the continued functioning and revitalization of the academic library. Perceiving the academic library as an extension of the classroom can assist in achieving such change (Freeman, 2005). Positive and effective integration of technologies into the pedagogical framework can aid in developing thriving and successful academic libraries that endeavor to achieve effective technological integration with all future campus endeavors.

CONCLUSION

Academic institutions must address the impact technological change can have on faculty teaching style, productivity, and research (DeMillo, 2011). The effect of technological change on the academic library's role to aid in accessing, distributing, and disseminating information in support of the faculty's information needs is wide-ranging and still changing (Duderstadt, 2009). But the academic library must continue to participate in discussions and planning on how to traverse the technological landscape and navigate a course that stays true to its mission while just changing the tools and means to reach the future. "New directions for supporting research and study can be taken using the library and its librarian's experience, skills and knowledge" (Barner, 2011).

Libraries are transforming into information commons, technology-rich collaborative and interactive learning spaces, yet students still desire quiet and solitary study space free from distraction, noise, and commotion. Finding the correct balance between collaborative and solitary spaces will be a challenge that goes hand-in-hand with flexibly incorporating technological changes into the academic library of the future (Sullivan, 2010). Libraries can be positioned strategically to take advantage of any sudden change in the technology playing field and incorporate new tools and technologies into useful library services appropriate to the academic environment and its users, keeping in mind that colleges and universities are just one supplier among many for learning, ideas, and knowledge. The opportunity to embrace new channels for delivery of learning can make the difference between survival and success for the academic community. Colleges and universities can reassert themselves as the principal source for learning, and academic libraries can support this effort through a commitment to continue to provide support to the academic community through the demonstrated implementation of technologies in the library's learning and teaching service provision.

REFERENCES

Adelsberger, H. H., Collis, B., & Pawlowski, J. M. (2008). *Handbook on information technologies for education and training*. Berlin: Springer. doi:10.1007/978-3-540-74155-8.

Adner, R. (2002). When are technologies disruptive? A demand-based view of the emergence of competition. *Strategic Management Journal*, *23*(8), 667–688. doi:10.1002/smj.246.

Agostinho, S., Lefoe, G., & Hedberg, J. (1997). *Online collaboration for effective learning: A case study of a postgraduate university course (New South Wales, Southern Cross University)*. Retrieved from http://ausweb.scu.edu.au/proceedings/agostinho/paper.html

Ahn, G., Shehab, M., & Squicciarini, A. (2011). Security and privacy in social networks. *IEEE Internet Computing*, *15*(3), 10–12. doi:10.1109/MIC.2011.66.

Ally, M. (2007). Guest editorial: mobile learning. *International Review of Research in Open and Distance Learning*, *8*(2). Retrieved from http://www.irrodl.org/index.php/irrodl/article/viewArticle/451/918/.

Ally, M. (2009). *Mobile learning: Transforming the delivery of education and training*. Edmonton: AU Press.

Andersen, M. H. (2011). The world is my school: Welcome to the era of personalized learning. *The Futurist*, *45*(1), 12–17.

Arp, L., Woodard, B. S., Lindstrom, J., & Shonrock, D. D. (2006). Faculty-librarian collaboration to achieve integration of information literacy. *Reference and User Services Quarterly*, *46*(1), 18–23.

Ashford, R. (2010). QR codes and academic libraries: Reaching mobile users. *College & Research Libraries News*, *71*(10), 526–530.

Averett, P. (2001). People: the human side of systems technology. *Journal for Quality and Participation*, *24*(2), 34–37.

Barley, S. (2010). The next multicoloured wave of computer interaction. *New Scientist*, *206*(2763), 18–19. doi:10.1016/S0262-4079(10)61366-8.

Barner, K. (2011). The library is a growing organism: Ranganathan's Fifth Law of Library Science and the academic library in the digital era. *Library Philosophy and Practice*. Retrieved from http://digitalcommons.unl.edu/cgi/viewcontent.cgi?article=1579&context=libphilprac

Beldarrain, Y. (2006). Distance education trends: Integrating new technologies to foster student interaction and collaboration. *Distance Education*, *27*(2), 139–153. doi:10.1080/01587910600789498.

Blazer, E. (2010). Do clickers click in the class-room? In *NABET, Northeastern Association Of Business, Economics, and Technology, 2010 Proceedings*. Retrieved from http://www.nabet.us/Archives/2010/NABET%20Proceedings%202010.pdf#page=20

Bower, B. L. (2001). Distance education: Facing the faculty challenge. *Online Journal of Distance Learning Administration, 4*(2). Retrieved from http://www.westga.edu/~distance/ojdla/summer42/bower42.pdf.

Bower, J., & Christensen, M. (1996). Disruptive technologies: Catching the wave. *Journal of Product Innovation Management, 13*(1), 75–76.

Bower, S. L., & Mee, S. A. (2010). Virtual delivery of electronic resources and services to off-campus users: A multifaceted approach. *Journal of Library Administration, 50*(5/6), 468–483. doi:10.1080/01930826.2010.488593.

Boyle, J. T., & Nicol, D. J. (2003). Using class-room communication systems to support interaction and discussion in large class settings. *Alt-J: Research in Learning Technology, 11*(3), 43–57. doi:10.1080/0968776030110305.

Brumfield, E. (2008). Using online tutorials to reduce uncertainty in information seeking behavior. *Journal of Library Administration, 48*(3/4), 365–377. doi:10.1080/01930820802289417.

Buhay, D., Best, L. A., & McGuire, K. (2010). The effectiveness of library instruction: Do student response systems (clickers) enhance learning? *The Canadian Journal for the Scholarship of Teaching and Learning, 1*(1). Retrieved from http://ir.lib.uwo.ca/cgi/viewcontent.cgi?article=1006&context=cjsotl_rcacea doi:10.5206/cjsotl-rcacea.2010.1.5.

Campbell, J., & Mayer, R. E. (2009). Questioning as an instructional method: Does it affect learning from lectures? *Applied Cognitive Psychology, 23*(6), 747–759. doi:10.1002/acp.1513.

Carmigniani, J., Furht, B., Anisetti, M., Ceravolo, P., Damiani, E., & Ivkovic, M. (2011). Augmented reality technologies, systems and applications. *Multimedia Tools and Applications, 51*(1), 341–377. doi:10.1007/s11042-010-0660-6.

Carmody, L. E. (2009). Clayton M. Christensen, Michael B. Horn, and Curtis W. Johnson: Disrupting class: How disruptive innovation will change the way the world learns. *Educational Technology Research and Development, 57*(2), 267–269. doi:10.1007/s11423-009-9113-1.

Chan, E. K., & Knight, L. A. (2010). Clicking with your audience: Evaluating the use of personal response systems in library instruction. *Communications in Information Literacy, 4*(2), 192–201.

Chandler, D., & Munday, R. (2011). Embedded reporters. In *A dictionary of media and communication*. Oxford: Oxford University Press. Retrieved from www.oxfordreference.com/views/ENTRY.html?subview=Main&entry=t326.e848

Chen, B. (2009). Barriers to adoption of technology-mediated distance education in higher-education institutions. *Quarterly Review of Distance Education, 10*(4), 333-338,399.

Christensen, C. M. (1997). *The innovator's dilemma: When new technologies cause great firms to fail.* Boston, MA: Harvard Business School Press.

Christensen, C. M., Horn, M. B., & Johnson, C. W. (2011). *Disrupting class: How disruptive innovation will change the way the world learns.* New York: McGraw-Hill.

Churchill, W., & Langworth, R. M. (2008). *Churchill by himself: The definitive collection of quotations.* New York: Public Affairs.

Clark, B. R. (1983). *The higher education system.* Berkeley: University of California Press.

Collins, B. L., Tedford, R., & Womack, H. D. (2008). 'Debating' the merits of clickers in an academic library. *North Carolina Libraries*, *66*(1), 20–24.

Connolly, M., Cosgrave, T., & Krkoska, B. B. (2011). Mobilizing the library's web presence and services: A student-library collaboration to create the library's mobile site and iPhone application. *The Reference Librarian*, *52*(1/2), 27–35.

Dabbagh, N., & Kitsantas, A. (2012). Personal learning environments, social media, and self-regulated learning: A natural formula for connecting formal and informal learning. *The Internet and Higher Education*, *15*, 3–8. doi:10.1016/j.iheduc.2011.06.002.

Dangel, H. L., & Wang, C. X. (2008). Student response systems in higher education: Moving beyond linear teaching and surface learning. *Journal of Educational Technology Development and Exchange*, *1*(1), 93–104.

Dede, C. (2005). Planning for neomillennial learning styles. *EDUCAUSE Quarterly*, *28*(1), 7–12.

Deleo, P. A., Eichenholtz, S., & Sosin, A. A. (2009). Bridging the information literacy gap with clickers. *Journal of Academic Librarianship*, *35*(5), 438–444. doi:10.1016/j.acalib.2009.06.004.

DeMillo, R. A. (2011). *Abelard to Apple: The fate of American colleges and universities*. Cambridge, MA: MIT Press.

Dempsey, J. V., Fisher, S. F., Wright, D. E., & Anderton, E. K. (2008). Training and support, obstacles, and library impacts on elearning activities. *College Student Journal*, *42*(2), 630–636.

Dempsey, L. (2006). The (digital) library environment: Ten years after. *Ariadne*, *46*. Retrieved from http://www.ariadne.ac.uk/issue46/dempsey/.

Dewey, B. I. (2004). The embedded librarian: Strategic campus collaborations. *Resource Sharing & Information Networks*, *17*(1/2), 5–17. doi:10.1300/J121v17n01_02.

Duderstadt, J. J. (2009). Possible futures for the research library in the 21st Century. *Journal of Library Administration*, *49*(3), 217–225. doi:10.1080/01930820902784770.

Ehlers, U., & Schneckenberg, D. (Eds.). (2010). *Changing cultures in higher education: Moving ahead to future learning*. Berlin: Springer. doi:10.1007/978-3-642-03582-1.

Forsyth, E. (2011). Ar U feeling happy? Augmented reality, apps and mobile access to local studies information. *Aplis*, *24*(3), 125–132.

Freeman, G. T. (2005). The library as place: Changes in learning patterns, collections, technology, and use. In *Library as place: Rethinking roles, rethinking space*. Washington, DC: Council on Library and Information Resources. Retrieved from http://www.clir.org/pubs/reports/pub129/freeman.html.

Gibbons, S. (2007). *The academic library and the net gen student: Making the connections*. Chicago: American Library Association.

Godwin, M. (2010). Disruptive technology: What is it? How can it work for professional writing? *Writing Instructor*, *1*, Retrieved from http://www.writinginstructor.com/godwin2.

Gonzalez, A. C., & Westbrock, T. (2010). Reaching out with LibGuides: Establishing a working web of best practices. *Journal of Library Administration*, *50*(5/6), 638–656. doi:10.1080/01930826.2010.488941.

Hanna, D. E., & Johnson, M. J. (2006). The challenges and opportunities of technology in higher education. *Effective Practices for Academic Leaders*, *1*(6), 1–16.

Hannafin, M. J., & Kim, M. C. (2003). In search of a future: A critical analysis of research on web-based teaching and learning. *Instructional Science*, *31*, 347–351. doi:10.1023/A:1024646328974.

Hinchliffe, L., & Wong, M. (2010). From services-centered to student-centered: A "Wellness Wheel" approach to developing the library as an integrative learning commons. *College & Undergraduate Libraries*, *17*(2/3), 213–224. doi:10.1080/10691 316.2010.490772.

Hoffman, C., & Goodwin, S. (2006). A clicker for your thoughts: Technology for active learning. *New Library World*, *107*(9/10), 422–433. doi:10.1108/03074800610702606.

Hoffman, S. (2011). Embedded academic librarian experiences in online courses. *Library Management*, *32*(6), 444–456. doi:10.1108/01435121111158583.

Hricko, M. (2010). Using microblogging tools for library services. *Journal of Library Administration*, *50*(5/6), 684–692. doi:10.1080/01930826.2 010.488951.

Hudson, M., McGowan, L., & Smith, C. (2011). Technology and learner motivation in library instruction: a study of personal response systems. *Indiana Libraries*, *30*(1), 20–27.

Johnson, L., Levine, A., Smith, R., & Stone, S. (2010). *The 2010 Horizon Report*. Austin, Texas: The New Media Consortium. Retrieved from http://wp.nmc.org/horizon2010/

Junco, R., & Cole-Avent, G. A. (2008). An introduction to technologies commonly used by college students. *New Directions for Student Services*, (124): 3–17. doi:10.1002/ss.292.

Kargbo, J. A. A. (2009). Automation: Whither academic libraries? *Information Technology for Development*, *15*(1), 43–51. doi:10.1002/ itdj.20078.

Koszalka, T., & Ntloedibe-Kuswani, G. S. (2010). Literature on the safe and disruptive learning potential of mobile technologies. *Distance Education*, *31*(2), 139–157. doi:10.1080/01587919. 2010.498082.

Kreber, C., & Cranton, P. A. (2000). Exploring the scholarship of teaching. *The Journal of Higher Education*, *71*(4), 476–495. doi:10.2307/2649149.

Lee, H. W., Lim, K. Y., & Grabowski, B. L. (2010). Improving self-regulation, learning strategy use, and achievement with metacognitive feedback. *Educational Technology Research and Development*, *58*(6), 629–648. doi:10.1007/s11423-010-9153-6.

Legris, P., Ingham, J., & Collerette, P. (2003). Why do people use information technology? A critical review of the technology acceptance model. *Information & Management*, *40*(3), 191–204. doi:10.1016/S0378-7206(01)00143-4.

Lin, P., Chen, K., & Chang, S. (2010). Before there was a place called library – library space as an Invisible factor affecting students' learning. *Libri*, *60*(4), 339–351. doi:10.1515/libr.2010.029.

Lockwood, F. (2001). Leadership, disruptive technologies, lessons learned and changes in learning and teaching. *Open Learning*, *17*(3), 199–202. doi:10.1080/0268051022000048200.

Mahmood, M. A., Tariq, M., & Javed, S. (2011). Strategies for active learning: An alternative to passive learning. *Academic Research International*, *1*(3), 193–198.

Marshall, S. (2010). Change, technology and higher education: Are universities capable of organisational change? *ALT-J. Research in Learning Technology*, *18*(3), 179–192. doi:10.3402/rlt. v18i3.10762.

Mayer, R. E., Stull, A., DeLeeuw, K., Almeroth, K., Bimber, B., Chun, D., & Zhang, H. (2009). Clickers in college classrooms: Fostering learning with questioning methods in large lecture classes. *Contemporary Educational Psychology, 34*(1), 51–57. doi:10.1016/j.cedpsych.2008.04.002.

McBride, K. (2010). Leadership in higher education: Handling faculty resistance to technology through strategic planning. *Academic Leadership Journal, 8*(4).

McConatha, D., Praul, M., & Lynch, M. J. (2008). Mobile learning in higher education: An empirical assessment of a new educational tool. *Turkish Online Journal of Educational Technology, 7*(3), 15-21. Retrieved from http://www.tojet.net/articles/v7i3/732.pdf

Meyer, K. A. (2010). The role of disruptive technology in the future of higher education. *EDUCAUSE Quarterly, 33*(1), 6. Retrieved from http://www.educause.edu/EDUCAUSE+Quarterly/EDUCAUSEQuarterlyMagazineVolum/TheRoleofDisruptiveTechnologyi/199378.

Oakey, R. (2007). Are disruptive technologies disruptive [disruptive technologies]. *Engineering Management, 17*(2), 10. doi:10.1049/em:20070201.

Oblinger, D. G. (2005). Learners, learning and technology: The Educause learning initiative. *EDUCAUSE Review, 40*(5). Retrieved from http://connect.educause.edu/Library/EDUCAUSE+Review/EDUCAUSEReviewMagazineVol/40585.

Osguthorpe, R. T., & Graham, C. R. (2003). Blended learning environments. *Quarterly Review Of Distance Education, 4*(3), 227–233.

Owusu-Ansah, E. K. (2004). Information literacy and higher education: Placing the academic library in the center of a comprehensive solution. *Journal of Academic Librarianship, 30*(1), 3–16. doi:10.1016/j.jal.2003.11.002.

Palloff, R., & Pratt, K. (2001). Lessons from the cyberspace classroom. *17TH Annual Conference on Distance Teaching and Learning.* Retrieved from http://www.uwex.edu/disted/conference/resource_library/proceedings/01_20.pdf

QR. (2011). In *OED online.* Oxford: Oxford University Press. Retrieved from www.oed.com/view/Entry/155604?redirectedFrom=qr%20code#eid246543457

Ranjan, J. (2008). Impact of information technology in academia. *International Journal of Educational Management, 22*(5), 442–455. doi:10.1108/09513540810883177.

Rizzuto, T. E., & Reeves, J. (2007). A multidisciplinary meta-analysis of human barriers to technology implementation. *Consulting Psychology Journal: Practice and Research, 59*(3), 226–240. doi:10.1037/1065-9293.59.3.226.

Schrum, L., & Hong, S. (2002). Dimensions and strategies for online success: Voices from experienced education. *Journal of Asynchronous Learning Networks, 6*(1), 57–67.

Scott, G. (2003). Effective change management in higher education. *EDUCAUSE Review, 38*(6), 64–80.

Selim, H. (2007). Critical success factors for e-learning acceptance: Confirmatory factor models. *Computers & Education, 49*(2), 396–413. doi:10.1016/j.compedu.2005.09.004.

Shaffer, D., & Collura, M. (2009). Evaluating the effectiveness of a personal response system in the classroom. *Teaching of Psychology, 36*(4), 273–277. doi:10.1080/00986280903175749.

Shih, Y., & Mills, D. (2007). Setting the new standard with mobile computing in online learning. *International Review of Research in Open and Distance Learning, 8*(2), 1–16.

Shumaker, D. (2009). Who let the librarians out? *Reference and User Services Quarterly, 48*(3), 239–242.

Shumaker, D., & Makins, A. (2012). Lessons from successful embedded librarians. *Information Outlook, 16*(3), 10–12.

Smith, A., Rainie, L., & Zickuhr, K. (2011, July 19). *College students and technology* (Rep.). Retrieved from http://pewinternet.org/Reports/2011/College-students-and-technology/Report.aspx

Smith, A. G. (1997). Testing the surf: criteria for evaluating Internet information resources. *The Public-Access Systems Review, 8*(3), 5-23. Retrieved from http://journals.tdl.org/pacsr/article/viewFile/6016/5645

Staley, D. J., & Malenfant, K. J. (2010). Futures thinking for academic librarians: Higher education in 2025. *Information Services & Use, 30*(1/2), 57–90.

Stoffle, C. J., & Cuillier, C. (2011). From surviving to thriving. *Journal of Library Administration, 51*(1), 130–155. doi:10.1080/01930826.2011.531645.

Su, S., & Kuo, J. (2010). Design and development of Web-based information literacy tutorials. *Journal of Academic Librarianship, 36*(4), 320–328. doi:10.1016/j.acalib.2010.05.006.

Sullivan, R. (2010). Common knowledge: Learning spaces in academic libraries. *College & Undergraduate Libraries, 17*(2), 130–148. doi:10.1080/10691316.2010.481608.

Tang, F. (2010). Reference tools in Second Life®: Implications for real life libraries. *New Library World, 111*(11/12), 513–525. doi:10.1108/03074801011094886.

Traxler, J. (2005). Defining mobile learning. In *IADIS International Conference Mobile Learning Proceedings* (pp. 261-266). International Association for the Development of the Information Society.

Trees, A., & Jackson, M. (2007). The learning environment in clicker classrooms: Student processes of learning and involvement in large university-level courses using student response systems. *Learning, Media and Technology, 32*(1), 21–40. doi:10.1080/17439880601141179.

Vaughan, N. (2007). Perspectives on blended learning in higher education. *International Journal on E-Learning, 6*(1), 81–94.

Viggiano, R. G. (2004). Online tutorials as instruction for distance students. *Internet Reference Services Quarterly, 9*(1-2), 37–54. doi:10.1300/J136v09n01_04.

Yan, H. (2010). On the clouds: A new way of computing. *Information Technology & Libraries, 29*(2), 87–92.

ADDITIONAL READING

Brasley, S. S. (2008). Effective librarian and discipline faculty collaboration models for integrating information literacy into the fabric of an academic institution. *New Directions for Teaching and Learning*, (114): 71–88. doi:10.1002/tl.318.

Budd, J. M. (2005). *The changing academic library: Operations, culture, environments (ACRL Publications in Librarianship, no. 56)*. Chicago, IL: Association of College and Research Libraries.

Dresselhaus, A., & Shrode, F. (2012). Mobile technologies & academics: Do students use mobile technologies in their academic lives and are librarians ready to meet this challenge? *Information Technology and Libraries, 31*(2), 82–101. doi:10.6017/ital.v31i2.2166.

Duncan, V., & Gerrard, A. (2011). All together now! Integrating virtual reference in the academic library. *Reference and User Services Quarterly, 50*(3), 280–292.

Elton, L. (2003). Dissemination of innovations in higher education: A change theory approach. *Tertiary Education and Management, 9*(3), 199–214. doi:10.1080/13583883.2003.9967104.

Garrison, D., & Kanuka, H. (2004). Blended learning: Uncovering its transformative potential in higher education. *The Internet and Higher Education, 7*(2), 95–105. doi:10.1016/j.iheduc.2004.02.001.

Han, N. (2012). Managing a 21st-century library collection. *The Serials Librarian, 63*(2), 158–169. doi:10.1080/0361526X.2012.700781.

Harvey, T. R., & Broyles, E. A. (2010). *Resistance to change: A guide to harnessing its positive power.* Lanham, MD: Rowman & Littlefield Education.

Howard, H. (2012). Looking to the future: Developing an academic skills strategy to ensure information literacy survives in a changing higher education world. *Journal of Information Literacy, 6*(1), 72–81. Retrieved from http://ojs.lboro.ac.uk/ojs/index.php/JIL/article/view/LLC-V6-I1-2012-2 doi:10.11645/6.1.1677.

Kamenetz, A. (2010). *DIY U: Edupunks, edupreneurs, and the coming transformation of higher education.* White River Junction, VT: Chelsea Green Publishing.

Kemp, J. (2006). Isn't being a librarian enough?—Librarians as classroom teachers. *College & Undergraduate Libraries, 13*(3), 3–23. doi:10.1300/J106v13n03_02.

Kraemer, E. W., Lombardo, S. V., & Lepkowski, F. J. (2007). The librarian, the machine, or a little of both: A comparative study of three information literacy pedagogies at Oakland University. *College & Research Libraries, 68*(4), 330–342.

Latimer, K. (2011). Collections to connections: Changing spaces and new challenges in academic library buildings. *Library Trends, 60*(1), 112–133. doi:10.1353/lib.2011.0035.

Lippincott, J. K. (2010). A mobile future for academic libraries. *RSR. Reference Services Review, 38*(2), 205–213. doi:10.1108/00907321011044981.

Machin, A., Harding, A., & Derbyshire, J. (2009). Enhancing the student experience through effective collaboration: A case study. *New Review of Academic Librarianship, 15*(2), 145–159. doi:10.1080/13614530903240437.

Martinez-Torres, M. R., Toral Marin, S., Garcia, F. B., Vazquez, S. G., Oliva, M. A., & Torres, T. (2008). A technological acceptance of e-learning tools used in practical and laboratory teaching, according to the European higher education area. *Behaviour & Information Technology, 27*(6), 495–505. doi:10.1080/01449290600958965.

McNish, M. (2001). Guidelines for managing change: A study of their effects on the implementation of new information technology projects in organisations. *Journal of Change Management, 2*(3), 201–211. doi:10.1080/738552754.

Michalak, S. C. (2012). This changes everything: transforming the academic library. *Journal of Library Administration, 52*(5), 411–423. doi:10.1080/01930826.2012.700801.

Owens, R., & Bozeman, D. (2009). Toward a faculty-librarian collaboration: Enhancement of online teaching and learning. *Journal of Library & Information Services in Distance Learning, 3*(1), 31–38. doi:10.1080/15332900902794898.

Peacock, J. (2005). Information literacy education in practice. In Levy, P., & Roberts, S. (Eds.), *Developing the new learning environment* (pp. 153–180). London: Facet.

Rushby, N. (2010). Editorial: Topics in learning technologies. *British Journal of Educational Technology*, *41*(3), 343–348. doi:10.1111/j.1467-8535.2010.01063.x.

Russell, C. (2009). A systemic framework for managing e-learning adoption in campus universities: Individual strategies in context. *Alt-J*, *17*(1), 3–19. doi:10.1080/09687760802649871.

Salmon, G. (2003). *E-moderating: The key to teaching and learning online* (2nd ed.). London: RoutledgeFalmer.

Shumaker, D., & Makins, A. (2012). Lessons from successful embedded librarians. *Information Outlook*, *16*(3), 10–12.

Thomsett-Scott, B., & May, F. (2009). How may we help you? Online education faculty tell us what they need from libraries and librarians. *Journal of Library Administration*, *49*(1), 111–135. doi:10.1080/01930820802312888.

KEY TERMS AND DEFINITIONS

Disruptive Technology: A technology that does new things for new users, such as distance education or cloud computing.

Embedded Librarians: A librarian assigned to a specific distance education course to provide information literacy instruction and research support for the duration of the course.

Geo-Social Networking: Social networking enhanced with geographic and location specific services and capabilities.

Gesture-Based Computing: Input and interaction with computing devices without keyboard or mouse but rather through taps, swipes, and other body movements.

M-learning: Mobile learning; distance learning through the use of mobile devices.

Personal Learning Environments (PLE): Systems that allow a learner to take control of the learning interaction, allowing for the learners the independence to set their own learning goals and communicate with other learners.

Personal Learning Networks (PLN): The people a learner is connected to, interacts with, and gains knowledge from in a personal learning environment.

QR Codes: Quick response codes; codes that are readable from a cell phone with a QR code reader, these graphic representations of information can include a web page link, text, and geo-location information.

Student Response Systems: Interactive keypad systems used by students in the classroom to interact or respond to questions, polls, or surveys; sometimes called clickers.

Sustaining Technology: A technology that replaces traditional activities and improves an established technology, such as color TV replacing back & white TV.

Chapter 11
Lending and Borrowing Library Materials:
Automation in the Changing Technology Landscape

Regina H. Gong
Lansing Community College, USA

Dao Rong Gong
Michigan State University, USA

ABSTRACT

The first application of robotic technologies in libraries is in the area of storage and retrieval of library materials. This chapter discusses past, present, and future developments in robotic technologies in the area of library circulation. Issues and challenges libraries face in light of rapid developments in the electronic realm are discussed in relation to circulation. This chapter also highlights future trends and technologies for library lending, as well as possibilities for advancement with the increasing shift towards electronic content in libraries.

BACKGROUND

The ability of patrons to borrow materials is a core service provided by libraries. Patrons want to borrow materials, and libraries want to lend. For this reason, borrowing and lending were among the first operations to become automated when integrated library systems were introduced in the market. Circulation is considered to be the most mature and stable automated process performed in libraries. However, until recently, there has been a dearth of information about these advances in circulation technologies. Not much has been written about libraries employing high-tech circulation technologies, nor has much attention been given

DOI: 10.4018/978-1-4666-3938-6.ch011

to this service compared to other areas within library operations.

Today, library circulation services are at a crossroads. The rapid development of online, digital, and multimedia technologies coupled with the shift in patrons' information seeking and retrieval pushes us to rethink how circulation services can better fit today's needs as well as patrons' expectations.

Circulation or library lending typically encompasses activities such as charging (checkout), returning (check in), routing items for internal use or handling, renewing, billing and fines transactions, holds management, shelving, inventory, and interlibrary or consortia lending (Aswal, 2006). Circulation can be viewed as a transaction-oriented system and serves as a direct interface between user and the library (Burns, 1975). Circulation can also be described as "the flow of materials within a library" (Dempsey, 2006). Essentially, circulation services provide the most basic library function of making materials available to its users and allowing access to these resources.

The technology development in circulation can be seen in two areas. First is the automated library catalog system that handles the work flow of circulation, and second, the automated mechanical system, or the robots that physically process library circulation-related tasks. Automated library systems and library robots work for both library staff and the library users. Their aim is to reduce human intervention during the circulation process, extend service coverage, and increase efficiency and accuracy of circulation.

OVERVIEW OF LIBRARY AUTOMATION

Much of the literature in circulation focuses on technology developments in storage and retrieval of printed materials. To understand how this came about, it is useful to examine advances in library automation and library information technology

side by side. Efficiency and productivity were the impetus for automation in libraries. As computer systems became more widely available, libraries began a slow and gradual shift from focusing on collections to focusing on information access (Borgman, 1997).

Lynch (2000) provides a framework for understanding the changes in the library automation industry and its effect on organizations based on the three-phase process introduced by Richard West and Peter Lyman:

Modernization (doing what you are already doing, though more efficiently); innovation (experimenting with new capabilities that the technology makes possible); and transformation (fundamentally altering the nature of the organization through these capabilities) (Lynch, 2000, p. 60).

The first phase of the automation age occurred in the 1960s and 1970s when technology advanced in the personal computing arena. Due to this development, libraries felt the need to introduce efficiency in their internal operations, or to bring the benefit of automation to library staff. The earliest library automation projects computerized routine and core library operations in the areas of circulation, acquisitions, cataloging, and serials management mainly to improve internal workflow and processes (Borgman, 1997). Circulation control at this stage was based on batch processing techniques using keypunch or key-to-tape devices in order to convert information about individual circulation transactions to computer-readable form (Saffady, 1989). Workflow typically involved batch processing of circulation data, which was then uploaded, maintained on magnetic tapes, and printed so that checked-out materials could be viewed by library staff. Some circulation transactions were not real-time and because there was a time lag in between batch processing and printing, circulation data were difficult to track. Nonetheless, automated circulation processes showed great potentials in a computer-aided environment.

Another important development during this stage was the introduction of shared copy-cataloging systems that paved the way for cooperation and collaboration between libraries. The idea for this system was to minimize, if not eliminate, the duplication of cataloging for printed materials, and to enable libraries to view holdings of a particular item across multiple libraries or consortia.

The second stage expanded on these developments with benefits for library users. As library systems sped up ordering, processing, and circulation of library materials, public access to local library resources became the focus. From the mid-1970s to early 1980s, the manual card catalog was replaced by "the online public access catalog," or OPAC. It was considered revolutionary, enabling users to search for library materials through a computer terminal, making the collection more accessible and discoverable. Online catalogs brought significant improvement in materials access with their ability to perform keyword, Boolean searching, and material type limiting. With integrated library systems in place, the online catalog made it possible to determine whether an item was owned, available, or on order (Borgman, 1997).

Circulation at this stage was characterized by the emergence of online, real-time systems that processed circulation transactions as they occurred (Saffady, 1989). The ability of the library system to reflect circulation transactions in real time enabled the library and its patrons to accurately view the status of an item and eliminated the paper printout of circulation data. In addition, libraries now had the ability to set up their own circulation parameters, such as loan periods, borrower types, and restrictions on circulation for certain categories of materials, which made for greater flexibility and control (Saffady, 1989).

The growing popularity of the online catalog among library users created an expectation for improved delivery as well as improved access. Once users could find the location of needed items online, a demand arose for delivering the actual content online. In response, vendors developed abstracting and indexing services in database form, so that individual articles from journals could be accessed. At the same time, academic libraries made considerable investments in resource sharing with the advent of shared national union catalogs and computer-assisted interlibrary loan systems. While these developments were successful in providing improved delivery of library materials and journal articles across libraries, resource sharing became very limited because of copyright and license agreement limitations (Lynch, 2000). Introducing machines to aid physical library operation resulted in the need for an inventory system which the library automated system has made possible.

The third stage of automation was the shift from print content to electronic. Online catalogs created a demand for the actual content in digital format. Coupled with rapid advances in network connectivity and the emergence of the Internet, materials could be digitized and made available across the Web. Consequently, "publishers and aggregators …began to offer this material to libraries" and users developed a preference for electronic versions of journal articles, even if the print material was available in their library (Lynch, 2000). At this stage, shift from print to electronic was primarily centered around journals rather than books. This period saw the development of the many hundreds of databases to which academic libraries subscribe for access to electronic journal contents. This in turn contributed to the significant increase of academic libraries' collection budget. Along with this shift came the beginnings of digitization in research libraries for their specialized materials which they made available to students and researchers.

Over the past years, circulation technology development has been centered on handling physical materials – until very recently, physical materials were the only materials actually circulated by libraries. (Access was provided to electronic materials, but outside the parameters of a circulation transaction.) Over time, circulation formed a distinctive library operation for technology to grow and mature. In that sense, lending books to patrons or borrowing books from a library became very much settled. However, this started to change in recent years as the growing number of electronic books came as an alternative option for users. The beginning of the 21st century ushered in the transformation phase. It is what Lynch calls the "networked information revolution" and defines it as "vast constellation of digital content and services that were accessible through the network at any time, from any place, could be used and reused, navigated and integrated, and tailored to the needs and objectives of each user" (Lynch, 2000). Now more than ever, information is being transformed in a digital and networked environment into various iterations (multimedia, video, images, audio, and other learning objects) and the ever-changing needs of library users for information retrieval, access, and distribution cannot be ignored. The academic library has to evolve in response to these challenges if it is to remain as a vital player in the teaching, learning, and scholarly activity of the university.

CIRCULATION TECHNOLOGY: REVIEW OF CURRENT APPLICATIONS

Traditionally, circulation service is the library operation most heavily involved with handling of library books on the shelf. This made circulation a distinctive and self-contained operation compared with other library services. Although at present, circulation operations still depend largely on library staff, there have been solid technology developments in this area, mainly centered around machinery used for physical materials handling. Today, these technologies are reaching maturity. Many are considered as state-of-the-art and cutting-edge technologies.

Library Materials Tracking and Identification: Barcodes and RFID

The technology for library materials detection dates back to the early 1970s. It was deployed to address a very old dilemma: protecting books from theft while offering open shelf access to users. Material detection systems rely on sensitive magnetic strips which are attached (often inconspicuously) to books and other materials. Some strips can be desensitized during the check-out process; if still sensitized, they will trigger a warning alarm at the detection gate. A variation involves strips which are permanently sensitized; in that case library staff hand checked-out materials to the user after he or she has passed through the detection gate. Over the years, the magnetic strip system has been widely adopted: its precision, flexibility, and relatively low cost have made it a standard in many library environments.

Many libraries today use scanners to read barcodes on library materials. These barcode readers are input devices for a computer system to read coded information on the surface of library materials by image recognition. This technology has evolved in the past decade since the encoding matrix has extended from linear barcode into 2D barcodes, but the basic function is the same. Barcodes are a low-cost, low-maintenance method for quickly identifying library materials. However, it is considered "contact identification," meaning that the operator must identify the location of barcode on an item and place it in close proximity to a scanner. Books on shelves have to be pulled out one by one to check barcodes – a time-consuming process, and even more so if the barcode has been placed inside the book.

In recent years, Radio Frequency Identification technology (RFID) has become a competing option for identifying library materials. RFID is a standard that conceptualizes a two-way communication method for devices to transfer and receive information from each other, within close range. In this method, a computer chip (often called a tag or smart label) is embedded on library materials. The chip stores information about that particular item and can be read from a nearby device via radio frequency signals. RFID chips can store barcode information and also function as the magnetic security strip, thus accomplishing both book tracking and identification. RFID chips can be read at greater distance than barcodes, so that it is possible to read the chips from each volume in a stack of books without separating them, even if the chips are affixed inside the books. This allows for an increased efficiency in retrieving identification information from library materials. Moreover, RFID systems allow information to be exchanged between the physical item, the library system, and the library staff who operate it. The following sections will review other library technologies based on RFID systems.

Self-Check-In and Check-Out

The circulation function of borrowing and returning library materials was the perfect testing ground for patrons to perform self-service. This service became possible with the development of self-check systems that facilitate self-check-out (charging) and self-check-in (return) with minimal staff intervention, if any. It is also increasingly being used in conjunction with automated document requests and delivery as well as online viewing and updating of patron records (Vaughan & Fabbi, 2008). Unlike other existing circulation technologies, self-check units are mature systems that have been in existence since 1992. Initial investment for these machines can be very high but with increased efficiency in workflows and expedited transactions for patrons, it can be cost effective in the long run.

The self-service checkout stations for patrons are very straightforward. The normal work flow for checking out library materials consists of a checkout station which can identify patron's library card, read information from the items to be checked out, flag each item in system as being

checked out, and desensitize the book's security protection so it won't trigger the alarm when carried out of the library. The self-check-in follows a similar pattern such that when an item is brought back by patron, the check-in station identifies the item, flags the item as returned in the integrated library system, and then sensitizes the security protection so the book can be re-shelved.

Standards have been established to support interoperability and communication between library catalog software and the hardware or peripherals used in automated circulation systems. The Standard Interchange Protocol (SIP) introduced by 3M in 1993 became the primary standard and protocol for communication between the Integrated Library Systems (ILS) applications and self-service devices. Another standard at play is the NISO (National Information Standards Organization) Circulation Interchange Protocol (NCIP). NCIP establishes a standard between automated systems to support effective communication in the following four areas: patron self-services, library circulation and interlibrary loan, consortial borrowing, and access to electronic resources (Needleman, Bodfish, O'Brien, Rush, & Stevens, 2001).

Automated Storage Systems and Materials Sorting

Robotic technology has long been utilized and adopted in the fields of manufacturing, retail, industrial warehousing, medicine, and engineering. While circulation control was the earliest and most widely automated library operation, use of robots and machines that effectively eliminate human intervention in academic library operations have only been in place in recent years.

Perhaps the most successful showcase of robotic technology application in libraries was the development of the Automated Storage and Retrieval System (ASRS) and the automated sorting system. These technologies are essentially a combination of equipment and controls that handle, store and retrieve materials as needed with precision, accuracy, and speed under a defined degree of automation. These systems vary from relatively simple, manually controlled order-picking machines operating in small storage structures to extremely large, computer-controlled storage and retrieval systems totally integrated into a manufacturing and distribution process.

Libraries soon found out how application of this technology would eventually solve the problem of overcrowding shelf space and limited library footprint to house printed materials. The ASRS system works parallel to the idea of a warehouse. The process is initiated when patrons select the items they want and submits the request through the library catalog with which the system interfaces. This will trigger the robotic retrieval system, usually a robotic arm or crane, to go into the environmentally enclosed area and retrieve the storage bin that contains the book being requested. In most cases, the books contain barcodes that are mapped to their bin locations in the ASRS system. A circulation staff member is normally at the end of the automated process to receive the book bin, select the item, and bring it to a designated area for the patron. The total turnaround time is just under five minutes from request, to retrieval of the item, and finally into the hands of the patron.

The California State University Northridge (CSUN) was the first library that implemented an ASRS back in 1991. Two decades later, more than two dozen academic libraries in North America, Japan, Europe, Australia, and the United Kingdom have implemented this technology. The newest soon-to-be-completed building, with probably the biggest storage capacity for an ASRS system, will be the Hunt Library at the North Carolina State University (NCSU), which is slated for completion in fall 2012. The building will have a robot-driven automated delivery system called bookBot, able to accommodate up to two million volumes with items ready for delivery in just under five minutes.

It is not hard to see why this technology has been a phenomenal success despite the huge initial investment by these academic libraries. It is also not uncommon to see academic libraries which are in the process of rebuilding or redesigning their building and facilities incorporate automated storage and retrieval systems into their plan. ASRS allows libraries high-density use of their space, sometimes eliminating the need for off-site storage. In addition, it provides a climate-controlled environment conducive to collection preservation, not to mention the added convenience of having materials accessible to patrons in a fast and efficient manner.

Another technology now revolutionizing materials handling in libraries is the automated return/sorting system. This machine utilizes robots to sort items that are transported by a conveyor belt from the self-check machines so that staff can put it back on the shelf. It accepts returned books into a sorting bed or conveyor where the RFID tag or barcode from the book is scanned by a reader. Based on the information from the item, the system interfaces with the library system and determines the sorting destination of the item and transfers it to a book bin, trolley, or a tote so that it can be transported back to the correct library location.

Public libraries – especially those with large, multi-branch locations – typically use automated sorting systems. This is mainly due to their high volume of circulation and returns. The most important benefit of return/sorting systems is the considerable savings in operating and labor costs because it eliminates the need for library staff to perform these functions. Staff will not have to empty the book bins or drops, move the materials inside the library, check them in and activate their security tags. All of these routine functions can be performed in an automated way with minimal staff intervention. These benefits translate to improved delivery of services to patrons by having items quickly available for re-shelving thereby increasing availability of materials. Lastly, libraries that employ return/sorting systems minimize the risk for their staff to suffer repetitive motion injuries from performing these manual tasks.

Automated Materials Vending

Originally used in retail and food service industries, vending technology has now penetrated libraries in the form of book and media vending machines. For libraries, vending technology is a new way of offering outreach services and expanding the library's presence in the community. It is also a marketing tool, showing how the library is responding to changing demands and lifestyles of its users. For the public, it offers a convenient way to collect items reserved online and better access to the library collection since they are no longer restricted by the hours the library is open (Monley, 2010). Automated vending machines bring the library to the people and extend its presence in the community. They offer a way to reach out to certain niches and segments of the population, and allow patrons to experience their libraries and its services in an unexpected way.

A growing number of public libraries and some academic libraries in Europe, Asia, United States, and Canada are now implementing vending machines where library materials such as books and DVDs can be borrowed by any user who has a library account. Imagine a RedBox kiosk but instead of renting DVDs at a cost, patrons can borrow DVDs for free, courtesy of their public library. These Automated Teller Machine (ATM) type models are stand-alone, mobile units that perform a variety of functions, depending on the manufacturer and model chosen by the library. Some machines can only dispense books, while others allow robotic delivery from internal storage or serve as an automated library branch with full service capabilities such as placing holds and requests and even visual browsing.

Libraries in Scandinavia and the Netherlands have the Bokomaten and Mediamaten and have been using these machines for a number of years. They are freestanding machines that store and vend books, DVDs and CDs, and handle loans and returns automatically. They have a touchpad similar to an ATM screen where users can select

from about 500 items that are delivered through a slot in front of the unit (Havens & Storey, 2009). They are integrated with the library system and can handle multiple types of media other than books such as DVD, CD, and VHS. The popularity of these kiosks have inspired other libraries to follow suit. For example, the Contra Costa County Library in California has book vending machines called Library a-Go-Go, strategically located in commuter transit stations across the Bay Area. Depending on the model and manufacturer selected, these vending machines can hold as many as 400 to 750 items. In Ohio and Colorado, the Cuyahoga County Library and Arapahoe District Library both have "RedBox" type machines, which vend DVDs for free to patrons with library cards. Patrons can check out DVDs and return them using the machine. In Spain, book vending machines called LibroExpress are used to reach out to the reading public, especially those using public transport who do not have time to go to libraries. Aside from the ability to borrow and return books, users can also suggest titles for purchase by their library. At the Shenzhen Library in China, an automated librarian machine debuted in 2008 that is powered by RFID technology and integrates seamlessly with the Shenzhen Library database. Patrons can reserve books either through the library catalog or the machine itself. A patron will also receive a text message indicating the book is available, which will then be delivered to the self-service machine closest to the patron's location.

Robotic Technologies

Compared to the standard circulation automation system provided by manufacturers, we have seen some interesting initiatives by scientists and engineers on the deployment and application of robot technologies in libraries. These projects represent pioneering research based on robotic and sensing technologies with specific applications to assist circulation and other library tasks. Although some are still in the prototype stage, these experiments are creative and forward-looking projects that explore new directions in library automation.

At Tsukuba University in Japan, researchers created a mobile robot able to help humans browse books located in a library from a remote location via the Internet. This remote book browsing system enables a human to have a substitute, in this case a mobile robot, browse books on their behalf. The robot retrieves the book selected by the remote user from the shelves, then opens the page and sends the image of the opened page to the user (Tomizawa, Ohya, & Yuta, 2003). In addition, researchers at Tsukuba University also developed a librarian robot that has the basic function of a librarian. This robot is able to greet library users much like a receptionist and programmed with a "human behavior prediction function using a laser range finder" (Mikawa, Morimoto, & Tanaka, 2010). It is also designed to talk to a library user using natural language, find library resources from the online catalog or databases, and present the results of the search to the user through a monitor. Also in Japan, at the National Institute of Advanced Industrial Science and Technology, researchers have developed a librarian robot using ambient intelligence with a basic goal of arranging books on shelves (Kim, Ohara, Kitagaki, & Ohba, 2009). The robot is able to organize books by removing them from the shelf, and replacing them in proper order using RFID tags.

Another example of robotic technology design for libraries is Jaume, the first librarian robot developed by a research team from the Robotic Intelligence Lab at the Universitat Jaume-I (UJI) in Spain. This robot is designed to search and retrieve a book and then take it to the user. Manipulation through multi-sensory-based grasping techniques using a robotic arm is still being perfected by these researchers before they launch this in a real library scenario (Ramos-Garijo, Prats, Sanz, & Del Pobil, 2003). Still under development is a mobile robotic library assistant, named LUCAS

(Limerick University Computerised Assistive System) designed to assist elderly individuals with mild cognitive or physical impairments within a library environment (Behan & O'Keefe, 2005). It has a graphical interface that displays a human-like animated character and serves as a guide to users in a library.

In the United States, the CAPM (Comprehensive Access to Printed Materials) project by researchers from the Johns Hopkins University aims to introduce robotics technology to not only retrieve items but to digitally scan materials. In this project, a prototype robotic system was developed for an off-site shelving facility. The robot can be controlled remotely via a Web interface to retrieve books from the shelves and bring them to a scan station for scanning (Suthakorn et al., 2002).

Integrated Library Systems

Library circulation technology cannot be discussed without mentioning the integrated library system or the ILS. Often the term ILS is used interchangeably with library automation since both describe the system design that automates all library operations. In a narrower sense, the ILS refers to a package of software solutions, or what we normally call the library catalog system. Compared to the hardware solutions that we have discussed so far, ILS has a special place in this discussion as it is the centerpiece, or the intelligent brain, in an automated library environment. Libraries have been using ILS systems since their beginning in the 1960s. The fundamental functions of the circulation module within an ILS have experienced stable development and wide adoption. Competing software solutions by commercial vendors and open source alternatives have provided libraries with a variety of choices.

Circulation is a focal point of library services. It has done a fairly good job in serving library staff and patrons on the first line of library services. Much like the mechanical robotics that we have looked at, the integrated library system for library circulation also has two service targets. One is the library patron, and second is the library staff. As most interaction between library staff and patrons occur at this stage, the ILS plays a central role in supporting both groups in an automated way that allows for increased efficiency. Thus, instead of direct human to human interaction which must take place in the library, ILS provides automated alternatives taking place online, without the presence of library staff.

Normally, there are two types of automated processes: the patron-initiated process and system-initiated process. Patron-initiated services are not new technology. They follow the business model of online transactions that are widely adopted today. Patron-initiated processes are triggered by actions performed by patrons from the access point, often in the public catalog interface. Patrons interacting with the system initiate and complete tasks online, for instance, to complete library registration, renew checked out books, place a hold or recall request for a book out to another patron, or pay library fines. In a system-initiated service, the library system acts upon a set of predefined conditions based on circulation status, and when that condition is met, the system automatically takes action against the appropriate service target, mostly library patrons. These system-initiated services also include patron notices for book pick-up, reminder of past due items, or information regarding a new book request. When such event happens, the automation process would alert patrons or library staff systematically by email, telephone, text messaging or even via Web feeds.

The contributions of the integrated library system in circulation cannot be ignored because they provide the efficiencies needed to perform automated circulation processes. The next step – and one that is already happening – is to veer away from the stand-alone ILS system and instead perform circulation functions which are Web-based or Web-scale based in the "cloud." This will be discussed in the succeeding sections.

CIRCULATION TECHNOLOGY: WHAT LIES AHEAD

Electronic Books and Digital Lending

Most of the circulation technologies mentioned in this chapter focus on storage, access and retrieval of physical or printed materials so they can be made more accessible to library users. However, the next stage for radical change in circulation is in the electronic frontier. This is already happening now that library collections are increasingly shifting to electronic. A new model of circulation technology is taking place in response to this changing landscape.

The proliferation and wide adoption of e-books in libraries greatly affected how circulation is now being carried out. According to the Library Journal/School Library Journal 2011 survey of e-book penetration, 82 percent of public libraries and 95 percent of academic libraries offer e-books. In the same report, there has been a staggering 184 percent increase in the number of e-books offered by public libraries while academic libraries had an impressive 93 percent increase in 2011. The Pew Internet and American Life Survey on e-reading reports that 21 percent or one-fifth of Americans have read an e-book and 29 percent of Americans age 18 and older own at least one e-book reading device (Rainie, 2012). Meanwhile, the 2012 Pearson Foundation Survey on Students and Tablets shows that 70 percent of college students have read digital textbooks and 58 percent prefer reading books in electronic format. Data from the Association of Research Libraries (ARL) Statistics, 2008-2009, shows a 27 percent drop in initial circulations in academic libraries since 1991 (Kyrillidou & Morris, 2011). This trend already represents an increasing change in attitude among college students and the general reading public in terms of their format preferences for books as they lean more towards reading e-books, whether for fun or as part of their coursework. Most would surmise that the decline in circulation of print materials has something to do with the increased allocation for electronic materials acquisition in libraries and their growing adoption of e-books. The reasons are as varied as they are complex. But there is little doubt that the shift in user behavior and the ubiquitous Web have greatly impacted the declining levels of circulation in academic libraries.

Library circulation has now crossed the barrier of physical lending and transitioned to digital lending or what can be termed as "circulation in the cloud." This was brought about by the introduction of electronic reading devices that enable library users to access, download, and borrow e-books. Aggregators, publishers, and vendor platforms that mediate between publishers and libraries also enable digital lending, where, most of the time, no mediation from the library is required other than authentication of users. The multitude of digital formats – each of which may be either compatible or incompatible with each different reading platform – make e-book circulation more complex and daunting for the average user. Restrictions imposed by publishers such as digital rights management (DRM) mechanisms allow them to dictate and control what can be done with their e-books – how the book must be accessed, how many people can access it, and how long can the book be accessed. In the digital lending world, the boundaries of space and time are blurred. Whereas with print materials, library users go to the physical library to check-out books, as it is now, users enter their virtual library online to discover the book. Once the desired item is discovered or identified, a few clicks then take the user to the platform or portal where the e-book is located. Everything happens seamlessly without obvious library staff intervention. This is patron self-service at its best because once the book loan period expires, the digital copy is automatically erased from the reader's device, thus eliminating the need for users to return the book to the library.

As more books become available in electronic format, the way they will be circulated to library users will continue to evolve. The situation is certainly not perfect at this time and causes frustration and turn-away for some, so it is imperative that libraries work with publishers, vendors, and aggregators to mitigate the negative aspects of e-book lending in order to gain buy-in and increase adoption, especially for those who have yet to join the e-lending bandwagon.

Circulation in the Digital Age

The past years have seen the adoption of mechanical robots and systems that handle physical materials in the library to enable fast and efficient retrieval and access. And, the digital age and the availability of e-books have reshaped the way circulation is carried out.

The advent of the digital age and rise of electronic reading has resulted in new ways of defining the properties of a book. From the readers' viewpoint, the fundamental difference between a printed book and its electronic version is the medium that carries the content of the book. Through the years, the function of a printed book has never really changed. The contents are printed on paper which a user can check-out or borrow from the library to read. The physical book contains all that is needed for a reader to access the information. There is nothing that stands between a book and its reader. However, in the electronic age, we are seeing an electronically created book accessible through a device. Much like any other digital resources, this adds an extra layer between the contents of the book and its reader. The layer is the technological tool or the reading device that transforms the digital contents of a book to a human-readable format. Under this new way of reading, the Internet, computers, and reading devices all become part of the e-book technology which users must have access to in order to read.

From the circulation service point of view, the result of this digital technology is disruptive. It brings circulation into a new frontier – from service taking place in a physical library location, to service taking place in the electronic landscape, or the cloud. Access to library owned e-books are increasingly through the cloud where the library does not need to have any direct involvement. These e-books are mostly accessible via content providers or content hosting services and have substantially contributed to a reduced overhead in technical processing and handling costs during the circulation process. For electronic materials, access management becomes a crucial part of circulation, extending to new areas such as copyright, licensing, and technical platform issues. The checking out process is being replaced to some extent by patron-initiated transaction and computer-mediated automation. With this change, the library portal or online catalog is no longer a place for users to search and find references to information, but a discovery tool to access and retrieve materials at the point of need. Thus, this virtual library shelf becomes the alternative access point to the library, providing a 24/7 solution for the users who "expect to be able to work, learn, and study whenever and wherever they want to" (Johnson, Adams, & Cummins, 2012).

The co-existence of print and digital as two parallel properties of the book is a distinctive view in today's technology landscape and it will stay this way for the foreseeable future (Young, 2007). Libraries are adjusting and coping with the changes by offering more electronic materials while at the same time maintaining print materials. We have seen various acquisition models such as approval plans or patron-initiated selection that are geared toward addressing the needs of users for print and electronic materials. In academic libraries, the merging of digital to print services is evident with the introduction of the Espresso Book Machine (EBM) that allows for an automated system for printing and binding books on-the-fly based on an electronic version of a document. On the opposite end, there are services offered by aggregators and publishers that allow downloading and printing of the entire e-book so that it can be packaged as if it was a

printed book. These two seemingly competing frameworks – digital to print and print to digital – clearly illustrate the co-existence of the two formats, which will be part of the library landscape for years to come. Given the state of current digital technologies, there is no doubt that the changes libraries are dealing with will impact the way circulation service is carried out. The emergence of electronic materials has changed the way libraries handle circulation and how information is accessed and delivered. Electronic circulation is already underway with its new set of service models and options for libraries. The point of maturity has not yet been achieved, and further development and refinement need to occur for this to happen.

The Future of Circulation

"The future has already arrived. It's just not evenly distributed yet."

—*William Gibson*

Predicting future scenarios in the area of library circulation is fraught with uncertainty. Even with advances in the handling of print materials, their circulation will continue to decrease steadily, if not rapidly, in the coming years. Digital lending as it is developing now will be the new face of circulation. But the future of electronic circulation will have limitations and this hinges on how digital technology will evolve.

The library as a place where borrowing and lending takes place will move in great degree to the cloud and patrons will have control over this experience. It does not even matter whether an item is print or electronic. The lines between formats will be blurred to the point where the user experience is simply the seamless delivery of content to a preferred device. Twist (2004) provides a glimpse of this future library circulation scenario:

Some years down the line, someone in the UK could go online and request a book in a US library at 3am in the morning. A robot in the US library could fetch the book and, as directed by the Web user, turn to the correct pages and scan the text and images (Twist, 2004, para. 29).

We do not see any library staff taking part in the transaction described in this scenario. Instead, a robot will have to perform the required tasks to fulfill and deliver the user's request. In this model, the traditional role of robots in libraries – where they retrieve print materials from a set of vaults or bins – would have an added layer digitizing the item as well as retrieving it. Bits and pieces of this technology are currently happening already with mass digitization projects such as Project Gutenberg, Google books, Hathi Trust, and the Internet Archive. However, the idea presented above for book digitization on demand and delivery anywhere and in any device might still be a long way in the future – complex issues of copyright, licensing, and digital rights management must also be defined and resolved.

Another issue is that the information landscape is changing, and lending is not just an institutional library phenomenon anymore (Spalding, 2010). In today's electronic environment, vendors, aggregators, publishers, and shared digital repositories all provide platforms for lending materials. Libraries must still pay for ownership or subscription fees for these electronic materials, regardless of where they are available from. But to the library user, there is no cost involved for borrowing. In this setup, the library becomes just a conduit and a middleman between third party vendors and its users. Imagine a time when the library is excluded from this equation: electronic circulation would continue but not in the library's turf anymore. However, since libraries will still continue to operate in a hybrid print and electronic environment, they will have to deal with both formats for a very long time (Lugg, 2010). In this scenario, print circulation from within the confines of the

library will still remain but as mentioned earlier, circulation rates will steadily decrease as levels of print book purchasing decline and as more materials become available in electronic format.

The future of library circulation is thought provoking but given the technology development in this area, robotics in itself is not remotely futuristic. As we have seen in the past, the existence of technology does not always guarantee its immediate adoption. Technology is an enabler of change but not change in itself. When it comes to predicting the future, technology has to react to social and economic impacts in its implementation and advancement. The future of library circulation automation is not determined by technology. In fact, the future has less to do with technology and more to do with the leadership guiding the strategic use of technology.

CONCLUSION

Over the past years, circulation technology development was centered on the storage, retrieval, and handling of library physical materials. For a long period until recently, the printed materials are predominantly the majority of what the library circulates. Over time, this formed a perfect environment for circulation technology to grow and mature. We saw huge investments made by large research libraries to manage, house, and preserve their extensive print material collections and make them available to users. The online library systems and automated robotic technologies worked in tandem to make these materials easily discoverable and retrievable. But developments in robotic technologies have not gotten as far as addressing the growing emphasis of libraries in electronic content and resource acquisition. The role robotic technologies will play in the electronic arena is fraught with uncertainty since it was largely developed to augment physical materials delivery for items that are held in large storage facilities. As more academic libraries are questioning whether their low-use print collections will be driven to obsolescence because of the massive and easily accessible digital book collections, the need for huge investments in robotic technologies will be lessened.

The research undertaken for this chapter indicates that many of the efforts by scientists towards applying robotic technologies lie in an attempt to simulate the human librarian or staff member so that retrieval of physical items is accomplished quickly and efficiently. While this has proven to be a timesaving solution for large public and academic libraries, it seems that this is a wasted effort. A machine cannot replace the human touch, intelligence, and empathy a librarian can offer. Instead, efforts should be directed towards creating mechanisms to blur the lines between local circulation, consortial borrowing, and interlibrary loans (Breeding, 2008). By doing this, materials can be put on the hands of borrowers regardless of format, location, ownership, and source with the option to read it in any device they choose to. This is happening now to some extent, but in a restrictive environment where digital rights management still prevails. It would be ideal and surely to the benefit of all, whether one is a consumer or producer of information, if a DRM-free world of e-books can happen and becomes the norm rather than the exception. Some publishers have gone this route and are surviving, if not thriving, but it takes a reversal of mindset for this change to go across the board because after all, the bottom line is always the top priority.

There is plenty of room for improvement in library circulation services, especially in the digital arena. Libraries alone cannot accomplish this. It will take a concerted, unified, and determined effort between librarians, publishers, vendors, and authors to bring improvements in the system that ultimately will benefit the library user. The technology to handle the challenges of an increasingly electronic world is already available to us. The question for librarians is how to take these technologies to the next level.

REFERENCES

Aswal, R. S. (2006). *Library automation for 21st century*. New Delhi: Ess Ess Publications.

Behan, J., & O'Keeffe, D. T. (2005). *LUCAS: The library assistant robot, implementation and localization*. Paper presented at the International Conference on Computational Intelligence for Modelling, Control, and Automation, and International Conference on Intelligent Agents, Web Technologies, and Internet Commerce. New York, NY.

Borgman, C. (1997). From acting locally to thinking globally: A brief history of library automation. *Library Quarterly, 67*(3), 215-249. doi: http://www.jstor.org/stable/40039721

Burns, R. W. (1975). *Library performance measures as seen in the descriptive statistics generated by a computer managed circulation system*. Unpublished manuscript. Fort Collins, CO: Colorado State University Libraries.

Dempsey, L. (2006). Libraries and the long tail: Some thoughts about libraries in a network age. *D-Lib Magazine, 12*(4). Retrieved from http://www.dlib.org/dlib/april06/dempsey/04dempsey.html doi:10.1045/april2006-dempsey.

DiFilipo, S. (2011). Connecting the dots to the future of technology in higher education. *EDUCAUSE Review, 46*(4), 58–59.

Havens, A., & Storey, T. (2009). Making innovation personal. *Nextspace, 11*, 6–11.

Johnson, L., Adams, S., & Cummins, M. (2012). *The NMC horizon report: 2012 higher education*. Austin, TX: The New Media Consortium.

Kinner, L., & Rigda, C. (2009). The integrated library system: From daring to dinosaur? *Journal of Library Administration, 49*(4), 401–417. doi:10.1080/01930820902832546.

Kyrillidou, M., & Morris, S. (2011). *ARL statistics 2008-2009*. Washington, DC: Association of Research Libraries.

Lugg, R. (2010). *Natural selection: E-books, libraries, and vendors*. Retrieved from http://www.harrassowitz.de/documents/HARRASSOWITZ_Natural_SelectionEBooks_Libraries_and_Vendors.pdf

Lynch, C. (2000). From automation to transformation: Forty years of libraries and information technology in higher education. *EDUCAUSE Review*, 60–68. Retrieved from http://net.educause.edu/apps/er/erm00/pp060068.pdf.

Mikawa, M., Morimoto, Y., & Tanaka, K. (2010). *Guidance method using laser pointer and gestures for librarian robot*. Paper presented at the 19th IEEE International Symposium on Robot and Human Interactive Communication. New York, NY.

Monley, B. (2010). *Vending machine collection dispensers in libraries: Discussion paper*. Retrieved from http://www.slq.qld.gov.au/info/publib/research/?a=183480

Needleman, M., Bodfish, J., O'Brien, T., Rush, J., & Stevens, P. (2001). The NISO circulation interchange protocol (NCIP) - An XML based standard. *Library Hi Tech*, *19*(3), 223–230. doi:10.1108/07378830110405526.

Ramos-Garijo, R., Prats, M., Sanz, P. J., & Del Pobil, A. P. (2003). *An autonomous assistant robot for book manipulation in a library*. Paper presented at the IEEE International Conference on Systems, Man, and Cybernetics. New York, NY.

Saffady, W. (1989). Library automation: An overview. *Library Trends*, *37*(3), 269–281.

Spalding, T. (2010). Why are you for killing libraries? *Thing-ology blog*. Retrieved from http://www.librarything.com/blogs/thingology/2010/02/why-are-you-for-killing-libraries/

Suthakorn, J., Lee, S., Zhou, Y., Thomas, R., Choudhury, S., & Chirikjian, G. S. (2003). *A robotic library system for an off-site shelving facility*. Retrieved from http://custer.lcsr.jhu.edu/wiki/images/c/c9/Suthakorn02_a.pdf

Tomizawa, T., Ohya, A., & Yuta, S. (2003). *Remote book browsing system using a mobile manipulator*. Paper presented at the IEEE International Conference on Robotics and Automation. New York, NY.

Twist, J. (2004). *Robots get bookish in libraries*. Retrieved from http://news.bbc.co.uk/2/hi/technology/3897583.stm

Vaughan, J., & Fabbi, J. (2008). The helping you buy series: Self-check systems. *Computers in Libraries*, *28*(2), 29–35.

Young, S. (2007). *The book is dead, long live the book*. Sydney, Australia: University of New South Wales Press.

ADDITIONAL READING

Anderson, R. (2008). Future-proofing the library: Strategies for acquisitions, cataloging, and collection development. *The Serials Librarian*, *55*(4), 560–567. doi:10.1080/03615260802399908.

Anderson, R. (2011). Print on the margins. *Library Journal*, *136*(11), 38–39.

Ayre, L. B. (2012). *Top tech trends in materials handling*. Retrieved from http://www.galecia.com/sites/default/files/Ayre_Materials_Handling_Trends_SpeakerNotes.pdf

Boss, R. (2003). RFID technology for libraries. *Library Technology Reports*, *39*(6), 1–58.

Coyle, K. (2006). Technology and the return on investment. *Journal of Academic Librarianship*, *32*(5), 537-539. doi: http://dx.doi.org/10.1016/j.acalib.2006.06.007

Holt, G. E., Larsen, J. I., & Vlimmeren, T. (2002). *Customer self service in the hybrid library.* Retrieved from http://www.public-libraries.net/html/x_media/pdf/selfservice_engl_mit_fotos.pdf

Kirsh, S. E. (1999). *Automated storage and retrieval: the next generation: How Northridge's success is spurring a revolution in library storage and circulation.* Retrieved from http://www.ala.org/acrl/sites/ala.org.acrl/files/content/conferences/pdf/kirsch99.pdf

Lougee, W. P. (2002). *Diffuse libraries: Emergent roles for the research library in the digital age.* Retrieved from http://www.clir.org/pubs/reports/pub108/pub108.pdf

O'Connor, S., & Au, L. (2009). Steering a future through scenarios: Into the academic library of the future. *Journal of Academic Librarianship, 35*(1), 57-64. doi: http://dx.doi.org/10.1016/j.acalib.2008.11.001

Ramzan, M., & Singh, D. (2010). Factors affecting librarians' attitudes toward IT application in libraries. *The Electronic Library, 28*(2), 334–344. doi:10.1108/02640471011033675.

KEY TERMS AND DEFINITIONS

Automated Materials Handling: Includes conveyor and sorting systems that can move library materials from one location and sort them according to call numbers. Once materials are sorted, they are deposited in bins for shelving.

Automated Materials Vending: These are freestanding machines that store and vend books, DVDs, and CDs for borrowing by library patrons. It operates like commercial vending machines except it is free and patrons only have to scan their library cards to obtain library materials.

Circulation: Also known as library lending that includes activities such as charging (checkout), returning (check in), routing items for internal use or handling, renewing, billing and fines transactions, holds management, shelving, inventory, and interlibrary or consortia lending.

Digital Lending: Usually refers to making electronic books available for borrowing in a networked or Internet-based environment.

Digital Rights Management: Technologies that are used by publishers, vendors, and copyright holders that limit use and access of digital materials such as e-books, music, or movies. It is essentially a "lock" that protects certain materials from copyright infringement and ensures that only those who paid for the content are able to use it.

Integrated Library Systems: Software developed to handle basic housekeeping functions of a library. It is considered integrated because it shares a common database needed to perform all the basic functions in a library.

Library Automation: The use and application of computer technologies to automate library functions and patron services. It replaces manual tasks in a library so that services can be provided in a timely and efficient manner.

Radio Frequency Identification: Refers to a system that uses radio frequency technology in combination with microchip technology. It is typically used as a security and theft detection system in libraries including inventory management.

Robotic Technologies: Refers to the use of mechanized robots in retrieving items on the shelf and delivering it to the users in a fraction of the time.

Compilation of References

Adams, C. (2010, September). *New managers of automated library storage: What to know*. Presented at the 25th annual Material Handling and Logistics Conference, Document Management Track, Park City, UT. Retrieved April 20, 2012, from http://www.hkplanet.net/resources-mhlc-presentations.cfm?year=2010

Adelsberger, H. H., Collis, B., & Pawlowski, J. M. (2008). *Handbook on information technologies for education and training*. Berlin: Springer. doi:10.1007/978-3-540-74155-8.

Adner, R. (2002). When are technologies disruptive? A demand-based view of the emergence of competition. *Strategic Management Journal, 23*(8), 667–688. doi:10.1002/smj.246.

Agostinho, S., Lefoe, G., & Hedberg, J. (1997). *Online collaboration for effective learning: A case study of a postgraduate university course (New South Wales, Southern Cross University)*. Retrieved from http://ausweb.scu.edu.au/proceedings/agostinho/paper.html

Ahn, G., Shehab, M., & Squicciarini, A. (2011). Security and privacy in social networks. *IEEE Internet Computing, 15*(3), 10–12. doi:10.1109/MIC.2011.66.

Alan, R., Chrzastowski, T. E., German, L., & Wiley, L. (2010). Approval Plan Profile Assessment in Two Large ARL Libraries. *Library Resources & Technical Services, 54*(4), 179–179.

Ally, M. (2007). Guest editorial: mobile learning. *International Review of Research in Open and Distance Learning, 8*(2). Retrieved from http://www.irrodl.org/index.php/irrodl/article/viewArticle/451/918/.

Ally, M. (2009). *Mobile learning: Transforming the delivery of education and training*. Edmonton: AU Press.

American Library Association. (2012). Academic libraries. *State of America's Libraries Report 2012*. Retrieved April 21, 2012, from http://www.ala.org/news/mediapresscenter/americaslibraries/soal2012/academic-libraries

Andersen, M. H. (2011). The world is my school: Welcome to the era of personalized learning. *The Futurist, 45*(1), 12–17.

Anderson, R. (2011, November 16). *It is about us, or is it about them? Libraries and collections in a patron-driven world*. Retrieved April 21, 2012 from http://www.alaeditions.org/blog/categories/acquisitions-and-collection-development

Anzolin, H. H. (2009). Rede Pergamum: história, evolução e perspectivas. *Revista ACB: Biblioteconomia em Santa Catarina, 14*(2), 493-512. Retrieved March 30, 2012, from http://revista.acbsc.org.br/index.php/racb/article/view/640

Arp, L., Woodard, B. S., Lindstrom, J., & Shonrock, D. D. (2006). Faculty-librarian collaboration to achieve integration of information literacy. *Reference and User Services Quarterly, 46*(1), 18–23.

Ashford, R. (2010). QR codes and academic libraries: Reaching mobile users. *College & Research Libraries News, 71*(10), 526–530.

Automation. (2012). In *Merriam-Webster.com*. Retrieved March 20, 2012, from http://www.merriam-webster.com/dictionary/automation

Averett, P. (2001). People: the human side of systems technology. *Journal for Quality and Participation, 24*(2), 34–37.

Azolin, B. R. (1999). O futuro da informática no brasil. Retrieved March 30, 2012, from http://www-usr.inf.ufsm.br/~cacau/elc202/futuro.html

Bailey, A., & Back, G. (2006). LibX - A Firefox extension for enhanced library access. *Library Hi Tech, 24*(2), 290–304. doi:10.1108/07378830610669646.

Bailey, A., & Back, G. (2007). Retrieving known items with LibX. *The Serials Librarian, 53*(4), 125–140. doi:10.1300/J123v53n04_09.

Baker, N. E. (2007). *LibX IE. University Libraries.* Blacksburg, Va.: Virginia Polytechnic Institute and State University.

Balby, C. N. (2002). *Estudos de uso de catálogos on-line (OPACs): revisão metodológica e aplicação da técnica de análise de log de transações a um OPAC de biblioteca universitária brasileira.* (Unpublished Doctoral dissertation). Universidade de São Paulo, São Paulo, SP.

Banks, J., & Pracht, C. (2008). Reference desk staffing trends: A survey.[from *MasterFILE Premier.* EBSCO. http://web.ebscohost.com.proxy.oplin.org]. *Reference and User Services Quarterly, 48*(1), 54. Retrieved March 8, 2011

Barcelos, M. E., & Gomes, M. L. (2004, October). *Preparando sua biblioteca para avaliação do MEC.* Paper presented at the Seminário Nacional de Bibliotecas Universitárias, 13. Retrieved March 30, 2012, from http://repositorio.cfb.org.br/handle/123456789/495

Barley, S. (2010). The next multicoloured wave of computer interaction. *New Scientist, 206*(2763), 18–19. doi:10.1016/S0262-4079(10)61366-8.

Barner, K. (2011). The library is a growing organism: Ranganathan's Fifth Law of Library Science and the academic library in the digital era. *Library Philosophy and Practice.* Retrieved from http://digitalcommons.unl.edu/cgi/viewcontent.cgi?article=1579&context=libphilprac

Barsotti, R. (1990). *A informática na biblioteconomia e na documentação.* São Paulo: Polis, APB.

Bavousett, R. (2012) *Koha migration toolbox.* Retrieved on March 15, 2012, from https://www.gitorious.org/koha-toolbox/koha-migration-toolbox

Behan, J., & O'Keeffe, D. T. (2005, 28-30 November). *LUCAS: The library assistant robot, implementation and localization.* Paper presented at the International Conference on Computational Intelligence for Modelling, Control and Automation, and International Conference on Intelligent Agents, Web Technologies and Internet Commerce. Retrieved April 10, 2012, from IEEE Xplore.

Beldarrain, Y. (2006). Distance education trends: Integrating new technologies to foster student interaction and collaboration. *Distance Education, 27*(2), 139–153. doi:10.1080/01587910600789498.

Blazer, E. (2010). Do clickers click in the classroom? In *NABET, Northeastern Association Of Business, Economics, and Technology, 2010 Proceedings.* Retrieved from http://www.nabet.us/Archives/2010/NABET%20Proceedings%202010.pdf#page=20

Bohle, S. (2010). The Neil A. Armstrong Library and Archives: That's one small step for a virtual world library, one giant leap for education! *Journal of Virtual Worlds Research, 3*(1). Retrieved January 23, 2012, from http://journals.tdl.org/jvwr/article/view/1600

Borgman, C. (1997). From acting locally to thinking globally: A brief history of library automation. *The Library Quarterly, 67*(3), 215–249. doi:10.1086/629950.

Bower, B. L. (2001). Distance education: Facing the faculty challenge. *Online Journal of Distance Learning Administration, 4*(2). Retrieved from http://www.westga.edu/~distance/ojdla/summer42/bower42.pdf.

Bower, J., & Christensen, M. (1996). Disruptive technologies: Catching the wave. *Journal of Product Innovation Management, 13*(1), 75–76.

Bower, S. L., & Mee, S. A. (2010). Virtual delivery of electronic resources and services to off-campus users: A multifaceted approach. *Journal of Library Administration, 50*(5/6), 468–483. doi:10.1080/01930826.2010.488593.

Bowersox, T., Oberlander, C., Sullivan, M., & Black, M. (2011, March). GIST: Getting it System Toolkit: a Remix of Acquisitions, Collection Development, Discovery, Interlibrary Loan, and Technical Services. Paper presented at the OCLC ILLiad International Conference, Virginia Beach, VA. Retrieved April 14, 2012 from http://www.atlas-sys.com/2011-illiad-international-conference-session-archive/

Boyle, J. T., & Nicol, D. J. (2003). Using classroom communication systems to support interaction and discussion in large class settings. *Alt-J: Research in Learning Technology, 11*(3), 43–57. doi:10.1080/0968776030110305.

Brasil. Ministério da Educação. (2004). SINAES – Sistema Nacional de Avaliação da Educação Superior: da concepção à regulamentação. Retrieved March 30, 2012, from http://www.abem-educmed.org.br/pdf/sinaes.pdf

Brasil. Ministério da Educação. (2010). Instrumento de avaliação para credenciamento de Instituição de Educação Superior (Faculdade). Retrieved March 30, 2012, from http://download.inep.gov.br/download/superior/institucional/2010/instrumento_avaliacao_para_credenciamento_IES.pdf

Brasil. Ministério da Educação. (2010). *Instrumento de avaliação institucional externa*. Retrieved March 30, 2012, from http://download.inep.gov.br/download/superior/institucional/2010/instrumento_avaliacao_institucional_externa_recredenciamento.pdf

Brasil. Ministério da Educação. (2012). Sistema e-MEC: instituições de educação superior e cursos cadastrados. Retrieved March 30, 2012, from http://emec.mec.gov.br

Brasil. Ministério da Educação. (2012). Instrumento de Avaliação de Cursos de Graduação presencial e a distância. Retrieved March 30, 2012, from http://download.inep.gov.br/educacao_superior/avaliacao_cursos_graduacao/instrumentos/2012/instrumento_retificado_fevereiro_2012.pdf

Breeding, M. (2011, August). Serials solutions to build web-scale management solution. *Smart Libraries Newsletter, 31*(8).

Breeding, M. (2011, August). The beginning of the end of the ILS in academic libraries. *Smart Libraries Newsletter, 31*(8).

Breeding, M. (2011, December 5). OCLC WorldShare platform: OCLC brands and strengthens its Webscale strategy. *Information Today*. Retrieved from http://newsbreaks.infotoday.com/NewsBreaks/OCLC-WorldShare-Platform-OCLC-Brands-and-Strengthens-its-Webscale-Strategy-79208.asp

Breeding, M. (2011, January). Ex Libris marks progress in developing URM. *Smart Libraries Newsletter, 31*(1).

Breeding, M. (2011, May). Innovative interfaces to launch Sierra: A new generation automation platform. *Smart Libraries Newsletter, 31*(5).

Breeding, M. (2011, September). *The new frontier: Libraries seek new technology platforms for and end-user discovery, collection management, and preservation*. Paper presented at the Conference of the International Group of Ex Libris Users. Retrieved March 30, 2012, from http://igelu.org/conferences/haifa-2011/archive-of-presentations

Breeding, M. (2012). *Lib-Web-Cats*. Retrieved on March 3, 2012, from http://www.librarytechnology.org/libwebcats/

Breeding, M. (2012, April). Innovative interfaces joins the private equity club. *Smart Libraries Newsletter, 32*(4).

Breeding, M. (2012, June). OCLC navigates the realm of open linked data. *Smart Libraries Newsletter, 32*(6).

Breeding, M. (2012, March). Ex Libris: Alma, Aleph, and Primo. *Smart Libraries Newsletter, 32*(3).

Breeding, M. (2012, May 1). *Knowledge Base and Link Resolver Study: General Findings*. Prepared for the National Library of Sweden. Retrieved from http://www.kb.se/dokument/Knowledgebase_linkresolver_study.pdf

Breeding, M. (2010). *Next-gen library catalogs*. New York, NY: Neal-Schuman Publishers.

Breeding, M. (2010, June). Access articles through encore synergy. *Smart Libraries Newsletter, 30*(6), 2–4.

Breeding, M. (2011, November). Progress in the new generation of library service platforms. *Smart Libraries Newsletter, 31*(11), 3–4.

Breeding, M. (2012, April 1). Automation marketplace 2012: Agents of change. *Library Journal, 137*(5).

Breeding, M. (2012, August). Crossing the threshold: Boston College places Alma into production. *Smart Libraries Newsletter, 32*(8), 3–5.

Breeding, M. (2012, March). Looking forward to the next generation of Discovery Services. *Computers in Libraries, 32*(2), 28–32.

Breeding, M. (2012, March). OCLC Launches the WorldShare brand and wins an ARL member. *Smart Libraries Newsletter, 32*(3), 6–7.

Breeding, M. (2012, September). The Orbis Cascade Alliance selects Alma for consolidated automation. *Smart Libraries Newsletter, 32*(8), 6.

Broadus, R. N. (1991). The History of Collection Development. In C. Osburn & R. Atkinson (Eds.). Collection Management: A New Treatise. (Vol. 26. Part A, pp. 3-28). Greenwich, CT: JAI Press.

Brown-Syed, C. (2011). *Parents of invention the development of library automation systems in the late 20th century*. Santa Barbara: ABC-CLIO.

Brumfield, E. (2008). Using online tutorials to reduce uncertainty in information seeking behavior. *Journal of Library Administration, 48*(3/4), 365–377. doi:10.1080/01930820802289417.

Brynjolfsson, E., & McAfee, A. (2011). *Race against the machine: How the digital revolution is accelerating innovation, driving productivity, and irreversibly transforming employment and the economy*. Digital Frontier Press.

Buhay, D., Best, L. A., & McGuire, K. (2010). The effectiveness of library instruction: Do student response systems (clickers) enhance learning? *The Canadian Journal for the Scholarship of Teaching and Learning, 1*(1). Retrieved from http://ir.lib.uwo.ca/cgi/viewcontent.cgi?article=1006&context=cjsotl_rcacea doi:10.5206/cjsotl-rcacea.2010.1.5.

Bullard, R., & Wrosch, J. (2009). Eastern Michigan University's automated storage and retrieval system: 10 years later. *Journal of Access Services, 6*(3), 388–395. doi:10.1080/15367960902894187.

Burin, C. K., Lucas, E. R. D. O., & Hoffmann, S. G. (2004, October). *Informatizar por quê? a experiência das bibliotecas universitárias informatizadas da região sul*. Paper presented at the Seminário Nacional de Bibliotecas Universitárias, 13. Retrieved March 30, 2012, from http://www.pergamum.pucpr.br/redepergamum/trabs/Camila_K_Burin-Informatizar_por_que.pdf

Burke, J. J. (2009). *Neal-Schuman library technology companion: A basic guide for library staff*. Neal-Schuman Publishers.

Burnett, M., Cook, C., & Rothermel, G. (2004). End-user software engineering. *Communications of the ACM, 47*(9), 53–58. doi:10.1145/1015864.1015889.

Burns, R. W. (1975). *Library performance measures as seen in the descriptive statistics generated by a computer managed circulation system*. (Unpublished manuscript). Colorado State University Libraries, Fort Collins, CO.

Campbell, J., & Mayer, R. E. (2009). Questioning as an instructional method: Does it affect learning from lectures? *Applied Cognitive Psychology, 23*(6), 747–759. doi:10.1002/acp.1513.

Carmigniani, J., Furht, B., Anisetti, M., Ceravolo, P., Damiani, E., & Ivkovic, M. (2011). Augmented reality technologies, systems and applications. *Multimedia Tools and Applications, 51*(1), 341–377. doi:10.1007/s11042-010-0660-6.

Carmody, L. E. (2009). Clayton M. Christensen, Michael B. Horn, and Curtis W. Johnson: Disrupting class: How disruptive innovation will change the way the world learns. *Educational Technology Research and Development, 57*(2), 267–269. doi:10.1007/s11423-009-9113-1.

Carroll, D., & Cummings, J. (2010). Data Driven Collection Assessment Using a Serial Decision Database. *Serials Review, 36*(4), 227–239. doi:10.1016/j.serrev.2010.09.001.

Carvalho, C. H. A. D. (2004). Agenda neoliberal e a política pública para o ensino superior nos anos 90. Retrieved March 30, 2012, from http://www.anped.org.br/reunioes/27/gt11/t114.pdf

Carvalho, I. C. L. (1997, July). *Bibliotecas universitárias federais: o cenário da informatização*. Paper presented at the Congresso Brasileiro de Biblioteconomia e Documentação.

Castro, A. D., & Barboza, T. L. (2011, August). *Família ISIS: do Microisis ao ABCD*. Paper presented at the Congresso Brasileiro de Biblioteconomia, Documentação e Ciência da Informação, 24. Retrieved March 30, 2012, from http://www.febab.org.br/congressos/index.php/cbbd/xxiv/paper/view/510

Cataloging, U. C., & the Metadata Common Interest Group. (2009). Change in cataloging policy for government document monographs. Retrieved April 21, 2012, from http://libraries.universityofcalifornia.edu/hots/camcig/GovDocMonoCatChangePolicy.pdf

CDL Shared Collections and Services sets up Acquisitions at UC San Diego. (1998, June 10). *CDLINFO* News. Retrieved April 21, 2012, from http://www.cdlib.org/cdlinfo/1998/06/10/cdl-shared-collections-and-services-sets-up-acquisitions-at-uc-san-diego/

Chandler, D., & Munday, R. (2011). Embedded reporters. In *A dictionary of media and communication*. Oxford: Oxford University Press. Retrieved from www.oxfordreference.com/views/ENTRY.html?subview=Main&entry=t326.e848

Chan, E. K., & Knight, L. A. (2010). Clicking with your audience: Evaluating the use of personal response systems in library instruction. *Communications in Information Literacy, 4*(2), 192–201.

Chen, B. (2009). Barriers to adoption of technology-mediated distance education in higher-education institutions. *Quarterly Review of Distance Education, 10*(4), 333-338,399.

Cheng, R., Bischof, S., & Nathanson, A. J. (2002). Data collection for user-oriented library services: Wesleyan University Library's experience. *OCLC Systems & Services, 18*(4), 195–204. doi:10.1108/10650750210450130.

Chen, H., & Cheng, R. (2007). *ZK: AJAX without the JavaScript framework*. Apress.

Christensen, A. (2007, August 28). *A trend from Germany: Library chatbots in digital reference*. Paper presented at the International Ticer School, Digital Libraries à la Carte, Module 2: Technological Developments: Threats and Opportunities for libraries. Tilburg, Netherlands. Retrieved March 2, 2012 from http://www.slideshare.net/xenzen/a-trend-from-germany-library-chatbots-in-electronic-reference-presentation

Christensen, C. M. (1997). *The innovator's dilemma: When new technologies cause great firms to fail*. Boston, MA: Harvard Business School Press.

Christensen, C. M., Horn, M. B., & Johnson, C. W. (2011). *Disrupting class: How disruptive innovation will change the way the world learns*. New York: McGraw-Hill.

Chudnov, D. (2006). Coins for the link trail. *Library Journal, 131*, 8–10.

Churchill, W., & Langworth, R. M. (2008). *Churchill by himself: The definitive collection of quotations*. New York: Public Affairs.

Clark, B. R. (1983). *The higher education system*. Berkeley: University of California Press.

Clinton, D. (2008). *OpenSearch 1.1 (Draft 4)*. Retrieved August 13, 2009, from http://www.opensearch.org/Specifications/OpenSearch/1.1

Collins, B. L., Tedford, R., & Womack, H. D. (2008). 'Debating' the merits of clickers in an academic library. *North Carolina Libraries, 66*(1), 20–24.

Connolly, M., Cosgrave, T., & Krkoska, B. B. (2011). Mobilizing the library's web presence and services: A student-library collaboration to create the library's mobile site and iPhone application. *The Reference Librarian, 52*(1/2), 27–35.

Côrte, A. R. E., Almeida, I. M. D., Pellegrini, A. E., Lopes, I. O., Saenger, J. C., Esmeraldo, M. B. P., et al. (1999). Automação de bibliotecas e centros de documentação: o processo de avaliação e seleção de softwares. *Ciência da Informação, 28*(3), 241-256. Retrieved March 30, 2012, from http://dx.doi.org/10.1590/S0100-19651999000300002; doi: 10.1590/S0100-19651999000300002

Côrte, A. R. E., Almeida, I. M. D., Rocha, E. G., & Lago, W. G. D. (2002). Avaliação de softwares para bibliotecas e arquivos: uma visão do cenário nacional (2.ed. rev. e ampl. ed.). São Paulo: Polis.

Coyle, K. (2012, January 11). *Bibliographic framework: RDF and linked data*. Retrieved April 21, 2012 from http://kcoyle.blogspot.com/search/label/linked%20data

Coyle, K. (2006). Identifiers: Unique, persistent, global. *Journal of Academic Librarianship, 32*(4), 428–431. doi:10.1016/j.acalib.2006.04.004.

Creaghe, N., & Davis, D. A. (1986). Hard copy in transition: An automated storage and retrieval facility for low-use library materials. *College & Research Libraries, 47*(5), 495–499.

Cundiff, M., & Trail, N. (2007, June 25). *Using METS and MODS to create XML standards-based digital library applications*. Retrieved April 21, 2012, from http://www.loc.gov/standards/mods/presentations/mets-mods-morgan-ala07/

Cunha, M. B. D. (1985). A informática e a biblioteconomia: uma união de muito futuro. *Revista de Biblioteconomia de Brasília, 13*(1), 1-7. Retrieved March 30, 2012, from http://www.brapci.ufpr.br/documento.php?dd0=0000001822&dd1=fa22d

Dabbagh, N., & Kitsantas, A. (2012). Personal learning environments, social media, and self-regulated learning: A natural formula for connecting formal and informal learning. *The Internet and Higher Education, 15*, 3–8. doi:10.1016/j.iheduc.2011.06.002.

Dangel, H. L., & Wang, C. X. (2008). Student response systems in higher education: Moving beyond linear teaching and surface learning. *Journal of Educational Technology Development and Exchange, 1*(1), 93–104.

De Angeli, A., Johnson, G. I., & Coventry, L. (2001). The unfriendly user: Exploring social reactions to chatterbots. In M.G. Helander, H.M. Kalid, & T. Ming Po (Eds.), *Proceedings of the International Conference on Affective Human Factor Design, London: ASEA*. Retrieved from http://disi.unitn.it/~deangeli/homepage/lib/exe/fetch.php?media=references:deangeli:theunfriendlyuser.pdf

De Fino, M., & Lo, M. L. (2011). New Roads for Patron-Driven E-Books: Collection Development and Technical Services Implications of a Patron-Driven Acquisitions Pilot at Rutgers. *Journal of Electronic Resources Librarianship, 23*(4), 327–338. doi:10.1080/1941126X.2011.627043.

Dede, C. (2005). Planning for neomillennial learning styles. *EDUCAUSE Quarterly, 28*(1), 7–12.

Deleo, P. A., Eichenholtz, S., & Sosin, A. A. (2009). Bridging the information literacy gap with clickers. *Journal of Academic Librarianship, 35*(5), 438–444. doi:10.1016/j.acalib.2009.06.004.

Demas, S., & Miller, M. (2012). Curating Collective Collections — What's Your Plan? Writing Collection Management Plans. *Against the Grain, 24*(1), 65-68.

DeMillo, R. A. (2011). *Abelard to Apple: The fate of American colleges and universities*. Cambridge, MA: MIT Press.

Dempsey, J. V., Fisher, S. F., Wright, D. E., & Anderton, E. K. (2008). Training and support, obstacles, and library impacts on elearning activities. *College Student Journal, 42*(2), 630–636.

Dempsey, L. (2006). Libraries and the long tail: Some thoughts about libraries in a network age. *D-Lib Magazine, 12*(4). Retrieved from http://www.dlib.org/dlib/april06/dempsey/04dempsey.html doi:10.1045/april2006-dempsey.

Dempsey, L. (2006). The (digital) library environment: Ten years after. *Ariadne, 46*, Retrieved from http://www.ariadne.ac.uk/issue46/dempsey/.

Dempsy, L. (n.d.). *Big data.big trend*. Lorcan Dempsey's Weblog. Retrieved April 4, 2012, from http://orweblog.oclc.org/archives/002196.html

Dewey, B. I. (2004). The embedded librarian: Strategic campus collaborations. *Resource Sharing & Information Networks, 17*(1/2), 5–17. doi:10.1300/J121v17n01_02.

Drueke, J. (2001). St. Osmund's New Legacy: The Scriptorium Informs Electronic Text. *Libraries & Culture, 36*(4), 506–517. doi:10.1353/lac.2001.0066.

Dublin Core Metatdata Initiative. (2011). *User guide*. Retrieved April 17, 2012, from http://wiki.dublincore.org/index.php/User_Guide

Duderstadt, J. J. (2009). Possible futures for the research library in the 21st Century. *Journal of Library Administration, 49*(3), 217–225. doi:10.1080/01930820902784770.

Ehlers, U., & Schneckenberg, D. (Eds.). (2010). *Changing cultures in higher education: Moving ahead to future learning*. Berlin: Springer. doi:10.1007/978-3-642-03582-1.

Evans, G. E. (2000). *Developing Library and Information Center Collections*. Englewood, CO: Libraries Unlimited.

Fast, K., & Campbell, G. (2004). 'I still like Google': University student perceptions of searching OPACs and the web. *Proceedings of the American Society for Information Science and Technology, 41*(1), 138–146. doi:10.1002/meet.1450410116.

Figueiredo, N. M. D. (1986). Aplicação de computadores em bibliotecas: estudo comparativo entre países desenvolvidos e o Brasil. *Revista de Biblioteconomia de Brasília, 14*(2), 227–244.

Fischer, G. (2009). End-user development and meta-design: Foundations for cultures of participation. End-user development, 5435, 3-14). Berlin: Springer.

Forster, G., & Bell, A. (2006). The Subscription Libraries and Their Members. In Leedham-Green, E., & Webber, T. (Eds.), *The Cambridge History of Libraries in Britain and Ireland (Vol. III*, pp. 147–168). Cambridge, England: Cambridge University Press. doi:10.1017/CHOL9780521780971.014.

Forsyth, E. (2011). Ar U feeling happy? Augmented reality, apps and mobile access to local studies information. *Aplis, 24*(3), 125–132.

Franklin, B. (2005). Managing the Electronic Collection With Cost Per Use Data. *IFLA Journal, 31*(3), 241–248. doi:10.1177/0340035205058809.

Franklin, B., & Plum, T. (2008). Assessing the Value and Impact of Digital Content. *Journal of Library Administration, 48*(1), 41–57. doi:10.1080/01930820802029334.

Free Software Foundation. (2012). *GNU General Public License, version 2*. Retrieved on March 3, 2012, from http://www.gnu.org/licenses/gpl-2.0.html

Freeman, G. T. (2005). The library as place: Changes in learning patterns, collections, technology, and use. In *Library as place: Rethinking roles, rethinking space*. Washington, DC: Council on Library and Information Resources. Retrieved from http://www.clir.org/pubs/reports/pub129/freeman.html.

French, P. S., Culbertson, R., & Hsiung, L.-Y. (2002). One for nine: The shared cataloging program of the California Digital Library. *Serials Review, 28*(1), 4–12. doi:10.1016/S0098-7913(01)00169-1.

Frey, T. (2010). *The future of libraries: Beginning the great transformation*. Louisville, CO: DaVinci Institute. Retrieved April 16, 2012 from http://www.davinciinstitute.com/papers/the-future-of-libraries/

Gaat, T. (2008). *The LibX edition builder. University Libraries*. Blacksburg, Va.: Virginia Polytechnic Institute and State University.

Gaffney, M. (2011, March). *Item Shipped!: Purchase on Demand and ILLiad 8 Addons*. Paper presented at the ILLiad conference, Virginia Beach, VA. Retrieved April 14, 2012 from http://www.atlas-sys.com/ILLiadConf/Presentations/ItemShipped.pdf

Gallagher, K. (2010, August 4). HK systems to be acquired by Dematic. *Milwaukee, Wisconsin Journal Sentinel*. Retrieved April 23, 2012, from http://www.jsonline.com/business/99969954.html

Gelernter, D. (1985). Generative communication in Linda. *ACM Transactions on Programmable Language Systems, 7*(1), 80–112. doi:10.1145/2363.2433.

Gibbons, S. (2007). *The academic library and the net gen student: Making the connections*. Chicago: American Library Association.

Gilbert, S. W. (2002, February, 12). *The beauty of low threshold applications*. Retrieved April 17, 2012, from http://campustechnology.com/articles/2002/02/the-beauty-of-low-threshold-applications.aspx

Godwin, M. (2010). Disruptive technology: What is it? How can it work for professional writing? *Writing Instructor, 1*, Retrieved from http://www.writinginstructor.com/godwin2.

Gonzalez, A. C., & Westbrock, T. (2010). Reaching out with LibGuides: Establishing a working web of best practices. *Journal of Library Administration, 50*(5/6), 638–656. doi:10.1080/01930826.2010.488941.

Greene, M. A., & Meissner, D. E. (2005). More product, less process: Revamping traditional archival processing. *The American Archivist, 68*, 208–263.

Gregorio, J., & de Hora, B. (2007). *The atom publishing protocol*. RFC 5023. 2012. Retrieved from http://tools.ietf.org/html/rfc5023

Gregory, V. L. (2011). *Collection Development and Management for 21st Century Library Collections: An Introduction*. New York, NY: Neal-Schuman Publishers.

Guilherme, R. C. (2009). Introdução ao ABCD (Automação de Bibliotecas e Centros de Documentação). Retrieved March 4, 2012, from http://abcdisis.files.wordpress.com/2009/03/microsoft-word-apostila-abcd.pdf

Hanna, D. E., & Johnson, M. J. (2006). The challenges and opportunities of technology in higher education. *Effective Practices for Academic Leaders, 1*(6), 1–16.

Hannafin, M. J., & Kim, M. C. (2003). In search of a future: A critical analysis of research on web-based teaching and learning. *Instructional Science, 31*, 347–351. doi:10.1023/A:1024646328974.

Harrington, M., & Stovall, C. (2011, November). Contextualizing and Interpreting Cost Per Use for Electronic Journals. Paper presented at Charleston Conference, Charleston, SC. Retrieved April 23, 2011 from http://www.slideshare.net/group/2011-charleston-conference/slideshows/3

Harris, M. H. (1995). *History of Libraries in the Western World.* Metuchen, NJ: Scarecrow Press.

Haslam, M., Kwon, M., & Pearson, M. (2002). The automated storage and retrieval system (ASRS) in Lied Library. *Library Hi Tech, 20*(1), 71–89. doi:10.1108/07378830210420708.

Havens, A., & Storey, T. (2009). Making innovation personal. *Nextspace, 11*, 6–11.

Hazen, D. (1991). Selection: Function, Models, Theory. In Osburn, C., & Atkinson, R. (Eds.), *Collection Management: a New Treatise* (pp. 273–300). Greenwich, CT: JAI Press.

Hinchliffe, L., & Wong, M. (2010). From services-centered to student-centered: A "Wellness Wheel" approach to developing the library as an integrative learning commons. *College & Undergraduate Libraries, 17*(2/3), 213–224. doi:10.1080/10691316.2010.490772.

Hoffman, C., & Goodwin, S. (2006). A clicker for your thoughts: Technology for active learning. *New Library World, 107*(9/10), 422–433. doi:10.1108/03074800610702606.

Hoffman, S. (2011). Embedded academic librarian experiences in online courses. *Library Management, 32*(6), 444–456. doi:10.1108/01435121111158583.

Holley, B. (2012). Demise of Traditional Collection Development. *Against the Grain, 24*(1), 30-31.

Horava, T. (2010). Challenges and Possibilities for Collection Management in a Digital Age. *Library Resources & Technical Services, 54*(3), 142–152.

Hot new tool? ITSO CUL. (2004). *Backstory, 1*(1). Retrieved April 10, 2012 from http://www.library.cornell.edu/backstory/v1n1/itsofeature.htm

Hricko, M. (2010). Using microblogging tools for library services. *Journal of Library Administration, 50*(5/6), 684–692. doi:10.1080/01930826.2010.488951.

Huber, J. (2011). *Lean library management: Eleven strategies for reducing costs and improving services.* New York, NY: Neal-Schuman.

Hudson, M., McGowan, L., & Smith, C. (2011). Technology and learner motivation in library instruction: a study of personal response systems. *Indiana Libraries, 30*(1), 20–27.

International Records Management Trust & International Council on Archives. (1999). *Automating records services.* Retrieved April 22, 2012, from http://www.irmt.org/documents/educ_training/public_sector_rec/IRMT_automating_rec_serv.doc

Jensen, K. (2006). Universities and Colleges. In *The Cambridge History of Libraries in Britain and Ireland* (Vol. I, pp. 345–362). Cambridge, England: Cambridge University Press. doi:10.1017/CHOL9780521781947.016.

Johnson, L., Levine, A., Smith, R., & Stone, S. (2010). *The 2010 Horizon Report.* Austin, Texas: The New Media Consortium. Retrieved from http://wp.nmc.org/horizon2010/

Johnson, L., Adams, S., & Cummins, M. (2012). *The NMC horizon report: 2012 higher education edition.* Austin, Texas: The New Media Consortium.

Johnson, M. D. (2003). Turning Ph.D.'s into librarians. *The Chronicle of Higher Education, 50*(8), C4.

Jones, D. (2011). On-Demand Information Delivery: Integration of Patron-Driven Acquisition into a Comprehensive Information Delivery System. *Journal of Library Administration, 51*(7/8), 764–776. doi:10.1080/01930826.2011.601275.

Junco, R., & Cole-Avent, G. A. (2008). An introduction to technologies commonly used by college students. *New Directions for Student Services*, (124): 3–17. doi:10.1002/ss.292.

Kaplan, M. (2009). Library Automation. In Nof, S. J. (Ed.), *Springer Handbook of Automation* (pp. 1285–1298). Berlin: Springer. doi:10.1007/978-3-540-78831-7_72.

Kargbo, J. A. A. (2009). Automation: Whither academic libraries? *Information Technology for Development*, *15*(1), 43–51. doi:10.1002/itdj.20078.

Keiser, B. (2010). Library of the future -- Today! *Searcher*, *18*(8), 18-54. Retrieved March 30, 2012, from http://www.allbusiness.com/media-telecommunications/information-services-libraries/15180365-1.html

Kelly, K. (2010). *What technology wants*. New York: Viking.

Kinner, L., & Rigda, C. (2009). The integrated library system: From daring to dinosaur? *Journal of Library Administration*, *49*(4), 401–417. doi:10.1080/01930820902832546.

Koha Community. (2012). *Koha migration tools repository*. Retrieved on March 30, 2012, from http://git.koha-community.org/gitweb/?p=contrib/migration-tools.git;a=summary

Koha Community. (2012). Retrieved on March 3, 2012, from http://www.koha-community.org

Koszalka, T., & Ntloedibe-Kuswani, G. S. (2010). Literature on the safe and disruptive learning potential of mobile technologies. *Distance Education*, *31*(2), 139–157. doi:10.1080/01587919.2010.498082.

Kountz, J. (1987). Robots in the library: Automated storage and retrieval systems. *Library Journal*, *112*(20), 67.

Kreber, C., & Cranton, P. A. (2000). Exploring the scholarship of teaching. *The Journal of Higher Education*, *71*(4), 476–495. doi:10.2307/2649149.

Kyrillidou, M., & Morris, S. (2011). *ARL statistics 2008-2009*. Washington, DC: Association of Research Libraries.

Lage, Â. (1989, June). *Automação de bibliotecas universitárias do Brasil: tendências e perspectivas*. Paper presented at the Seminário Nacional de Bibliotecas Universitárias, 6. Retrieved March 30, 2012, from http://www.dominiopublico.gov.br/download/texto/me001650.pdf

Lamas, S. D. F. T. B. (2007). Automação de bibliotecas: do ISIS à biblioteca do futuro. *Pesquisa Brasileira em Ciência da Informação e Biblioteconomia*, *2*(1).

Lang, B., Gerz, M., Meyer, O., & Sim, D. (2008). *An enterprise architecture for the delivery of a modular interoperability solution*. Retrieved April 21, 2012, from http://ftp.rta.nato.int/public//PubFullText/RTO/MP%5CRTO-MP-IST-101///MP-IST-101-08.doc

Leedham-Green, E. S., & Webber, T. (Eds.). (2006). *The Cambridge History of Libraries in Britain and Ireland*. Cambridge, England: Cambridge University Press. doi:10.1017/CHOL9780521781947.

Lee, H. W., Lim, K. Y., & Grabowski, B. L. (2010). Improving self-regulation, learning strategy use, and achievement with metacognitive feedback. *Educational Technology Research and Development*, *58*(6), 629–648. doi:10.1007/s11423-010-9153-6.

Legris, P., Ingham, J., & Collerette, P. (2003). Why do people use information technology? A critical review of the technology acceptance model. *Information & Management*, *40*(3), 191–204. doi:10.1016/S0378-7206(01)00143-4.

Leonard, E. (2011). *Lost in translation: The emerging technology librarian and the new technology*. Retrieved April 21, 2012, from http://www.slideshare.net/eleonard/lost-in-translation-8440125

Lerner, F. A. (1998). *The Story of Libraries: From the Invention of Writing to the Computer Age*. New York, NY: Continuum.

Lerner, F. A. (1999). *Libraries Through the Ages*. New York, NY: Continuum.

Library of Congress. (2011, October 31). *A bibliographic framework for the digital age*. Retrieved April 21, 2012, from http://www.loc.gov/marc/transition/news/framework-103111.html

Lima, G. Â. B. (1999). Softwares para automação de bibliotecas e centros de documentação na literatura brasileira até 1998. *Ciência da Informação, 28*(3), 310-321. Retrieved March 30, 2012, from http://dx.doi.org/10.1590/S0100-19651999000300009; doi: 10.1590/S0100-19651999000300009

Lima, G. Â. B. D. O., & Mendonça, A. M. (1998). A Utilização do MicroISIS no Brasil. *Perspectiva em Ciência da Informação, 3*(1), 125-136. Retrieved March 30, 2012, from http://portaldeperiodicos.eci.ufmg.br/index.php/pci/article/view/601

Lin, P., Chen, K., & Chang, S. (2010). Before there was a place called library – library space as an Invisible factor affecting students' learning. *Libri, 60*(4), 339–351. doi:10.1515/libr.2010.029.

Lockwood, F. (2001). Leadership, disruptive technologies, lessons learned and changes in learning and teaching. *Open Learning, 17*(3), 199–202. doi:10.1080/0268051022000048200.

Lugg, R. (2011). Collection for the Moment: Patron-Driven Acquisitions as Disruptive Technology. In Swords, D. A. (Ed.), *Patron-Driven Acquisitions: History and Best Practices* (pp. 7–22). Berlin, Germany: De Gruyter Saur. doi:10.1515/9783110253030.7.

Lynch, C. (2000). From automation to transformation: Forty years of libraries and information technology in higher education. *EDUCAUSE Review*, 60–68. Retrieved from http://net.educause.edu/apps/er/erm00/pp060068.pdf.

Lynch, C. A. (1991). Evolution in action: The demise of the integrated library system and the rise of networked information resources. *Library Software Review, 10*(5), 336–337.

Lynn, V. A., FitzSimmons, M., & Robinson, C. K. (2011). Special report: Symposium on transoformational change in health sciences libraries: Space, collections, and roles. *Journal of Medical Library Association, 99*(1), 82-87. Retrieved April 16, 2012, from http://www.ncbi.nlm.nih.gov/pmc/articles/PMC3016656/

Mahmood, M. A., Tariq, M., & Javed, S. (2011). Strategies for active learning: An alternative to passive learning. *Academic Research International, 1*(3), 193–198.

Mann, L., & Chan, J. (Eds.). (2011). *Creativity and innovation in business and beyond: Social science perspectives and policy implications*. New York, NY: Taylor and Francis.

Marcelino, S. C. (2009). A contribuição da biblioteca para a construção e difusão do conhecimento no Instituto Nacional de Pesquisas Espaciais (Inpe). *Ciência da Informação, 38*(2), 80-95. Retrieved March 30, 2012, from http://revista.ibict.br/index.php/ciinf/article/view/1090

Marques, I. D. C. (2003). Minicomputadores brasileiros nos anos 1970: uma reserva de mercado democrática em meio ao autoritarismo. *História, Ciências, Saúde-Manguinhos, 10*(2), 657-681. Retrieved March 30, 2012, from http://www.scielo.br/scielo.php?script=sci_arttext&pid=S0104-59702003000200008

Marshall, S. (2010). Change, technology and higher education: Are universities capable of organisational change? *ALT-J. Research in Learning Technology, 18*(3), 179–192. doi:10.3402/rlt.v18i3.10762.

Mathews, B. (2012). Think like a startup: A white paper to inspire library entrepreneurialism. *VTechWorks Institutional Repository, Virginia Tech University*. Retrieved April 21, 2012, from http://hdl.handle.net/10919/18649

Mayer, R. E., Stull, A., DeLeeuw, K., Almeroth, K., Bimber, B., Chun, D., & Zhang, H. (2009). Clickers in college classrooms: Fostering learning with questioning methods in large lecture classes. *Contemporary Educational Psychology, 34*(1), 51–57. doi:10.1016/j.cedpsych.2008.04.002.

Mazzillo, C. A., Araujo, D. K. D., Viana, M. M. M., Crespo, I. M., & Naumann, P. (2011). Die Zentralbibliothek Irmão José Otão an der Päpstlich-Katholischen Universität von Rio Grande do Sul: Ein Beispiel für Innovation. *BIBLIOTHEK Forschung und Praxis, 35*(2), 231-235. Retrieved March 30, 2012, from http://www.degruyter.com/view/j/bfup.2011.35.issue-2/bfup.2011.032/bfup.2011.032.xml

McBride, K. (2010). Leadership in higher education: Handling faculty resistance to technology through strategic planning. *Academic Leadership Journal, 8*(4).

McCarthy, C. M. (1983). Library automation in Brazil: The state of the art. *Program, 17*(4), 233-240. Retrieved March 30, 2012, from http://www.emeraldinsight.com/journals.htm?articleid=1671037&show=abstract; doi:10.1108/eb046868

McCarthy, C. M., & Neves, F. I. (1990). Levantamento geral da automação de bibliotecas no Brasil. *Revista Biblioteconomia de Brasília, 18*(2), 51-57. Retrieved March 30, 2012, from http://www.brapci.ufpr.br/documento.php?dd0=0000002623&dd1=cd841

McCarthy, C. M. (1982). *The automation of libraries and bibliographic information systems in Brazil. Leicestershire.* Loughborough University of Technology.

McConatha, D., Praul, M., & Lynch, M. J. (2008). Mobile learning in higher education: An empirical assessment of a new educational tool. *Turkish Online Journal of Educational Technology, 7*(3), 15-21. Retrieved from http://www.tojet.net/articles/v7i3/732.pdf

McMurtrie, D. C. (1967). *The Book: The Story of Printing & Bookmaking.* New York, NY: Oxford University Press.

Mentor Public Library Board of Trustees. (2009, September 16). Board of trustees meeting minutes. Resolution #09-104.

Mesbah, A., & van Deursen, A. (2008). A component- and push-based architectural style for AJAX applications. *Journal of Systems and Software, 81*(12), 2194–2209. doi:10.1016/j.jss.2008.04.005.

Metz, P., & Cosgriff, J. (2000). Building a Comprehensive Serials Decision Database at Virginia Tech. *College & Research Libraries, 61*(4), 324.

Meyer, K. A. (2010). The role of disruptive technology in the future of higher education. *EDUCAUSE Quarterly, 33*(1), 6. Retrieved from http://www.educause.edu/EDUCAUSE+Quarterly/EDUCAUSEQuarterlyMagazineVolum/TheRoleofDisruptiveTechnologyi/199378.

Mikawa, M., Morimoto, Y., & Tanaka, K. (2010, 12-15 September). *Guidance method using laser pointer and gestures for librarian robot.* Paper presented at the 19th IEEE International Symposium on Robot and Human Interactive Communication. Retrieved April 11, 2012, from IEEE Xplore.

Modesto, F. (2006). O CDS/ISIS morreu? Viva o CDS/ISIS livre. Retrieved March 30, 2012, from http://www.ofaj.com.br/colunas_conteudo.php?cod=274

Monley, B. (2010). *Vending machine collection dispensers in libraries: Discussion paper.* Retrieved April 10, 2012, from http://www.slq.qld.gov.au/info/publib/research/?a=183480

Mosher, P. H. (1991). Reviewing for Preservation, Storage, and Weeding. In Osburn, C., & Atkinson, R. (Eds.), *Collection Management: a New Treatise* (pp. 373–391). Greenwich, CT: JAI Press.

Murray, P. (2008, June 20). *Riding the waves of content and change.* Retrieved April 16, 2012, from http://dltj.org/article/riding-the-waves/

Nelson, N. (1991). *Library technology 1970-1990: Shaping the library of the future: General session entitled "Mainstreets, landmarks, and cross roads: Mapping library technology."* 5th Annual conference on computers. Westport, CT. London: Meckler.

Nicholson, B. R. (2011). *Libx 2.0. University Libraries.* Blacksburg, Va.: Virginia Polytechnic Institute and State University.

NIH Data Sharing Information - Main Page. (n.d.). Retrieved August 29, 2012, from http://grants.nih.gov/grants/policy/data_sharing/

Norberg, B., Orcutt, D., & Vickery, J. (2011, November). New Tricks For Old Data Sources: Mashups, Visualizations and Questions Your ILS Has Been Afraid to Answer. Paper presented at the Charleston Conference, Charleston, SC. OCLC. *Creating the Conspectus.* Retrieved April 15, 2012, from http://www.oclc.org/research/activities/past/rlg/conspectus.htm

Novick, D., Elizalde, E., & Bean, N. (2007). *Toward a more accurate view of when and how people seek help with computer applications.* Paper presented at the SIGDOC '07: Proceedings of the 25th annual ACM international conference on Design of communication, El Paso, Texas, USA.

Oakey, R. (2007). Are disruptive technologies disruptive[disruptive technologies]. *Engineering Management, 17*(2), 10. doi:10.1049/em:20070201.

Oakleaf, M. (2010). *Value of academic libraries: A comprehensive research review and report.* Assoc. of College and Research Libraries.

Oblinger, D. G. (2005). Learners, learning and technology: The Educause learning initiative. *EDUCAUSE Review, 40*(5). Retrieved from http://connect.educause.edu/Library/EDUCAUSE+Review/ EDUCAUSEReviewMagazineVol/40585.

OCLC. (2003). *xISBN Web Service*. Retrieved April 22, 2012, from http://www.worldcat.org/affiliate/webservices/xisbn/app.jsp

OCLC. (2005). *Perceptions of libraries and information resources*. Retrieved April 22, 2012, from http://www.oclc.org/reports/2005perceptions.htm

OCLC. (2007). *xISSN* (Web Service). Retrieved April 22, 2012, from http://xissn.worldcat.org/xissnadmin/index.htm

OCLC. (2012). *WorldCat Registry*. Retrieved from http://www.oclc.org/registry/default.htm

Office of the Governor. State of Ohio. (2009). Governor Strickland's proposed framework to balance FYs 10/11. Retrieved from http://obm.ohio.gov/document.aspx?ID=d9f88041-a555-4840-a211-23f69bbf1829

Ohira, M. L. B. (1992). Automação de bibliotecas: Utilização do MicroISIS. *Ciência da Informação, 21*(3). Retrieved March 30, 2012, from http://revista.ibict.br/index.php/ciinf/article/viewArticle/1306

Ohira, M. L. B. (1994). Biblioinfo base de dados sobre automação em bibliotecas (informática documentária): 1986-1994. *Ciência da informação, 23*(3). Retrieved March 30, 2012, from http://revista.ibict.br/cienciadainformacao/index.php/ciinf/article/viewArticle/1159

Osguthorpe, R. T., & Graham, C. R. (2003). Blended learning environments. *Quarterly Review Of Distance Education, 4*(3), 227–233.

Owusu-Ansah, E. K. (2004). Information literacy and higher education: Placing the academic library in the center of a comprehensive solution. *Journal of Academic Librarianship, 30*(1), 3–16. doi:10.1016/j.jal.2003.11.002.

Palloff, R., & Pratt, K. (2001). Lessons from the cyberspace classroom. *17TH Annual Conference on Distance Teaching and Learning*. Retrieved from http://www.uwex.edu/disted/conference/resource_library/proceedings/01_20.pdf

Paulista, U. E. Assessoria de Comunicação e Imprensa. (2011). Unesp lança novo sistema de busca bibliográfica. *Unesp Informa, 2*(21). Retrieved March 30, 2012, from http://www.unesp.br/unespinforma/21/novo-sistema-de-busca-bibliografica

PerlMonks. (2012). Retrieved on March 28, 2012, from http://www.perlmonks.org

Petit, J. (2011). Twitter and Facebook for User Collection Requests. *Collection Management, 36*(4), 253–258. doi:10.1080/01462679.2011.605830.

Pilgrim, M. (2005). *Greasemonkey hacks: Tips & tools for remixing the Web with Firefox*. Sebastopol, CA: O'Reilly Media.

Pitcher, K., Bowersox, T., Oberlander, C., & Sullivan, M. (2010). Point-of-Need Collection Development: The Getting It System Toolkit (GIST) and a New System for Acquisitions and Interlibrary Loan Integrated Workflow and Collection Development. *Collection Management, 35*(3/4), 222–236. doi:10.1080/01462679.2010.486977.

Powell, B. B. (2009). *Writing: Theory and history of the technology of civilization*. Chichester, U.K. Malden, MA: Wiley-Blackwell.

Prado, N. S., & Abreu, J. D. (2002). Informatização das bibliotecas universitárias do estado de Santa Catarina: cenário. *Revista ACB, 7*(2). Retrieved March 30, 2012, from http://revista.acbsc.org.br/index.php/racb/article/view/394

Princeton University Cataloging. (2004, February 24). *Single-record approach for e-journals –Princeton practice, mid-1999+*. Retrieved April 21, 2012, from http://library.princeton.edu/departments/tsd/katmandu/electronic/single.html

Provider-Neutral E-Monograph Record Task Group. (2009, July 30). *Provider-neutral e-monograph record task group report*. Retrieved April 21, 2012, from http://www.loc.gov/catdir/pcc/bibco/PN-Final-Report.pdf

QR. (2011). In *OED online*. Oxford: Oxford University Press. Retrieved from www.oed.com/view/Entry/155604?redirectedFrom=qr%20code#eid246543457

Radford, M., et al. (2010, July). *Taking the library with you: VR going mobile*. Paper presented at the ALA Virtual Conference. Retrieved from http://www.amandaclaypowers.com/2010/06/28/taking-the-library-with-you-vr-going-mobile/

Ramos-Garijo, R., Prats, M., Sanz, P. J., & Del Pobil, A. P. (2003, 5-8 October). *An autonomous assistant robot for book manipulation in a library*. Paper presented at the IEEE International Conference on Systems, Man and Cybernetics. Retrieved April 11, 2012, from IEEE Xplore.

Ranasinghe, R. H. I. S. (2008). How Buddhism Influenced the Origin and Development of Libraries in Sri Lanka (Ceylon): From the Third Century BC to the Fifth Century AD. *Library History, 24*(4), 307–312. doi:10.1179/174581608X381602.

Ranjan, J. (2008). Impact of information technology in academia. *International Journal of Educational Management, 22*(5), 442–455. doi:10.1108/09513540810883177.

Rapp, D. (2011). Robot visions. *Library Journal, 136*(15), 20–24.

Rathemacher, A. J., Cerbo, M. A., II, & Li, Y. (2011). New England technical services librarians Spring 2011 Conference: 2020 vision: A new decade for technical services. *Technical Services Department Faculty Publications. Paper 42*. Retrieved April 16, 2012, from http://digitalcommons.uri.edu/lib_ts_pubs/42

Reese, T. (2012). *MARCEdit*. Retrieved on March 15, 2012, from http://people.oregonstate.edu/~reeset/marcedit/html/index.php

Ricketts, P. (2008, November 5). *Australian federal government and Web 2.0*. Retrieved April 21, 2012, from http://oracle-gtmi-anz.blogspot.com/search?q=little+choice

Rizzuto, T. E., & Reeves, J. (2007). A multidisciplinary meta-analysis of human barriers to technology implementation. *Consulting Psychology Journal: Practice and Research, 59*(3), 226–240. doi:10.1037/1065-9293.59.3.226.

Robredo, J. (1981). *Panorama dos planos e projetos de automação das bibliotecas universitárias brasileiras*. Paper presented at the Seminário Nacional de Bibliotecas Universitárias, 2.

Rowley, G. S., & Association of Research Libraries/Office of Management, S. (1995). *Organization of Collection Development*. United States of America: Association of Research Libraries. Office of Management Services.

Rowley, J. (1994). *Informática para bibliotecas*. Brasília: Briquet de Lemos/Livros.

Russo, M. (2010). *Fundamentos de Biblioteconomia e Ciência da Informação*. Rio de Janeiro: E-papers.

Saffady, W. (1989). Library automation: An overview. *Library Trends, 37*(3), 269–281.

Sambaquy, L. (1960). Article. *American Documentation (pre-1986), 11*(3), 205-205. Retrieved March 30, 2012, from http://search.proquest.com/docview/195446789

Sambaquy, L. D. Q. (1972). A Biblioteca do Futuro. *Revista da Escola de Biblioteconomia da UFMG, 1*(1), 62-68. Retrieved March 30, 2012, from http://www.brapci.ufpr.br/documento.php?dd0=0000001890&dd1=1b795

Santa Cruz, U. C. University Library. (2012). *Technical services mission statement*. Retrieved April 21, 2012, from http://library.ucsc.edu/content/technical-services-mission-statement

Saviani, D. (2010). A expansão do ensino superior no Brasil: Mudanças e continuidades. *Poíesis Pedagógica, 8*(2), 4-17. Retrieved March 30, 2012, from http://www.revistas.ufg.br/index.php/poiesis/article/view/14035

Sayão, L. F., Marcondes, C. H., Fernandes, C. C., & Medeiros, L. P. M. (1989). Avaliação dos processos de automação em bibliotecas universitárias. *Trans-in-informação, 1*(2), 233-254. Retrieved March 30, 2012, from http://www.brapci.ufpr.br/documento.php?dd0=0000000139&dd1=d6193

Sayers, E. (2010-). Entrez programming utilities help. Retrieved from http://www.ncbi.nlm.nih.gov/books/NBK25500/

Schmidt, A. (2011). Ready for a UX librarian? *Library Journal, 136*(18), 24.

Schonfeld, R. C., & Housewright, R. (2010) *Faculty Survey 2009: Strategic Insights for Libraries, Publishers, and Societies*. Retrieved April 21, 2012 from http://www.ithaka.org/ithaka-s-r/research/faculty-surveys-2000-2009/faculty-survey-2009

Schrum, L., & Hong, S. (2002). Dimensions and strategies for online success: Voices from experienced education. *Journal of Asynchronous Learning Networks, 6*(1), 57–67.

Scott, G. (2003). Effective change management in higher education. *EDUCAUSE Review, 38*(6), 64–80.

Selim, H. (2007). Critical success factors for e-learning acceptance: Confirmatory factor models. *Computers & Education, 49*(2), 396–413. doi:10.1016/j.compedu.2005.09.004.

Sennet, A. (2011). Ambiguity. In E. N. Zalta (Ed.), *The Stanford Encyclopedia of Philosophy*. Retrieved from http://plato.stanford.edu/archives/sum2011/entries/ambiguity/

Shaffer, D., & Collura, M. (2009). Evaluating the effectiveness of a personal response system in the classroom. *Teaching of Psychology, 36*(4), 273–277. doi:10.1080/00986280903175749.

Shanahan, M. (2009). The frame problem. In E. N. Zalta (Ed.), *The Stanford Encyclopedia of* Philosophy. Retrieved from http://plato.stanford.edu/archives/win2009/entries/frame-problem/

Sharpe, R. (2006). The Medieval Librarian. In Leedham-Green, E.S. & Webber. T. (Eds.). The Cambridge History of Libraries in Britain and Ireland (Vol. I, pp. 218-241). Cambridge, England: Cambridge University Press.

Shelburne, W. A. (2009). E-book usage in an academic library: User attitudes and behaviors. *Library Collections, Acquisitions & Technical Services, 33*(2-3), 58–72. doi:10.1016/j.lcats.2009.04.002.

Shen, L., Cassidy, E. D., Elmore, E., Griffin, G., Manolovitz, T., & Martinez, M. et al. (2011). Head First into the Patron-Driven Acquisition Pool: A Comparison of Librarian Selections Versus Patron Purchases. *Journal of Electronic Resources Librarianship, 23*(3), 203–218. doi:10.1080/1941126X.2011.601224.

Shih, Y., & Mills, D. (2007). Setting the new standard with mobile computing in online learning. *International Review of Research in Open and Distance Learning, 8*(2), 1–16.

Shirato, L., Cogan, S., & Yee, S. G. (2001). The impact of an automated storage and retrieval system on public services. *RSR. Reference Services Review, 29*(3), 253–261. doi:10.1108/EUM0000000006545.

Shumaker, D. (2009). Who let the librarians out? *Reference and User Services Quarterly, 48*(3), 239–242.

Shumaker, D., & Makins, A. (2012). Lessons from successful embedded librarians. *Information Outlook, 16*(3), 10–12.

Silva, F. C. C. D., & Favaretto, B. (2005). Uso de softwares para o gerenciamento de bibliotecas: Um estudo de caso da migração do sistema Aleph para o sistema Pergamum na Universidade de Santa Cruz do Sul. *Ciência da Informação, 34*(2), 105-111. Retrieved fromhttp://dx.doi.org/10.1590/S0100-19652005000200011; doi: 10.1590/S0100-19652005000200011

Silveira, A., Knoll, M. M. D. D. C., & Araújo, F. M. B. G. D. (1990). Mini-micro CDS/ISIS: Uma proposta de aplicação no ensino da informática na Biblioteconomia e Ciência da Informação. *Revista de Biblioteconomia de Brasília, 18*(2). Retrieved March 30, 2012, from http://www.brapci.ufpr.br/documento.php?dd0=0000002624&dd1=912cd

Silvis, G. A. (2012). *The impending demise of the local OPAC*. Retrieved April 21, 2012, from www.wils.wisc.edu/events/opac06/impending_demise.pdf

Sitepal. (2008, May). *Virtual meeting with Kathleen improved sales productivity by 50%*. Retrieved January 23, 2012 from http://www.sitepal.com/pdf/casestudy/loftusphotography.pdf

Sitepal. (2010, February). *Daughter nature increased online sales by 40%*. Retrieved January 23, 2012, from http://www.sitepal.com/ccs2/oddcast/casestudy/14.pdf

Slote, S. J. (1997). *Weeding Library Collections: Library Weeding Methods*. Englewood, CO: Libraries Unlimited.

Smith, A. G. (1997). Testing the surf: criteria for evaluating Internet information resources. *The Public-Access Systems Review, 8*(3), 5-23. Retrieved from http://journals.tdl.org/pacsr/article/viewFile/6016/5645

Smith, A., Rainie, L., & Zickuhr, K. (2011, July 19). *College students and technology* (Rep.). Retrieved from http://pewinternet.org/Reports/2011/College-students-and-technology/Report.aspx

Smith, S. (2011). The technical services-public services connection: Tips for managing change. *American Assoication of Law Libraries Spectrum, 16*(2). Retrieved April 16, 2012, from http://www.aallnet.org/main-menu/Publications/spectrum/Vol-16/No-2/tech.html

Sompel, H. V. D., & Beit-Arie, O. (2001). Open linking in the scholarly information environment using the OpenURL Framework. *D-Lib Magazine, 7*(3).

Spalding, T. (2010, February 5). *Why are you for killing libraries?* Thing-ology blog. Retrieved from http://www.librarything.com/blogs/thingology/2010/02/why-are-you-for-killing-libraries/

Stahl, R., & Correa-Morris, C. Hwang. S. (2012, April). Solving the Complexities of Ebook Record Management in Millennium. Paper presented at the Innovative Users Group Meeting, Chicago, IL.

Staley, D. J., & Malenfant, K. J. (2010). Futures thinking for academic librarians: Higher education in 2025. *Information Services & Use, 30*(1/2), 57–90.

Steele, C. R. (1987). From punched cards to robots: Our ascent into technology. *Wilson Library Bulletin, 62*(2), 29–32.

Stephens, J., & Presley, R. (1996). *EDI: Slow walk to fast forward.* Retrieved April 21, 2012, from http://www.ala.org/acrl/sites/ala.org.acrl/files/content/conferences/pdf/stephens99.pdf

Stoffle, C. J., & Cuillier, C. (2011). From surviving to thriving. *Journal of Library Administration, 51*(1), 130–155. doi:10.1080/01930826.2011.531645.

Sullivan, R. (2010). Common knowledge: Learning spaces in academic libraries. *College & Undergraduate Libraries, 17*(2), 130–148. doi:10.1080/10691316.2010.481608.

Sun News Staff. (2009, June 24). Governor Ted Strickland's proposed budget cuts endanger future of public libraries. *Sun News.* Retrieved March 13, 2011, from http://blog.cleveland.com/garfieldmaplesun/2009/06/governor_ted_stricklands_propo.html

Su, S., & Kuo, J. (2010). Design and development of Web-based information literacy tutorials. *Journal of Academic Librarianship, 36*(4), 320–328. doi:10.1016/j.acalib.2010.05.006.

Suthakorn, J., Lee, S., Zhou, Y., Thomas, R., Choudhury, S., & Chirikjian, G. S. (2003). *A robotic library system for an off-site shelving facility.* Retrieved April 10, 2012, from http://custer.lcsr.jhu.edu/wiki/images/c/c9/Suthakorn02_a.pdf

Swierczek, J. (2010). Using Web 2.0 applications in technical services: An ALCTS webcast. Retrieved April 16, 2012, from http://www.ala.org/alcts/confevents/upcoming/webinar/092910web

Tang, F. (2010). Reference tools in Second Life®: Implications for real life libraries. *New Library World, 111*(11/12), 513–525. doi:10.1108/03074801011094886.

Teixeira, L. A., Oliveira, L. R. V., Lapa, R. C., & Assunção, R. V. D. (2011, August). *Pergamum: Serviços web e auto-atendimento na Biblioteca da Universidade Federal de Mato Grosso do Sul (UFMS).* Paper presented at the Congresso Brasileiro de Biblioteconomia, Documentação e Ciência da Informação (p. 24). Retrieved March 30, 2012, from http://febab.org.br/congressos/index.php/cbbd/xxiv/paper/view/293

The Document Foundation. (2012). *LibreOffice.* Retrieved on March 12, 2012, from http://www.libreoffice.org

Thomsen, E. (2008). Welcome to the NOBLE SwapShop. Retrieved April 20, 2012, from http://www.noblenet.org/swapshop/

Tilevich, E., & Back, G. (2008). *'Program, enhance thyself!': Demand-driven pattern-oriented program enhancement.* Paper presented at the Proceedings of the 7th international Conference on Aspect-Oriented Software Development.

Tomizawa, T., Ohya, A., & Yuta, S. (2003, 14-19 September). *Remote book browsing system using a mobile manipulator.* Paper presented at the IEEE International Conference on Robotics and Automation. Retrieved April 10, 2012, from IEEE Xplore.

Traxler, J. (2005). Defining mobile learning. In *IADIS International Conference Mobile Learning Proceedings* (pp. 261-266). International Association for the Development of the Information Society.

Trees, A., & Jackson, M. (2007). The learning environment in clicker classrooms: Student processes of learning and involvement in large university-level courses using student response systems. *Learning, Media and Technology, 32*(1), 21–40. doi:10.1080/17439880601141179.

Tucker, C. (2009). Benchmarking Usage Statistics in Collection Management Decisions for Serials. *Journal of Electronic Resources Librarianship, 21*(1), 48–61. doi:10.1080/19411260902858581.

Twist, J. (2004, 21 July). *Robots get bookish in libraries*. Retrieved from http://news.bbc.co.uk/2/hi/technology/3897583.stm

UC Council of University Librarians. (2012). *Next-generation technical services (NGTS)*. Retrieved April 21, 2012, from http://libraries.universityofcalifornia.edu/about/uls/ngts/index.html

Vaughan, J., & Fabbi, J. (2008). The helping you buy series: Self-check systems. *Computers in Libraries*, *2*, 29–35.

Vaughan, N. (2007). Perspectives on blended learning in higher education. *International Journal on E-Learning*, *6*(1), 81–94.

Verzosa, F. A. (2010). *The changing library environment of technical services*. Retrieved April 16, 2012, from http://www.slideshare.net/verzosaf/the-changing-library-environment-of-technical-services

Vieira, A. D. S. (1972). A automação no currículo de Biblioteconomia. *Revista da Escola de Biblioteconomia da UFMG, 1*(1), 12-31. Retrieved March 30, 2012, from http://www.brapci.ufpr.br/documento.php?dd0=0000001886&dd1=58ac4

Vieira, D. K., & Souza, A. L. D. (2010). Bibliotecário X Indicadores: Biblioteca universitária nos resultados de avaliação do MEC. Retrieved March 30, 2012, from http://issuu.com/biblioteconomiafatea/docs/indicadores_avaliacaomec

Viggiano, R. G. (2004). Online tutorials as instruction for distance students. *Internet Reference Services Quarterly*, *9*(1-2), 37–54. doi:10.1300/J136v09n01_04.

VirtuOz Inc. (2011, May 10). VirtuOz/CCM benchmark group study predicts a 400 percent increase in ecommerce adoption of intelligent virtual agents by 2014. Retrieved January 23, 2012, from http://www.virtuoz.com/company/news-events/press-releases/virtuoz-ccm-benchmark-group-study-predicts-a-400-percent-increase-in-ecomme/

Wanderley, M. A. (1973). Utilização de processos de automação na Biblioteca Nacional. *Ciência da Informação*, *2*(1). Retrieved from http://revista.ibict.br/index.php/ciinf/article/view/1631

Ward, S. M., & Aagard, M. C. (2008). The Dark Side of Collection Management: Deselecting Serials from a Research Library's Storage Facility Using WorldCat Collection Analysis. *Collection Management*, *33*(4), 272–287. doi:10.1080/01462670802368638.

Warschauer, M., & Healey, D. (1998). Computers and language learning: An overview. *Language Teaching*, *31*, 57–71. doi:10.1017/S0261444800012970.

Warzala, M. (1994). The Evolution of Approval Services. *Library Trends*, *42*(3), 514–514.

Wilde, M., & Level, A. (2011). How to Drink From a Fire Hose Without Drowning: Collection Assessment in a Numbers-Driven Environment. *Collection Management*, *36*(4), 217–236. doi:10.1080/01462679.2011.604771.

Workshop on "Fuzzy is Scalable: Managing Huge Databases Using Fuzzy Methods" at the International Conference on Soft Computing as Transdisciplinary Science and Technology CSTST08. (2008). *Summary*. Retrieved April 21, 2012, from http://www.lirmm.fr/~laurent/FiS/

Wu, A., & Mitchell, A. M. (2010, July 1). Mass management of e-book catalog records: Approaches, challenges, and solutions. *Library Resources & Technical Services*, *54*(3), 164–174.

Yan, H. (2010). On the clouds: A new way of computing. *Information Technology & Libraries*, *29*(2), 87–92.

Young, S. (2007). *The book is dead, long live the book*. Sydney: University of New South Wales Press.

Zabel, D. (2005). Trends in reference and public services librarianship and the role of rusa part one.[from *MasterFILE Premier*. EBSCO. http://web.ebscohost.com.proxy.oplin.org]. *Reference and User Services Quarterly*, *45*(1), 7. Retrieved March 8, 2011

Zink, S. D. (2010, September). *A building for the post-Gutenberg era: The Mathewson-IGT Knowledge Center at the University of Nevada, Reno*. Presented at the 25th annual Material Handling and Logistics Conference, Document Management Track, Park City, UT. Retrieved April 20, 2012, from http://www.hkplanet.net/resources-mhlc-presentations.cfm?year=2010

About the Contributors

Edward Iglesias was born in Laredo, Texas and lived there much of his life. This environmental bias caused by living in a bilingual, bicultural society that has permanently affected his outlook on life. As a result he is drawn to subjects that don't easily fit description or are a blend of many things. His latest book *Robots in the Academic Library* is a good example as the field of library automation and technology is always in flux. After leaving Laredo Mr. Iglesias taught English at various colleges in Houston before deciding to venture into the world of libraries by getting his MLIS at the University of Texas. From there has quickly settled into academic libraries and has worked in the field ever since. Currently Mr. Iglesias is researching the role of maker spaces in libraries as a way for libraries to continue to be relevant and provide communities of creation for their users.

* * *

Carolyn Adams is a Reference and Instruction Librarian and Assistant Professor at Boise State University. Previously, she worked as the Head of Library Services at the Mathewson-IGT Knowledge Center at the University of Nevada, Reno where she managed the circulation, interlibrary loan, electronic reserves, stacks maintenance, and the automated storage departments. She began her library career at the Kresge Business Administration Library at the University of Michigan Ross School of Business. Carolyn earned a Master of Library and Information Science from Wayne State University and a Bachelor of Arts in Psychology from the University of Michigan. While Carolyn's research interests include management and employee performance, assessment, and the innovative use of technology in libraries, she has also published on various subjects related to access services.

Godmar Back is Associate Professor of Computer Science at Virginia Tech, where he has been doing research and teaching in Computer Science since 2004. Dr. Back obtained his PhD from the University of Utah and worked as a post-doctoral scholar at Stanford University. His research interests are diverse, including operating systems, virtualization, programming languages, scientific computing, and web technology, and library information systems. He is an active collaborator with librarians in the area of advancing library technology to ensure that modern technology can find its way into the library sphere. Since 2007, he has been involved in the LibX project, providing technical supervision and input.

Annette Bailey is currently the Assistant Director for Electronic Resources and Emerging Technology Services for the Jean Russell Quible Department of Collections and Technical Services at Virginia Tech. Bailey serves on the Program Planning Committee for the ER&L Conference. She co-developed the open source LibX plug-in, for which she received the 2007 LITA Brett Butler Entrepreneurship Award. She won a National Leadership Grant in 2006 for LibX and in 2008 for LibX 2.0. She has given several invited presentations on LibX at conferences and heads the LibX team.

Ruth Bavousett has spent most of the last two decades serving the IT needs of libraries, serving as a system administrator, ILS operations specialist, developer, project manager, and data wrangler. Her other professional work includes large IT project management for clinical lab and aerospace organizations. She has been a presenter at numerous conferences in the library technology arena. Outside of her work, Ruth is an avid reader of romance novels, blogging on series romance at http://numbersonthespines.com and about almost everything else at http://librarygeekgirl.net. She lives in Kansas with her daughter and two obnoxious cats. Recently, she attempted an experiment by clicking her heels together three times in downtown Lawrence. Nothing interesting happened. Conclusions will be presented in a future paper.

Marshall Breeding is an, author, speaker, and independent consultant specializing in the area of library technologies. He recently concluded a 27-year career at Vanderbilt University, most recently serving as Director for Innovative Technology and Research. He founded and maintains the Library Technology Guides website; he has authored the annual Library Journal "Automation Marketplace" feature since 2002, writes a monthly column for Computers and Libraries, and is editor and primary contributor to the Smart Libraries Newsletter published by ALA TechSource. He has written seven books, including Cloud Computing for Libraries and Next-Gen Library Catalogs. Breeding has authored over 500 articles, news stories, or technical reports published in professional and academic publications.

Denise A. Garofalo always wanted to be a librarian. She is currently the Systems and Catalog Services Librarian at Mount Saint Mary College in Newburgh, NY, but has worked in various special, school, academic, and public libraries in the northeastern United States. She has consulted with libraries on technology, information processing, and digitization projects, has served as an adjunct professor at the Department of Information Studies at the State University of New York at Albany, and is a trustee for the Marlboro (NY) Free Library. She has served on various library committees at the regional and state level and given presentations at various conferences. She writes a column for the Journal of Electronic Resources Librarianship and reviews for *Library Journal, School Library Journal, American Reference Books Annual* and *Technical Services Quarterly*. When not busy with technology in libraries, she enjoys science fiction, bicycling, and hiking with her husband and son.

Dao Rong Gong is Head of Library Catalog Services at Michigan State University (MSU) Libraries. He was the technical lead in implementing several projects at MSU including the library catalog merger with the Library of Michigan, incorporation of electronic resources management module and the system migration to a new catalog platform. He also led the library's initiative in planning and implementing the next generation discovery platform services. His research interests include library technology and automation, and has presented at a number of professional conferences. He co-authored a book on Management Information Systems, published in 2009. Prior to his stint in the academe, he worked as software developer and later as a service consultant for a library automation company. Dao obtained his MLIS from McGill University in Canada.

Regina H. Gong is the Head of Technical Services and Systems for the Lansing Community College (LCC) Library. Regina has over ten years' experience working for the Asian Development Bank, a multilateral financial institution based in the Philippines as reference librarian, cataloger, and head of acquisitions and serials. She moved to the U.S. to work for an integrated library system vendor as systems librarian. In her current position, Regina oversees the Technical Services Unit and provides overall leadership in the management of library resources and collection development. She also manages the integrated library system and Summon discovery service. She enjoys implementing technology solutions that allow LCC students and staff to discover and access a variety of resources for learning and teaching. Regina holds a Bachelor of Library Science from the University of the Philippines and a Master of Library and Information Science from Wayne State University.

Lai-Ying Hsiung has been in the academic library technical services field for over thirty years. She was the former Head of Technical Services at McHenry Library, University of California, Santa Cruz. Her major experience is in optimizing and streamlining technical services operations, including record loading and electronic resources, using various technological approaches, change management, and extensive cross-training.

Edward Lener. Having worked for many years in public services as College Librarian for the Sciences at Virginia Tech, he now serves as Associate Director of Collection Management. He currently is on the ALCTS Collection Management Section Publications Committee and is the university's representative to the Resources for User's Committee of the VIVA library consortium.

Michele McNeal is the Web Specialist at the Akron-Summit County Public Library. She is responsible for developing and managing the Library's Website, Intranet, and III WebOpac, as well as for assisting staff with blogs on Wordpress MU and digital collections using Content DM. She creates, maintains and articulates with these various websites a variety of databases and applications for the Library's staff and customers. She holds an MLIS from Kent State University, as well as a BA in French from the University of Akron and has completed an education certification program at that same institution. Prior to her current position she served as a Reference Librarian in the Science and Technology Division and as the Coordinator of Information Services at the Eisenhower National Clearinghouse for Mathematics and Science Education.

David Newyear is a reference librarian and cataloger at Lakeland Community College in Kirtland, Ohio. He and Michele McNeal have presented on artificial intelligence and library services at IUG 2011, ALA 2011, LITA 2011 National Forum, OLC, and Computers in Libraries 2012. He won the 2011 PLA Polaris John Iliff Innovation in Technology Award for "Emma the Catbot." David holds an MLS from Kent State University and a Bachelor of Music in Horn Performance from Northwestern University.

Leslie O'Brien is the director of collections and technical services at the University Libraries of Virginia Tech. She received her M.L.S. from the University of Maryland and began her library career as a cataloger in special libraries, including the National Wildlife Federation and the American Institute of Architects. Since 1991 she has held a variety of management positions at Virginia Tech. She has been active in ALA ALCTS, the Virginia Library Association, and the Virtual Library of Virginia. Her areas of professional interest are workflow analysis and collections assessment.

Connie Stovall is the Assistant Director for Collection Management at Virginia Tech in Blacksburg, Virginia, where much of her professional interests center on collection analysis. Prior to working in this position, she worked as a Humanities Reference and Instruction Librarian at Virginia Tech, and as student supervisor and student employee at The University of Alabama's Information Services Department. She has been working in libraries since 1999. As time permits, she indulges in her personal interests in observing flora and fauna, either by hiking trails or paddling trails, and in her fascination with the early cultural history of the Southeastern United States.

Michelângelo Mazzardo Marques Viana. Graduated in Library Science from Federal University of Rio Grande do Sul – UFRGS (1998) and graduated also in Business Administration and Systems Development from the Pontifical Catholic University of Rio Grande do Sul – PUCRS (2008). Works as Systems Librarian at the Pontifical Catholic University of Rio Grande do Sul since 1999. Currently holds the position of Coordinator of the Library Systems, in charge of the management of library automation and information retrieval systems, RFID and self-service equipments. He was editor of the Librarianship Guide, of the SobreSites Project (Brazil), since 2001. He also works on projects of regional and national levels as a consultant in the implementation of integrated systems for process automation in academic libraries and for information retrieval in hybrid environments. Participates in working groups of the International Group of Ex Libris Users (IGeLU).

Wei Wei is a Fellow of the Special Libraries Association (SLA) and former Engineering Librarian at the University of California, Santa Cruz. As an active SLA member since 1987, she has served as the chair of the Science and Technology Division, as well as the chairs of various committees at the association level. Wei is the editor and co-editor of *Scholarly Communication in Science and Engineering Research in Higher Education,* 2002; and *Leadership and Management Principles in Libraries in Developing Countries,* 2004. She was also a member of the editorial board of *Science and Technical Libraries,* a peer-reviewed journal for ten years.

Index

R

radio frequency identification technology (RFID) 6, 54, 61, 142-143, 150, 210-214, 220
Request for Proposal (RFP) 91
Request Tracker (RT) 163, 174
Resource Description and Access (RDA) 173
Resource Description Framework (RDF) 175
Rich Internet Application (RIA) 71, 81, 89

S

self-service 141-143, 150, 211, 213, 216
Serials Management 14-15, 21, 208
Serials Solutions 14, 16-18, 23, 26-30, 32, 34-35, 47, 49, 55
service adjustments 182
Shared Cataloging Program (SCP) 159, 170
short term loan (STL) 47
social networking 157, 195-196, 206
Society of Friends of the National Library (SABIN) 142
Staff reaction 104
Standard Interchange Protocol (SIP) 211
Statistical Method 79
storage and retrieval machines (SRMs) 119
Student-Centered Active Learning Environment for Undergraduate Programs (SCALE-UP) 54
student response systems 184-187, 200-201, 204, 206

Summon 17, 23, 26-27, 31, 53, 63, 65, 144
sustaining technologies 180, 183-184

T

Technical Services (TS) 157-158

U

Uniform Resource Locator (URL) 159

V

vendor management 4, 171
Virginia Tech Library System (VTLS) 141
virtual agents 102, 109, 111-113
virtual reference 4-5, 19, 112, 114, 182, 205

W

Wear Ur World (WUW) 197
Website Development 4
WorldCat Collection Analysis™ (WCA) 55
WorldCat Local (WCL) 159, 164
World Health Organization (WHO) 139
WorldShare Platform 14, 19-22, 35, 175

Y

Yankee Book Peddler (YBP) 53